Handbook of Optical Sensors

Handbook of Optical Sensors

Edited by Bob Tucker

CLANRYE INTERNATIONAL
www.clanryeinternational.com

Clanrye International,
750 Third Avenue, 9th Floor,
New York, NY 10017, USA

ISBN: 978-1-63240-781-8

Cataloging-in-Publication Data

Handbook of optical sensors / edited by Bob Tucker.
p. cm.
Includes bibliographical references and index.
ISBN 978-1-63240-781-8
1. Optical detectors. 2. Detectors. 3. Optoelectronic devices. I. Tucker, Bob.
TK8360.O67 H36 2019
681.25--dc23

For information on all Clanrye International publications visit our website at www.clanryeinternational.com

CLANRYE INTERNATIONAL

Printed in China.

Contents

Preface

Optical sensors are electronic detectors that function by the conversion of light into electrical signals. Such electro-optical sensors are used across industries for both simple and complex purposes as well as in devices like smartphones, smartwatches, medical devices, etc. There has been rapid progress in this field and its applications are finding their way across multiple fields like civil and transportation engineering, pipeline monitoring, medical diagnostics, etc. From theories to research to practical applications, case studies related to all contemporary topics of relevance to this field have been included in this book. It is compiled to provide a general overview of the subject while also providing details about the modern advancements in the field. It strives to be a valuable source of knowledge for students, researchers and experts.

This book has been the outcome of endless efforts put in by authors and researchers on various issues and topics within the field. The book is a comprehensive collection of significant researches that are addressed in a variety of chapters. It will surely enhance the knowledge of the field among readers across the globe.

It gives us an immense pleasure to thank our researchers and authors for their efforts to submit their piece of writing before the deadlines. Finally in the end, I would like to thank my family and colleagues who have been a great source of inspiration and support.

Editor

1

A Hybrid Φ/B-OTDR for Simultaneous Vibration and Strain Measurement

Fei PENG* and Xuli CAO

Key Laboratory of Optical Fiber Sensing & Communications (Ministry of Education), University of Electronic Science and Technology of China, Chengdu, 611731, China

*Corresponding author: Fei PENG E-mail: pengfei_uestc@126.com

Abstract: A hybrid phase-sensitive optical time domain reflectometry (Φ-OTDR) and Brillouin optical time domain reflectometry (B-OTDR) system which can realize simultaneous measurement of both dynamic vibration and static strain is proposed. Because the Rayleigh scattering light and spontaneous Brilliouin scattering light are naturally frequency-multiplexed, the heterodyne asynchronous demodulation of frequency shift keying (FSK) in optical fiber communications is utilized, and the demodulations of the two scattering signals are synchronized. In addition, the forward Raman amplification is introduced to the system, which not only makes up for the deficiency of spontaneous Brilliouin scattering based distributed fiber sensor, but also has the merit of the single end measurement of B-OTDR. The designed Φ/B-OTDR hybrid system has the sensing range of 49 km with 10 m spatial resolution. The vibration and strain experiments show that this hybrid system has great potential for use in long-distance structural health monitoring.

Keywords: Optical fiber sensor; Rayleigh scattering; Brillouin scattering; vibration; strain

1. Introduction

As the most attractive distributed fiber sensor (DFS), phase-sensitive optical time domain reflectometry (Φ-OTDR) is successfully employed in the safety monitoring of petroleum, transportation, and structures where the vibration/sound waves need to be measured [1−5]. Especially in recent years, the distributed acoustic sensor, which is developed from Φ-OTDR, is now being the most promising distributed fiber sensor [2]. To improve the sensing performance (including the sensitivity, sensing range, dynamic range, and bandwidth) of Φ-OTDR, researchers focused on increasing the power of backscattered Rayleigh scattering (RS) light for detecting the weak sensing signal RS with high signal noise ratio (SNR). No matter which kind of configuration being used, the nonlinear effects in the fiber have to be considered. The main nonlinear effect has been demonstrated to be modulation instability, which comes earlier than stimulated Brillouin scattering (BS) with the growing of the launching power of the probe pulse [3]. But BS still has to be considered even the stimulated BS does not occur, especially when the direct detection is employed (the spontaneous BS and RS are detected simultaneously) [4]. As to other aspect, in many applications, like structure health monitoring (SHM), after the vibration or acoustic wave generated by the structure transforming, the static strain (<0.1 Hz) of the structure usually changes accordingly. But it is difficult to cover the ultra-low frequency for the current Φ-OTDR or distributed acoustic sensor (DAS) [2]. In this article, by detecting spontaneous BS in Φ-OTDR, a hybrid Φ/B-OTDR (Brillouin optical time domain reflectometry) is proposed to simultaneously measure vibration (dynamic) and strain (static) in one sensing system. Therefore, a comprehensive analysis of vibration and strain can be realized by this hybrid system, offering more useful information for accurate estimation of structure health.

2. Principle of Φ/B-OTDR hybrid system

2.1 Wavelength relationship between the hybrid system and demodulator

Generally, the probe pulse of Φ-OTDR is modulated by the acoustic optical modulator (AOM) for high extinction ratio pulse light. In this configuration, the wavelength relationship of laser, probe pulse, RS, and BS can be shown as Fig. 1.

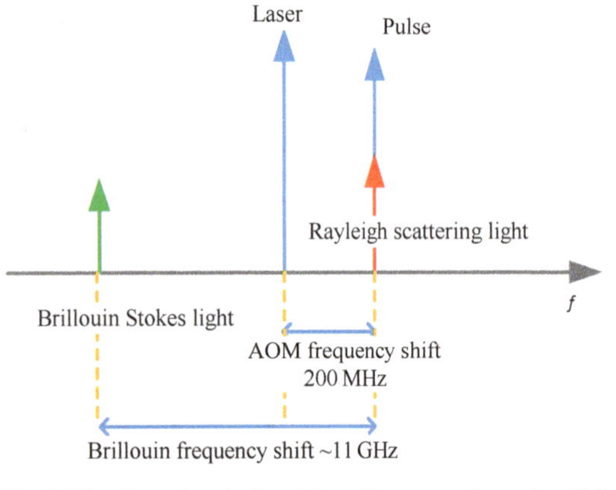

Fig. 1 Wavelength relationship of laser, probe pulse, RS, and BS in the hybrid system.

The frequency of pulse is upshifted by the AOM for 200 MHz (assuming the frequency shift of the AOM is 200 MHz). RS has the same frequency as the pulse and BS frequency downshift (~11 GHz). Utilizing the large frequency deviation of RS and BS and referring the principle of heterodyne asynchronous demodulation for frequency shift keying (FSK), we design the demodulator of the hybrid system, just as shown in Fig. 2. RS and BS first beat with the local oscillator light, and then the mixed light is detected by a wide bandwidth detector (>11 GHz). The radio frequency (RF) signal is divided by two branches. The upper branch demodulates the RS signal, and the lower branch demodulates the BS signal. The difference between this configuration and common heterodyne asynchronous demodulator for FSK is that the lower branch has to sweep the central frequency of the band pass filter (BPF) by a tunable BPF to fit the Brillouin spectrum of the fiber. After BPF, the envelopes of the mixed signals can be acquired by the envelope detectors. At last, the RS and BS curves can be sampled by the analog to digital (A/D) converter when the noise is further filtered by two low pass filters (LPFs).

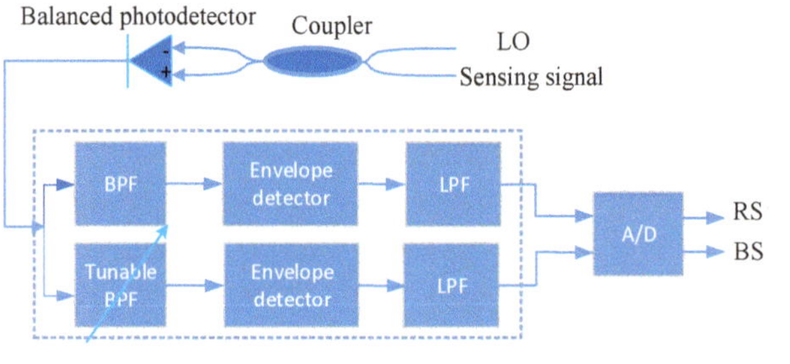

Fig. 2 Signal demodulation principle of the Φ/B-OTDR system.

2.2 Synchronization principle in the demodulation

To realize the simultaneous demodulation of RS and BS signals, we design the demodulating synchronization mechanism, just as shown in Fig. 3. Instead of the polarization scrambling method, the polarization diversity method is used to ensure the RS signal to keep the polarization state of probe pulse stable in a relatively long time [6, 7]. After $N{\cdot}T$ (T is the period of the light pulse, and N is the number of the light), the polarization of the local oscillator (LO) light is switched to polarization orthogonally to eliminate the polarization related gain of BS light by averaging the BS signal of the two LO polarizations. And the central frequency of the tunable BPF is changed around 11 GHz during each pulse repetition period to get the Brillouin spectrum of each position. The RS is processed within $N{\cdot}T$ (with the same LO polarization state during $N{\cdot}T$) to avoid the polarization-induced intensity fluctuation of RS.

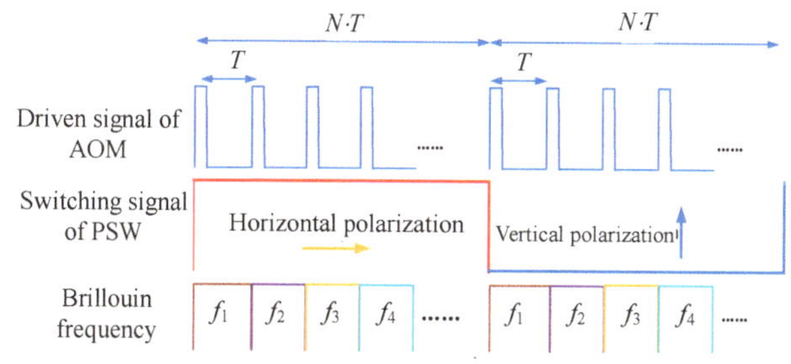

Fig. 3 Synchronization mechanism of the sensing signal demodulation.

3. Experiment

3.1 Experimental setup

The schematic diagram of the hybrid system is shown in Fig. 4. A narrow line width (~3 kHz) semiconductor laser at 1550 nm is used. The laser is divided into two portions by 50:50 Coupler 1. One portion is modulated to pulse with the 2 kHz repetition rate and 100 ns pulse width. 200 MHz frequency shift is induced by using an AOM. Then the pulse light is amplified by an erbium doped fiber amplifier (EDFA) and passes to the BPF to suppress the amplified spontaneous emission noise generated by the EDFA. Then the pulse light launches into the fiber via a circulator. A Raman fiber laser with the central wavelength of 1455 nm is used as the Raman pump. The probe pulses are distributionally amplified by forward Raman pumps, generating RS light and spontaneous BS light along the fiber. Another portion of light split by Coupler 1 is used as a local oscillator with a polarization switch (PSW), which is used to eliminate the polarization related gain of BS. The local oscillator light is mixed with the RS signal by Coupler 2. Then, the beat light is detected by a balanced photodetector with the 11 GHz bandwidth and analyzed by an electronic spectrum analyzer (ESA). The BS and RS curves are sampled by an A/D card with 50 MSample/s. Finally, the sampled signal is processed by a computer in real time. Using the demodulation method in Section 2, the vibration and strain information of the sensing fiber is updated with different speeds.

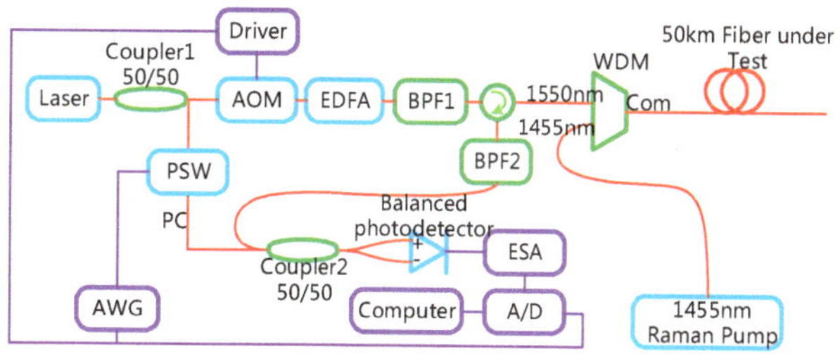

Fig. 4 Setup of the hybrid sensing system.

3.2 Vibration test

When the probe pulse and RS light are enhanced by the former Raman pump, the RS curve obtained is shown in Fig. 5. It can be found that the power of the RS signal over the whole 49 km keeps relatively even by optimizing the power of pulse light and forward Raman pump [4], and the power difference of the fitting curve between the strongest and the weakest positions is less than 3 dB.

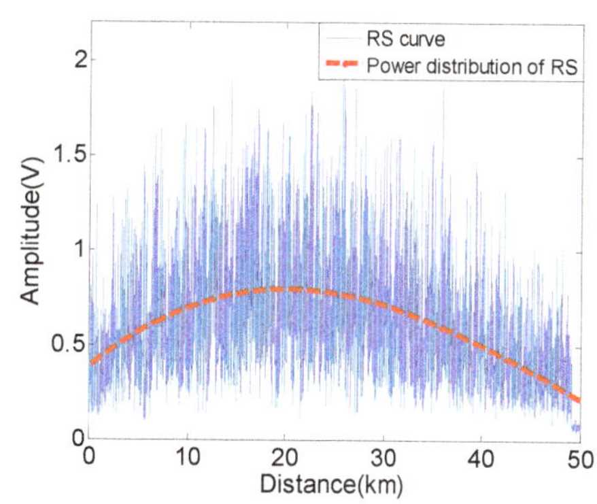

Fig. 5 RS curve acquired by the hybrid system.

To verify the vibration performance of this hybrid system, vibration measurement at the end of fiber is employed. We first spiral two-meter fiber at the distance of 49.03 km into a circle with the diameter of 10 cm and then paste it on the aluminum plate. The subtracted result of the adjacent two RS curves (16 times averaging) can be seen in Fig. 6. The maximum value of the subtracted curve appears at the distance of 49.03 km, where the ball drops.

We drop a ball at the location 49.03 km two times per second and storage the RS signals within one second. And then the vibration level of each position is calculated by root-mean-square of each position, just as shown in Fig. 7. It can be found that the SNR of the vibration location is very high.

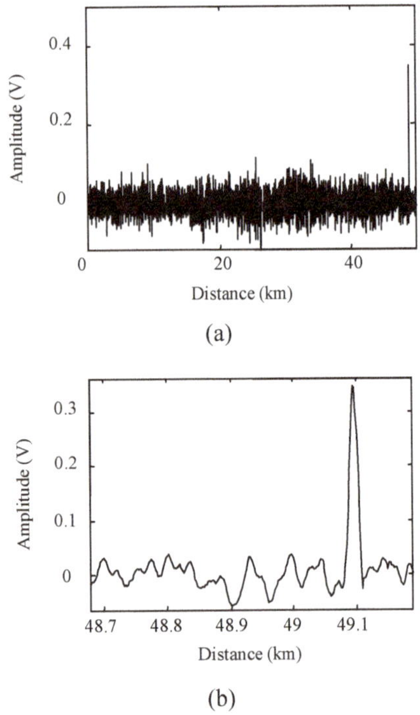

Fig. 6 Difference curve of RS of the falling ball (a) of all over the fiber and (b) near the vibration position when the falling ball experiment is employed.

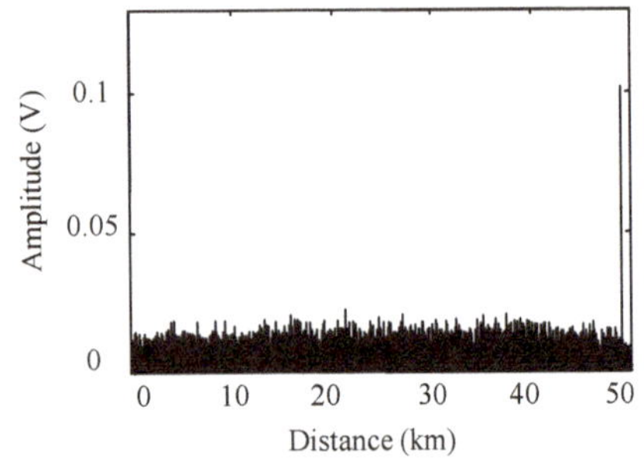

Fig. 7 Spatial distribution of vibration level when the falling ball experiment is employed.

3.3 Strain test

To verify the ability of measuring the static strain of the hybrid system, the strain experiment at the same position of vibration test, 49.03 km, is employed. The 10-m fiber is fixed on the micro-positioner. Then we adjust the strain imposed on the fiber from 200 $\mu\varepsilon$ to 700 $\mu\varepsilon$. When the strain is 700 $\mu\varepsilon$, the spatial distribution of the BS spectrum can be seen as shown in Fig. 8(a). Compared with the configuration without forward Raman amplification, the power of the BS spectrum of the whole fiber is relatively flat, ensuring the strain to be accurate at the end of fiber.

The BS spectra near the strain point are shown in Figs. 8(b) and 8(c). We can see the obvious frequency shift of the BS spectrum near the strain point, and the BS spectrum of all locations has no obvious broadening.

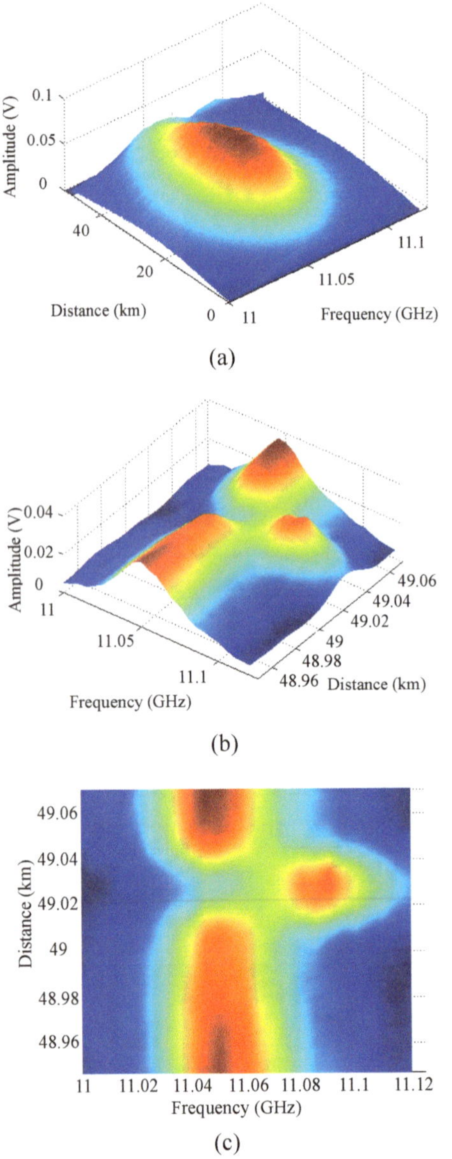

Fig. 8 BS spectra (a) of all over the fiber, (b) near the strain point, and (c) its vertical view.

By fitting the BS spectra of all positions, we can get Brillouin frequency shfit (BFS) of each location, as shown in Fig. 9.

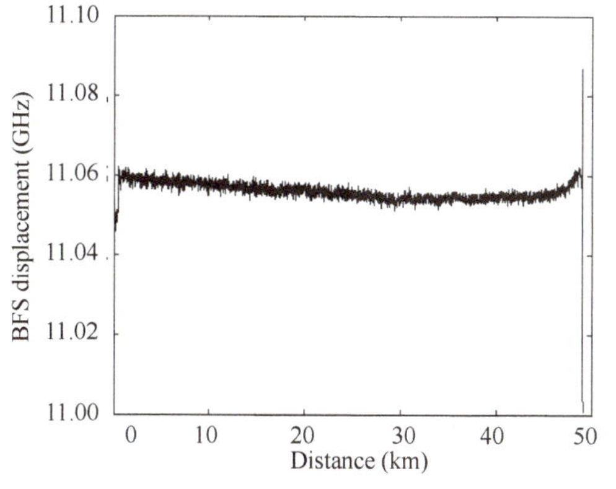

(a)

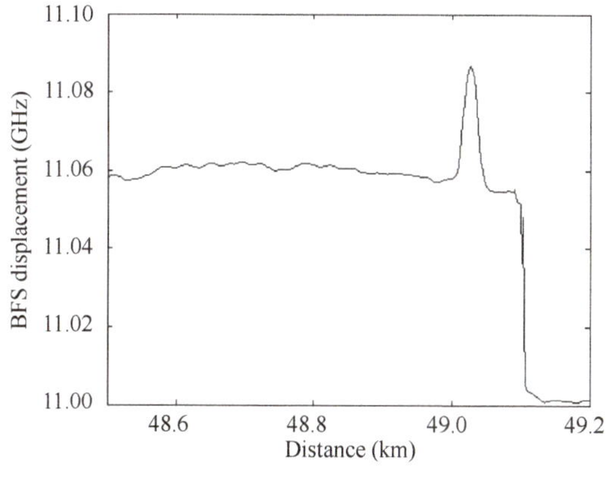

(b)

Fig. 9 BFS distributions (a) of all over the fiber and (b) near the strain point when the strain is 700με.

The spatial distribution of BFS under different strains can be seen in Fig. 10.

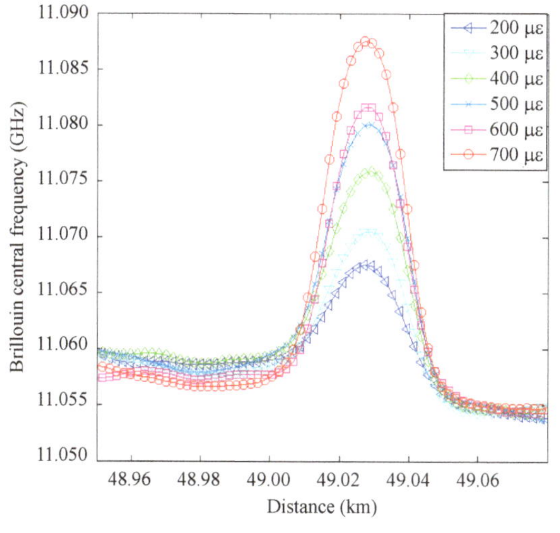

Fig. 10 Spatial distribution of BFS under different strains.

Because the strain experiment is not in constant temperature environment, we calibrate the peak value of Brillouin frequency shift under each strain by subtracting the peak value of Brillouin frequency shift at the location where fiber has natural conditions. Finally, we get the relationship between the strain and BFS of the strain point, as shown in Fig. 11. The strain and BFS have good linearity, and the norm of residual standard error is 1.2911 MHz.

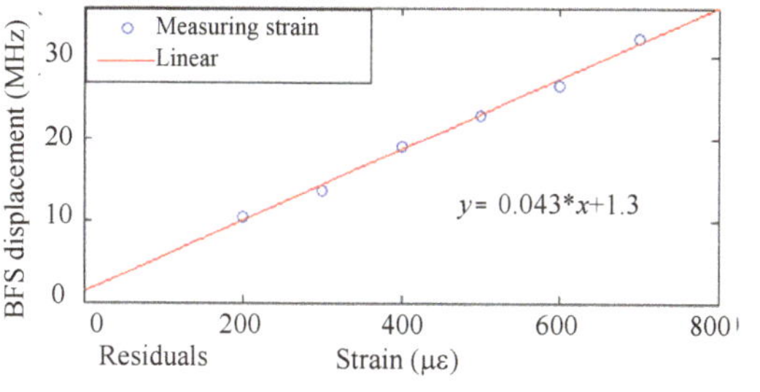

(a)

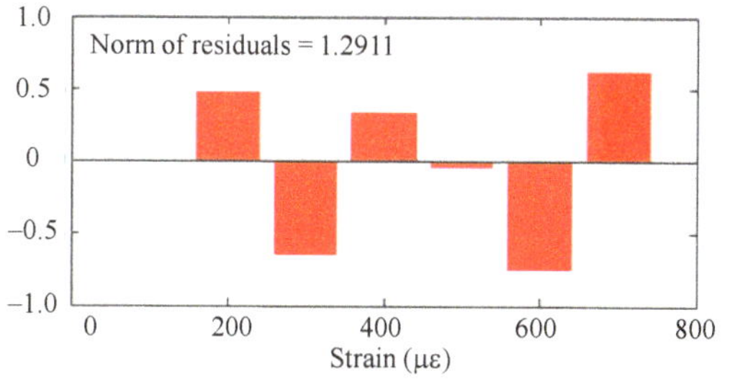

(b)

Fig. 11 Relationship between the strain and BFS displacement (a) and the residual of strain (b) when different strains are imposed.

4. Conclusions

In this article, a hybrid Φ/B-OTDR for simultaneous vibration and strain is proposed. Utilizing the heterodyne detection configuration of classical FSK communications, RS and BS signals in the fiber can be simultaneously detected by the designed synchronization mechanism. To keep the advantage of single-end measurement, we introduce the forward Raman amplification to the system. Not only the sensing range is extended, but also the measurement deficiency of the spontaneous-Brillouin-based DFS (distributed fiber sensor) compared with stimulated-Brillouin-based DFS is also made up for. To verify the sensing performance of the system, the vibration and strain at 49 km are measured successfully with 10 m spatial resolution.

This system has the advantages of long distance, single-end, and dual-parameter measurement, making it possible to be applied in long-distance SHM in future.

Acknowledgment

This work is supported by the National Nature Science Foundation of China (NSFC) under grants (61290312 and 61205048), the PCSIRT (IRT1218), and the 111 Project (B14039).

References

[1] J. C. Juarez and H. F. Taylor, "Field test of a distributed fiber-optic intrusion sensor system for long perimeters," *Applied Optics*, 2007, 46(46): 1968–1971.

[2] T. R. Parker, A. Gillies, S. V. Shatalin, and M. Farhadiroushan, "The intelligent distributed acoustic sensing," in *Proc. SPIE*, vol. 9157, pp. 91573Q-1–91573Q-4, 2014.

[3] H. F. Martins, S. Martin-Lopez, P. Corredera, P. Salgado, O. Frazão, and M. González-Herráez, "Modulation instability-induced fading in phase-sensitive optical time-domain reflectometry," *Optics Letters*, 2013, 38(6): 872–874.

[4] F. Peng, H. Wu, X. H. Jia, Y. J. Rao, Z. N. Wang, and Z. P. Peng, "Ultra-long high-sensitivity Φ-OTDR for high spatial resolution intrusion detection of pipelines," *Optics Express*, 2014, 22(11): 13804–13810.

[5] F. Peng, N. Duan, Y. Rao, and J. Li, "Real-time position and speed monitoring of trains using phase-sensitive OTDR," *IEEE Photonic Technology Letters*, 2014, 26(20): 2055–2057.

[6] X. Zhang, Y. Lu, F. Wang, H. Liang, and Y. Zhang, "Development of fully-distributed fiber sensors based on Brillouin scattering," *Photonic Sensors*, 2011, 1(1): 54–61.

[7] T. Horiguchi, K. Shimizu, T. Kuroshima, M. Tateda, and Y. Koyamada, "Development of a distributed sensing technique using Brillouin scattering," *Journal of Lightwave Technology*, 1995, 13(7): 1296–1302.

2

A Fiber Optic Temperature Sensor Based on the Combination of Epoxy and Glass Particles With Different Thermo-Optic Coefficients

Wolfgang WILDNER* and Dietmar DRUMMER

Institute of Polymer Technology, Friedrich-Alexander-University Erlangen-Nürnberg, Erlangen, Germany

*Corresponding author: Wolfgang WILDNER E-mail: wildner@lkt.uni-erlangen.de

Abstract: This paper describes the development and function of an optical fiber temperature sensor made out of a compound of epoxy and optical glass particles. Because of the different thermo-optic coefficients of these materials, this compound exhibits a strong wavelength and temperature dependent optical transmission, and it therefore can be employed for fiber optic temperature measurements. The temperature at the sensor, which is integrated into a polymer optical fiber (POF), is evaluated by the ratio of the transmitted intensity of two different light-emitting diodes (LED) with a wavelength of 460 nm and 650 nm. The material characterization and influences of different sensor lengths and two particle sizes on the measurement result are discussed. The temperature dependency of the transmission increases with smaller particles and with increasing sensor length. With glass particles with a diameter of 43 μm and a sensor length of 9.8 mm, the intensity ratio of the two LEDs decreases by 60% within a temperature change from 10℃ to 40℃.

Keywords: Temperature sensors; fiber optic sensors; epoxy; glass particles; thermo-optic coefficient

1. Introduction

Fiber optic temperature sensing enables monitoring in particular surroundings such as microwaves or explosion sensitivity [1–4]. Commercially, available fiber optic temperature sensors often use a semiconductor-chip, which has an absorption edge that moves with the temperature [5, 6]. Other sensors utilize Bragg gratings [7–9] or the temperature-dependent fluorescence intensity of certain materials [10] to measure the temperature. Another way is to analyze the Raman scattering in a glass fiber [11] or to analyze the temperature with a Fabry-Pérot interferometer (FPI) [12, 13] or with a Mach-Zehnder interferometer [14] integrated into an optic fiber. Also stimulated Brillouin scattering can be used for distributed temperature measurement [15, 16] whereby the scattered frequency depends on the temperature of the fiber.

Most temperature sensing applications use glass fibers for information transport and measurement. Nevertheless, the use of a polymer optical fiber (POF) has several advantages: POFs have low costs, high elastic strain limits, and high fracture toughness, [17] which enable sensing in deforming structures [18] amongst others.

Glass-filled polymers have a large field of applications in a wide variety of polymer parts. Glass as a filler is used to enhance the mechanical properties [19], to reduce the thermal expansion [20,

21] or to cause light scattering in backlight units of displays [22, 23].

Applied in transparent polymers, glass fillers with a similar refractive index (RI) to the polymer generally cause a RI difference dependent transmission [24–26]. The RI difference of a single material changes with the wavelength of the incident light (dispersion) and with the temperature (thermo-optic coefficient). The thermo-optic coefficient $\Delta n_{rel}/\Delta T$ of optical glass has a value between $-5.4 \cdot 10^{-6} K^{-1}$ and $32.2 \cdot 10^{-6} K^{-1}$ [27], while polymers have a negative thermo-optic coefficient, which is nearly two orders of magnitude higher [28]. This paper shows fiber optic temperature measurement with a compound comprised of glass particles and epoxy. The different thermo-optic coefficients and dispersions of the utilized glass particles and polymer cause a high temperature dependence of the transmission. The first experiments of this measuring principle with oil as the matrix have been reported before [29]. The described fiber-optic sensor is very compact in size and has the potential for low-cost production. It also enables the temperature evaluation by intensity measurements, therefore no spectrometer is needed.

2. Methodology

2.1 Sample preparation

For the investigations described in this paper, the transparent epoxy SpeciFix40 (Struers) was chosen as the matrix because of the high transparency and low shrinkage along with glass particles (N-BAK1, Schott) as the filler material. The SpeciFix40 system consisted of the resin and a curing agent which contains 3-aminomethyl-3,5,5-trimethy-lcyclohexylamine and benzyl alcohol. The RI of the glass was calculated with the Sellmeier Equation using the Sellmeier coefficients given from the manufacturer:

$$n(\lambda) = \sqrt{1 + \frac{B_1 \lambda^2}{\lambda^2 - C_1} + \frac{B_2 \lambda^2}{\lambda^2 - C_2} + \frac{B_2 \lambda^2}{\lambda^2 - C_3}} \quad (1)$$

where B_i and C_i are the Sellmeier coefficients, and λ is the wavelength in μm.

This glass was selected for its matching RI with the polymer matrix in the visible range which is necessary for the RI measurements described later. Table 1 summarizes some relevant properties of the employed glass and polymer.

The bulk glass was crushed and then sieved to produce particles in two different size ranges. The therefore employed two laboratory sieves have mesh widths of 63 μm and 90 μm. The resulting particles vary between 0–63 μm and 63 μm–90 μm. Afterwards, the glass particles, sieved with a mesh width between 63 μm and 90 μm, were washed in a supersonic bath to remove any residual dust particles.

Table 1 Relevant properties of the epoxy and glass.

Material property	Epoxy: Specifix40	Glass: N-BAK1
Density (g/cm³)	1.1 (1)	3.19 (1)
Total Transmission (%)	91.28 (2)	internal transmittance at 546 nm (10 mm): 99.8 (1)
Coefficient of thermal expansion [μm/(mK)]	77.9 (−10/+40℃) (3)	7.6 (−30/+70℃) (1)
Thermo-optic coefficient (K^{-1})	$-1.51 \cdot 10^{-4}$ (4)	$1.2 \cdot 10^{-6}$ (546 nm, +20/ +40℃) (1)
RI n_D (20℃) (-)	1.573 (1) 1.569 (4)	1.57250 (1)

(1) Manufacturer information;

(2) Spectrometer measurement with integrating sphere at 589 nm, sample thickness: 1 mm;

(3) TMA measurement;

(4) Measurement: see results section.

The sensor elements for temperature measurements were prepared via casting of the particle filled polymer in silicone molds. In order to obtain a high filling degree, the resin and curing agent in a ratio of 2.5 to 1 by weight and glass particles were mixed together in a test tube. Then it was waited for 30 minutes to let the particles settle down at the bottom of the tube. From there the mixture was filled into the silicone mold. A section of a polished POF (Hitronic POF Simplex) was attached to each end of the sensor element to connect the sensor to the light guides (Hitronic POF Simplex) during temperature sensing afterwards.

temperature during curing, and therefore, the curing properties [30, 31].

Figure 5 shows the temperature and the voltage U_M (for a wavelength of 460 nm and 650 nm) with respect to the time for sensor elements with different lengths and a particle size of $d_{50.3}$=43 μm. The change in the voltage per temperature increases for the two wavelengths with the length of the sensor element.

The percentage change in the intensity (voltage at 10℃=100%) was evaluated for different sensor lengths and two different particle sizes in Fig. 6. This result shows that longer sensors and smaller particles increase the dependency on the temperature. This is meant to be caused by the consequential larger scattering interface between the particles and the matrix [24, 32].

Figure 7(a) shows the temperature dependent voltage ratio (corresponding to the intensity ratio at the photodiode) of the sensor elements with a particle size of $d_{50.3}$=43 μm. The curves' slope increases with the length of the sensor elements due to the reasons described above. Additionally, the experiments show that the material combination almost exhibits a linear behavior of the intensity ratio in the investigated temperature range from 10℃ to 40℃.

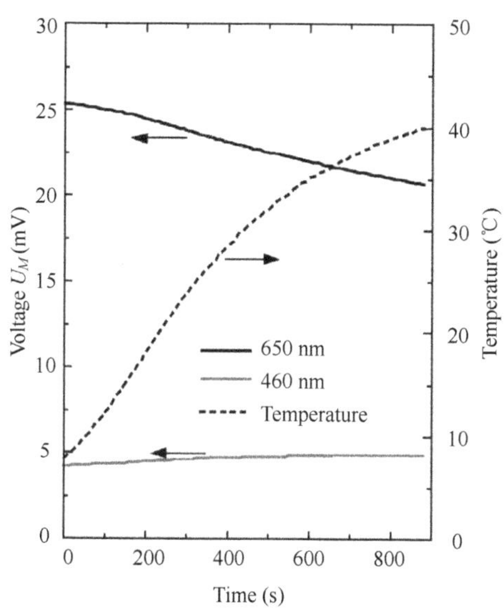

(a)

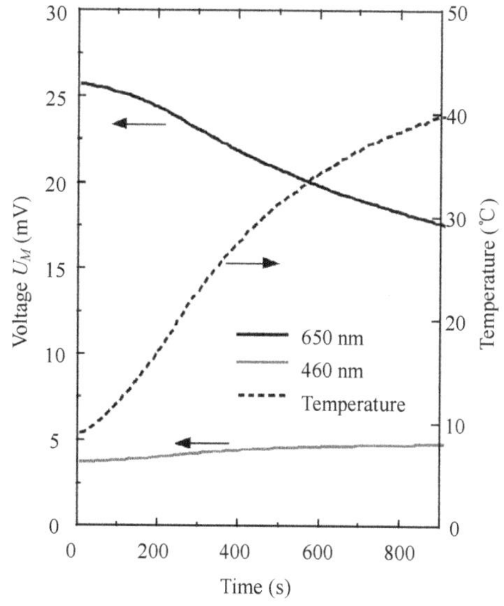

(b)

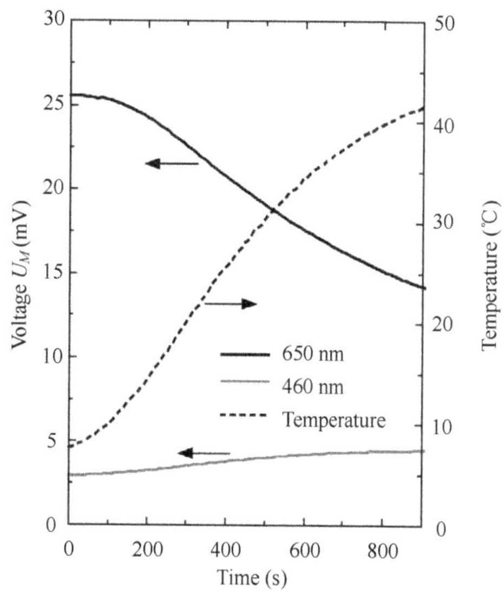

(c)

Fig. 5 Temperature and voltage at the photodiode for three different sensor elements with a particle size of $d_{50.3}$ = 43 μm; length from (a) 4.1 mm, (b) 7.3 mm, and (c) 9.8 mm.

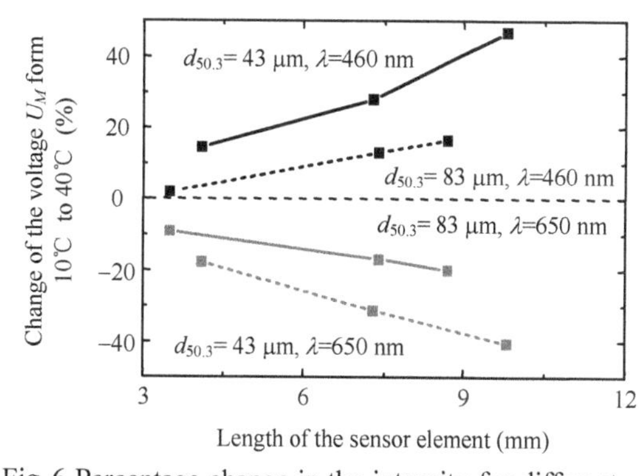

Fig. 6 Percentage change in the intensity for different sensor elements during the heating between 10℃ and 40℃ with regard to the intensity at a temperature of 10℃.

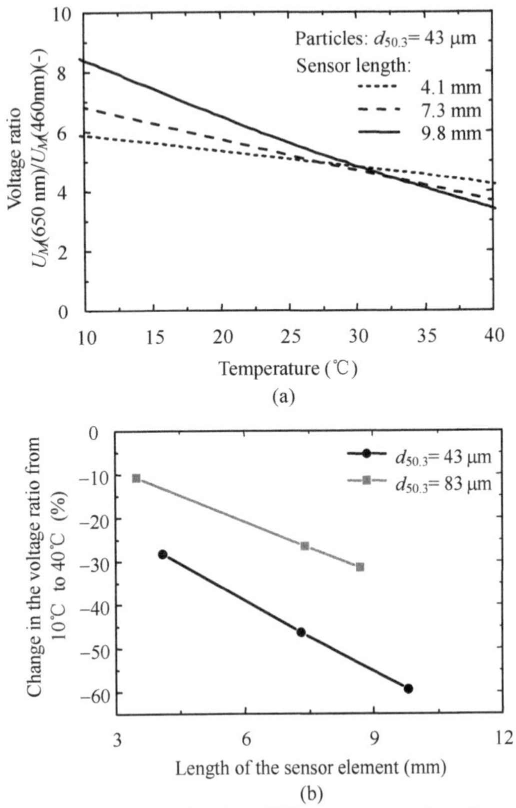

Fig. 7 Voltage ratio for different sensor lengths with a particle size of $d_{50.3}$ = 43 μm (a) and percentage change in the voltage ratio for different sensor elements between 10℃ and 40℃ (b).

The sensor element with a length of 9.8mm and particle size of $d_{50.3}$ = 43 μm shows the strongest decline in the intensity ratio of –60% for a temperature change of 30 ℃ (Fig. 7(b)). In comparison, the transmitted intensity of a sensor based on thermochromic material increases by 41% between 0 and 30℃ [33]. The intensity ratio of the Stokes line to the anti-Stokes line of a typical temperature Raman sensor increases by about 10 % (0.28 to 0.31), the Brillouin frequency (Brillouin sensor) increases about 0.3 % (12.78GHz to 12.82 GHz) and the Bragg wavelength of a typical Fiber Bragg Grating sensor increases by 0.02% (832.64 nm to 832.81nm) with a temperature shift from 0 to 30℃ [34].

Consequently, in order to build a high resolution sensor it is useful to employ materials with strong differences in the thermo-optic coefficients. As shown above, also small fillers and long scattering distances lead to more accurate measurements.

Nevertheless, potential negative influences on the transmission have to be taken into account. Several heating and cooling cycles may cause debonding of the particles and the matrix due to thermal stresses which might reduce the overall transmission. Also other disturbing influences on the measuring result, like yellowing because of the polymer's aging or RI changes because of the absorption of water or chemicals, have to be respected.

4. Conclusions

This paper shows the implementation of a fiber optic temperature sensor based on different thermo-optic coefficients of a polymer and glass particles. The wavelength dependent RI of the polymer was determined by the analyzation of the materials' matching RI, which led to a maximum transmission of the compound. The influences of the sensor length and two different particle sizes on the temperature measurement are described in the paper, whereas small particles and long sensor elements increase the temperature dependent scattering and therefore the sensor's potential accessible accuracy. With a particle size of $d_{50.3}$ = 43 μm and a sensor length of 9.8 mm, the intensity ratio of the two applied LEDs decreases by 60% within a temperature change of 30 ℃.

Acknowledgment

The authors would like to acknowledge the funding of the Deutsche Forschungsgemeinschaft (DFG) through the Cluster of Excellence Engineering of Advanced Materials and to thank Dr.-Ing. H. Kragl of Diemount GmbH for his help with the measurement setup.

References

[1] B. Lee, "Review of the present status of optical fiber sensors," *Optical Fiber Technology*, 2003, 9(2): 57–79.

[2] V. I. Busurin, A. S. Semenov, and N. P. Udalov, "Optical and fiber-optic sensors (review)," *Soviet Journal of Quantum Electronics*, 1985, 15(5): 595–621.

[3] K. T. V. Grattan and T. Sun, "Fiber optic sensor technology: an overview," *Sensors and Actuators, A: Physical*, 2000, 82(1): 40–61.

[4] A. D. Kersey, "A review of recent developments in fiber optic sensor technology," *Optical Fiber Technology*, 1996, 2(3): 291–317.

[5] C. Renschen, "Fi(e)be-thermometer, Prinzipien und Anwendungen der faseroptischen Temperature-messung," *Laser+Photonik*, 2004, 3(4): 38-40.

[6] K. Kyuma, S. Tai, T. Sawada, and M. Nunoshita, "Fiber-optic instrument for temperature measurement," *IEEE Journal of Quantum Electronics*, 1982, 18(4): 676–679.

[7] Y. J. Rao, "In-fiber Bragg grating sensors," *Measurement Science and Technology*, 1997, 8(4): 355–375.

[8] A. D. Kersey, M. A. Davis, H. J. Patrick, M. LeBlanc, K. P. Koo, C. G. Askins, *et al.*, "Fiber grating sensors," *Journal of Lightwave Technology*, 1997, 15(8): 1442–1462.

[9] J. Mandal, S. Pal, T. Sun, K. T. V. Grattan, A. T. Augousti, and S. A. Wade, "Bragg grating-based fiber-optic laser probe for temperature sensing," *IEEE Photonics Technology Letters*, 2004, 16(1): 218–220.

[10] E. Maurice, G. Monnom, B. Dussardier, A. Saissy, D. B. Ostrowsky, and G. W. Baxter, "Erbium-doped silica fibers for intrinsic fiber-optic temperature sensors," *Applied Optics*, 1995, 34(34): 8019–8025.

[11] D. Samiec, "Verteilte faseroptische Temperatur- und Dehnungsmessung mit sehr hoher Ortsauflösung," *Photonik*, 2011, 43(6): 34-37.

[12] W. H. Tsai and C. J. Lin, "A novel structure for the intrinsic Fabry-Perot fiber-optic temperature sensor," *Journal of Lightwave Technology*, 2001, 19(5): 682–686.

[13] C. Wu, H. Y. Fu, K. K. Qureshi, B. O. Guan, and H. Y. Tam, "High-pressure and high-temperature characteristics of a Fabry-Perot interferometer based on photonic crystal fiber," *Optics Letters*, 2011, 36(3): 412–414.

[14] Y. Wang, Y. Li, C. Liao, D. N. Wang, M. Yang, and P. Lu, "High-temperature sensing using miniaturized fiber in-line Mach-Zehnder interferometer," *IEEE Photonics Technology Letters*, 2010, 22(1): 39–41.

[15] D. Culverhouse, F. Farahi, C. N. Pannell, and D. A. Jackson, "Potential of stimulated Brillouin scattering as sensing mechanism for distributed temperature sensors," *Electronics Letters*, 1989, 25(14): 913–915.

[16] X. Bao and L. Chen, "Recent progress in Brillouin scattering based fiber sensors," *Sensors*, 2011, 11(4): 4152–4187.

[17] K. Peters, "Polymer optical fiber sensors – a review," *Smart Materials and Structures*, 2011, 20(1): 13002–17.

[18] W. R. Habel and K. Krebber, "Fiber-optic sensor applications in civil and geotechnical engineering," *Photonic Sensors*, 2011, 1(3): 268–280.

[19] B. D. Agarwal, L. J. Broutman, and K. Chandrashekhara, *Analysis and performance of fiber composites*. New York: Wiley, 2006.

[20] K. C. Yung, B. L. Zhu, T. M. Yue, and C. S. Xie, "Preparation and properties of hollow glass microsphere-filled epoxy-matrix composites," *Composites Science and Technology*, 2009, 69(2): 260–264.

[21] A. Chandra, W. H. Meyer, A. Best, A. Hanewald, and G. Wegner, "Modifying thermal expansion of polymer composites by blending with a negative thermal expansion material," *Macromolecular Materials and Engineering*, 2007, 292(3): 295–301.

[22] H. Q. Tang, Z. F. Tang, T. T. Zhao, W. L. Li, and Q. Ye, "Preparation and optical properties of polysiloxane microspheres/PMMA light scattering materials," *Guangzi Xuebao/Acta Photonica Sinica*, 2012, 41(6): 723–727.

[23] Y. Zhao, P. Ding, C. Ba, A. Tang, N. Song, and Y. Liu, "Preparation of TiO_2 coated silicate micro-spheres for enhancing the light diffusion property of polycarbonate composites," *Displays*, 2014, 35(4): 220–226.

[24] W. Wildner and D. Drummer, "Modelling haze and transmission of transparent filled systems in dependence of filler surface area, refractive index difference and wavelength," in *SPE ANTEC™ Orlando*, USA, 2015.

[25] W. Wildner and D. Drummer, "Wavelength dependent haze of transparent glass-particle filled poly(methyl methacrylate) composites," *International Scholarly Research Notice*, 2014, 2014: 1–5.

[26] H. Breuer and J. Grzesitza, "Trübungserscheinungen in zweiphasigen Polymersystemen (glasfaserverstarkte Polymere)," *Die Angewandte Makromolekulare Chemie*, 1975, 45(681): 1–19.

[27] *Optisches Glas: Datenblätter*: Schott AG, 2012.

[28] J. Graf, Entwicklung und Untersuchungen zur Herstellung verlustarmer passiver Wellenleiter und verstärkender Wellenleiter, Ph.D. dissertation, Universität des Saarlandes, Saarbrücken, Germany, 1999.

[29] W. Wildner and D. Drummer, "A fiber optic temperature sensor based on the combination of two

materials with different thermo-optic coefficients," *IEEE Sensors Journal*, 2015, 16(3): 688–692.

[30] A. C. Loos and G. S. Springer, "Curing of epoxy matrix composites," *Journal of Composite Materials*, 1983, 17(2): 135–169.

[31] J. H. Oh and D. G. Lee, "Cure cycle for thick glass/epoxy composite laminates," *Journal of Composite Materials*, 2002, 36(1): 19–45.

[32] T. Naganuma and Y. Kagawa, "Effect of total particle surface area on the light transmittance of glass particle-dispersed epoxy matrix optical composites," *Journal of Materials Research*, 2002, 17(12): 3237–3241.

[33] N. Díaz-Herrera, M. C. Navarrete, O. Esteban, and A. González-Cano, "A fiber-optic temperature sensor based on the deposition of a thermochromic material on an adiabatic taper," *Measurement Science and Technology*, 2004, 15(2): 353–358.

[34] F. Bernhard, *Handbuch der technischen Temperaturmessung*. Berlin Heidelberg: Springer, 2014.

3

Analysis of a Plasmonic Based Optical Fiber Optrode With Phase Interrogation

H. MOAYYED[1*], I. T. LEITE[1], L. COELHO[1], J. L. SANTOS[1], and D. VIEGAS[2]

[1]*CAP / INESC TEC - Technology and Science and Faculty of Sciences from the University of Porto, Rua do Campo Alegre 687, 4169-007 Porto, Portugal*

[2]*INL - International Iberian Nanotechnology Laboratory, Av. Mestre José Veiga, 4715-310 Braga, Portugal*

[*]Corresponding author: H. MOAYYED E-mail: hmoayyed@fc.up.pt

Abstract: Optical fiber optrodes are attractive sensing devices due to their ability to perform point measurement in remote locations. Mostly, they are oriented to biochemical sensing, quite often supported by fluorescent and spectroscopic techniques, but with the refractometric approach considered as well when the objective is of high measurement performance, particularly when the focus is on enhancing the measurand resolution. In this work, we address this subject, proposing and analyzing the characteristics of a fiber optic optrode relying on plasmonic interaction. A linearly tapered optical fiber tip is covered by a double overlay: the inner one – a silver thin film and over it – a dielectric layer, with this combination allowing to achieve, at a specific wavelength range, surface plasmonic resonance (SPR) interaction sensitive to the refractive index of the surrounding medium. Typically, the interrogation of the SPR sensing structures is performed, considering spectroscopic techniques, but in principle, a far better performance can be obtained, considering the reading of the phase of the light at a specific wavelength located within the spectral plasmonic resonance. This is the approach which is studied here in the context of the proposed optical fiber optrode configuration. The analysis performed shows the combination of a silver inner layer with a dielectric titanium oxide layer with tuned thicknesses enables sensitive phase reading and allows the operation of the fiber optic optrode sensor in the third telecommunication wavelength window.

Keywords: Optrode; fiber sensing; surface plasmon resonance; phase interrogation; metallic-oxide coupled layers

1. Introduction

Along with the years, a large number of fiber based sensing head layouts have been proposed to measure a broad range of parameters, relying on diverse interaction mechanisms of the measurand with the optical field [1]. One of the most attractive layouts is the reflective fiber tip configuration, where the sensor functionality occurs at the end of the optical fiber, therefore readily allowing point sensing at remote locations. These fiber sensing devices are commonly identified as fiber optrodes, a generalization of the initial meaning of the word optrode, and the combination of the words "optics" and "electrode" indicates it strictly refers to the modulation of the optical field by the electric characteristics of an electrode which interacts with the targeted measurand [2]. To a large degree, fiber optrodes include a membrane at the tip of the fiber with its fluorescent and spectroscopic properties changing with the variation of the measurand (typically a chemical or biochemical entity), inducing an intensity/spectral modulation of the light that returns along the fiber which is also the illumination channel [3]. However, other operation modes are possible, and the one associated with refractometry has the particular importance, where the sensing head is sensitive to the refractive index of the surrounding medium, which by itself can show a specific dependence on a certain chemical or biochemical parameter [4].

Since its first observation by Wood in 1902 [5], the physical phenomenon of plasmonics has been found its way into practical applications in sensitive detectors, capable of detecting sub-monomolecular coverage. Physical interpretation of the phenomenon was initiated by Lord Rayleigh [6] and further refined by Fano [7], but a complete explanation was not possible until 1968, when Otto [8] and in the same year Kretschmann and Raether [9] reported the excitation of surface plasmons. Nowadays, plasmonics is one of the most active research topics in photonics. The fundamental reason behind the ever-growing importance of plasmonics in today's scientific research lies on the subjacent physical phenomenon, which is an interaction process between the electromagnetic radiation and the conduction electrons at metallic interfaces or in small metallic nanostructures, leading to an enhanced optical near-field of the sub-wavelength dimension. Physically, this arises from the fact that the length scale involved in the mode coupling between photons and electrons has an effective value much smaller than the light's wavelength, essentially determined by the electron dynamics in the metal, with the light energy carried as a package of electron oscillations [10].

Owning to the well-known advantages over conventional sensors, surface plasmon resonance fiber optic sensors have attracted much attention in the past two decades. The importance of surface plasmonic resonance (SPR) fiber optic sensors in the field of chemical, biological, and environmental sensing has notably increased in the last years [11−12]. Indeed, ever since the introduction to various optical methods in the excitation of the SPR at a metal-dielectric interface [13], it has been widely recognized that such effect can be utilized to achieve sensing or monitoring of various interfacial phenomena with ultrahigh sensitivity. These include, for example, chemical and biological sensing [14, 15], film-thickness sensing [16], temperature sensing [17], and angular measurement [18]. Recently, it has been demonstrated that the SPR technique applied to chemical and biological sensing can achieve refractive index (RI) resolutions down to 10^{-9}, values not accessible to other optical sensing technologies [19].

The light polarization can be TE or TM polarization, although only the TM polarization will excite the plasmon in a waveguide, which is the condition achieved looking for the phase matching between a plasmon and a waveguide mode, mathematically meaning the equality of the modal propagation constants of the plasmon (k_{SP}) and the optical waveguide mode ($k_{waveguide}$), which corresponds to the equality of their effective refractive indices. In general, there are four approaches to interrogate an SPR sensing structure, namely (1) the change in the resonant angle ("angular interrogation"), (2) the change in reflectance at the fixed incidence angle ("intensity interrogation"), (3) the change in the resonant wavelength at the fixed incidence angle ("wavelength interrogation"), and (4) the phase difference between the p- and s-polarization states of the reflected light ("phase interrogation") [15]. The plasmonic based modulation in an optical fiber does not rely on the incidence angle of the light, as in the Otto or Kretschmann-Raether configurations, but instead on the modulation of the intensity, wavelength or phase of the light, i.e., the output signal can be obtained by taking into account the total power transmitted through the device, analyzing the spectral properties of the transmitted light or by reading and processing the phase of this light.

In phase interrogation of SPR sensors, there is a lack of comprehensive analysis. Surely, this methodology is somewhat more complex to implement than most traditional ones, requiring deeper know-how on optical components, modulation approaches, and signal processing techniques, but eventually, the main reason for this situation is the circumstance: the great majority of the R&D (research and development) communities which has been working until recently in SPR sensing have a biochemical background, therefore with a tendency to use the spectroscopic equipment. This is changing since plasmonics is moving to fields other than sensing, such as imaging and data storage, attracting the subject researchers with diverse backgrounds. A consequence of this dynamics is a clear improvement of the resolution values obtained when considering phase interrogation. Indeed, for the case of heterodyne interrogation (when the TE and TM waves are provided by two light sources with slightly different optical frequencies that are made to interfere, resulting in a beat signal with the frequency low enough to be resolved by the photodetection and amplification electronics), a value of 2.8×10^{-9} was reported in 2008 [19], which seems to be the highest refractive index resolution achieved so far; when considering polarimetric interrogation (a polarizer is used to obtain, from a single optical source, TE and TM components which interfere on the photodetector), and under the optimized conditions, a resolution of 3.7×10^{-8} was obtained [20]; applying the methodology of the reference wave propagating along a different spatial path, resulting in interfering with the TM light on the photodetector, a resolution down to 2.2×10^{-7} was reported in 2011 [21].

Considering optical fiber sensing platforms, phase interrogation is attractive since the SPR induced phase change is greatly amplified due to the multiple attenuated total reflections undergone by light in the fiber, with a consequent increase in measurand readout sensitivity. This factor certainly contributes to the interest this interrogation approach has been inducing in recent years [22].

There are several methodologies to identify the performance of an SPR optical fiber sensing system. One of the most common methods is by evaluating its intrinsic sensitivity [23, 24]. Focusing on this characteristic, several attempts have been considered, such as tapering optical fibers [25]. For instance, a tapered optical fiber SPR microsensor based on white light excitation [26] and the use of dual-tapered and tetra-tapered fiber SPR probes for gas and liquid sensing have been proposed [27]. These structures present a substantial variation in evanescent field penetration along the tapered sensing region length. The evanescent field can be reached by the narrowing of the fiber cross section in the tapered region. Another approach, compatible with the mentioned mechanisms, is based on modifying the geometry of the sensing head (tip layout) to optimize the dynamic range and sensitivity [28–30]. These configurations are supported by beveling the distal end of the fiber probe, such as a chisel-tipped probe, where the top and bottom of the probe are polished to the same angle.

In this work, the combination "optrode + refractometry associated with plasmonics + phase interrogation" is analytically explored. A linearly tapered fiber tip coated with two thin film layers is considered: the inner one with a metallic material as silver (Ag) and the external layer with a dielectric oxide. This combination permits to achieve plasmonic resonance at a specific wavelength window sensitive to refractive index variations of the external medium. Insights into the experimental implementation and characterization of this SPR sensing structure are also delivered.

2. Sensor layout and analysis

Figure 1 shows an optical fiber tip optrode with the linear profile, in which the conical surface is coated with a reflective thin film. The tip length is assumed to be L, ρ_0 and ρ_L denote the radii of the fiber core and tip end, respectively. A geometric ray methodology combined with the transfer-matrix formalism for stratified optical media is applied here, following the approach of Gupta and co-workers [23, 25].

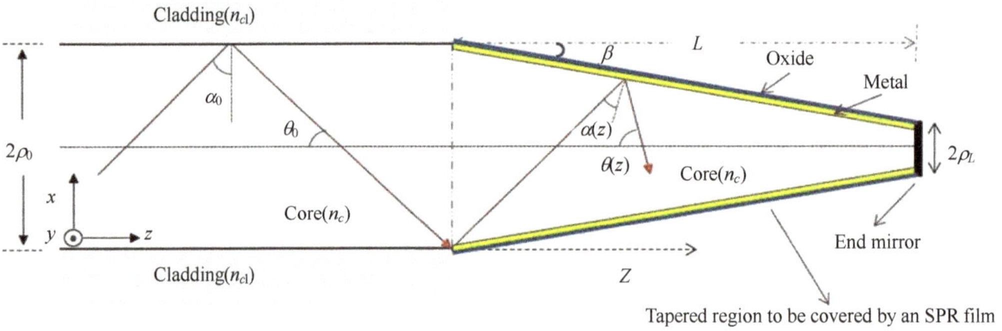

Fig. 1 Schematic illustration of the SPR fiber optrode under analysis (the top and bottom of the probe are polished to the same angle; the cladding of the optical fiber tip has been removed in the sensing region and replaced by a metallic film covered with an oxide overlayer).

If θ_0 represents the propagation angle that the optical ray makes with the axis of the fiber, it is transformed to an angle $\theta(z)$ inside the taper following the transformation relationship:

$$\rho_0 \sin\theta_0 = \rho(z)\sin\theta(z) \tag{1}$$

where $\rho(z)$ is the taper radius at a distance z from the input end of the taper (z=0). In the taper region, such optical ray reaches the interface with the metal-dielectric film at an angle of incidence, and $\alpha(z)$, at coordinate z, is related with the value α_0 in the non-tapered region as

$$\alpha(z) = \cos^{-1}\left[\frac{\rho_0\cos\alpha_0}{\rho(z)}\right] - \beta \tag{2}$$

where $\rho(z)=\rho_0-(z/L)(\rho_0-\rho_L)$ is the tip radius with the coordinate z, and $\beta=\tan^{-1}((\rho_0-\rho_L)/L)$ is the taper angle. Notice that as the radius $\rho(z)$ decreases, so does the angle $\alpha(z)$. Therefore, for a particular optical ray launching at the input end of the taper to undergo total internal reflection throughout the entire length of the taper, the condition $\alpha(z)>\alpha_c=\sin^{-1}(n_s/n_{co})$ must be satisfied for $0<z<L$ (n_s and n_{co} are the refractive indices of the surrounding medium and fiber core, respectively). Since $\alpha(z)$ becomes smaller along the propagation and considering (2), the above condition is fulfilled when

$$\cos^{-1}\left(\frac{\rho_0\cos\alpha_0}{\rho_L}\right) - \beta > \alpha_c\,. \tag{3}$$

This is the basic equation which permits to analyze the two relevant situations, namely for a fixed angle α_0 what is the maximum length of the optrode (taper) to keep the condition of total reflection along its length and the other way around, i.e., for a fixed taper length what is the minimum value for α_0.

Considering the first of these two situations, regarding (3), and assuming a fixed angle of incidence, the maximum taper length is

$$L > \frac{\rho_0 - \rho_L}{\tan\left[\cos^{-1}\left(\frac{\rho_0\cos\alpha_0}{\rho_L}\right) - \alpha_c\right]}\,. \tag{4}$$

On the other hand, when L is fixed, the minimum angle of incidence must satisfy

$$\alpha_0 > \cos^{-1}\left[\frac{\rho_L}{\rho_0}\cos[\beta + \alpha_c]\right]. \tag{5}$$

The angle range of the rays at the coordinate z alters to $[\alpha_1(z),\alpha_2]$ due to the variation of the fiber core diameter, where

$$\alpha_1(z) = \cos^{-1}\left[\frac{\rho_0\cos\alpha_{cr}}{\rho(z)}\right] - \beta \tag{6}$$

$$\alpha_2 = \frac{\pi}{2} - \beta \tag{7}$$

where $\theta_{cr}=\sin^{-1}(n_{cl}/n_{co})$ is the critical angle (n_{cl} and n_{co} are the refractive indices of the fiber cladding and core, respectively) [30].

In order to evaluate the optical response of the sensing structure, we use the multilayer transfer-matrix theory applied to a four-layer system comprising: (1) fiber-core, (2) inner-metal, (3) oxide-layer, and (4) probe-medium [23]. This methodology is applied to compute the propagation of the radiation through the layer system using Maxwell's equations subject to boundary conditions between two adjacent layers. From the basis of this formalism, the tangential components of the electric and magnetic field vectors (**E** and **H**, respectively) at the first and final boundaries, are related with [31]

$$\begin{bmatrix} U_0 \\ V_0 \end{bmatrix} = \begin{bmatrix} m_{11} & m_{12} \\ m_{21} & m_{22} \end{bmatrix} \begin{bmatrix} U_N \\ V_N \end{bmatrix} \tag{8}$$

where $(U,V)=(E_y,H_z)$ for the TE modes or $(U,V)=(H_y,-E_z)$ for the TM modes, and $[m_{ij}]$ is the transfer matrix of the multilayer structure comprised of N interfaces:

$$\begin{bmatrix} m_{11} & m_{12} \\ m_{21} & m_{22} \end{bmatrix} = \prod_{k=1}^{N-1} \begin{bmatrix} \cos\delta_k & \frac{-\mathrm{i}}{\eta_k}\sin\delta_k \\ -\mathrm{i}\eta_k \sin\delta_k & \cos\delta_k \end{bmatrix} \tag{9a}$$

with δ_k being the phase in the kth layer (k = 0 and k=N are the first and final media, respectively), given by

$$\delta_k = \frac{2\pi}{\lambda} n_k \cos\theta_k d_k = \frac{2\pi d_k}{\lambda}\sqrt{\varepsilon_k - n_0^2 \sin^2\theta_0} \tag{9b}$$

where d_k, n_k, and θ_k are, respectively, the thickness, refractive index, and propagation angle (with respect to the normal to the direction of stratification) of the kth layer, λ is the free-space wavelength of the light propagating in the system, and non-magnetic media are assumed (i.e., $\mu_k = 1$). The parameter η_k in (9a) is defined as a function of the polarization state as

$$\eta_k^s = \sqrt{\frac{\varepsilon_0 \varepsilon_k}{\mu_0 \mu_k}} \cos\theta_k \quad \text{(for TE waves)} \tag{10a}$$

and

$$\eta_k^p = \sqrt{\frac{\mu_0 \mu_k}{\varepsilon_0 \varepsilon_k}} \cos\theta_k = \frac{\mu_0 \mu_k}{\varepsilon_0 \varepsilon_k} \eta_k^s \quad \text{(for TM waves)} \tag{10b}$$

where ε_k and μ_k are the dielectric constant and relative magnetic permeability, respectively. The reflectance $R_{s,p}$ of the multilayer structure for each polarization state is obtained as

$$R_{s,p}(\lambda) = |r_{s,p}(\lambda)|^2 \tag{11}$$

where r_s and r_p are the Fresnel reflection coefficients which are calculated directly from the transfer-matrix elements in (8) and the parameters η of the incidence and transmission media.

$$r = \frac{(m_{11} + m_{12}\eta_N)\eta_0 - (m_{21} + m_{22}\eta_N)}{(m_{11} + m_{12}\eta_N)\eta_0 + (m_{21} + m_{22}\eta_N)}. \tag{12}$$

The normalized transmitted power through the sensing region of the fiber tip is given by [30]

$$P_{\text{trans}} = \frac{\int_0^L dz \int_{\alpha_1(z)}^{\alpha_2} \left[R_p\right]^{N_{ref}} I(\theta) d\theta}{\int_0^L dz \int_{\alpha_1(z)}^{\alpha_2} I(\theta) d\theta} \tag{13}$$

where

$$I(\theta) = \frac{n_0^2 \sin\theta \cos\theta}{(1 - n_0^2 \cos^2\theta)} \tag{14}$$

is the intensity distribution between the continua of guided modes (assuming meridional excitation). As shown in (11), R_p is the reflectivity of the multilayer structure, and N_{ref} is the number of reflections that the optical ray with the propagation angle θ undergoes in the sensing region.

In order to evaluate the number of reflections, we consider a length dL of the taper region around the coordinate z where the radius of the taper is $\rho(z)$, as illustrate in Fig. 2.

With regard to this figure, $\tan(\beta)=b/h$ and $\tan[\alpha(z)-\beta]=(h_0-h)/b$, therefore

$$h = \frac{h_0}{1 + \tan(\beta)\tan[\alpha(z) - \beta]}. \tag{15}$$

also

$$h_0 = 2\rho(z)\tan[\alpha(z) - \beta]. \tag{16}$$

With these relations, the equation for the longitudinal length advanced when the ray propagates from one side of the taper to the other is given by

$$h = \frac{2\rho(z)\tan[\alpha(z)-\beta]}{1+\tan\beta\tan[\alpha(z)-\beta]}. \tag{17}$$

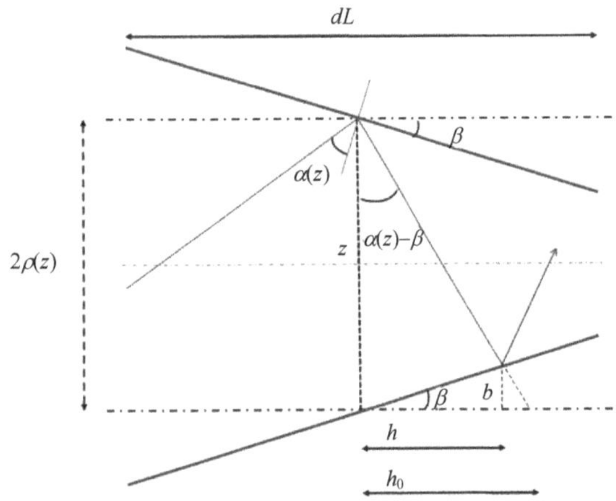

Fig. 2 Schematic of a length dL of the taper region around the coordinate z where the radius of the taper is $\rho(z)$.

The ray is reflected in the lower taper surface and propagates upwards towards the upper taper surface. For simplification, we can consider the same length h is advanced by the ray associated with this upwards propagation (actually, the real value is smaller than h). Therefore, in the distance $2h$, we have a reflection, which means in the distance dL the number of reflections is

$$\delta N = \frac{1+\tan(\beta)\tan[\alpha(z)-\beta]}{4\rho(z)\tan[\alpha(z)-\beta]}. \tag{18}$$

Considering the full taper length, the number of reflections can be calculated as

$$N_{\text{one-path}} = \int_0^L \frac{1+\tan(\beta)\tan[\alpha(z)-\beta]}{4\rho(z)\tan[\alpha(z)-\beta]}dz. \tag{19}$$

Since the configuration works in reflection, the total number of reflections is given by

$$N_{\text{ref}} = 2N_{\text{one-path}} = \int_0^L \frac{1+\tan(\beta)\tan[\alpha(z)-\beta]}{2\rho(z)\tan[\alpha(z)-\beta]}dz \tag{20}$$

with $\alpha(z)$ given by (2). In order to access the performance of phase interrogation of SPR sensors supported by fiber structures, it is convenient to express the amplitude reflection coefficients r_s and r_p in the polar form as

$$r_s = |r_s|e^{i\phi_s}, r_p = |r_p|e^{i\phi_p}. \tag{21}$$

If $(\delta\phi_{p,s})_j$=[$(\phi_p-\phi_s)_j$] is the phase shift of the light between the p and s polarizations for reflection j, the total phase shift accumulated by the light that propagates in the optrode is given by

$$(\phi_p-\phi_s)_{\text{total}} = 2\sum_{j=1}^{N_{\text{one-path}}}(\phi_p-\phi_s)_j. \tag{22}$$

The situation is somehow simplified if an average phase shift per reflection is determined. At the input of the taper, the angle of incidence can be considered to be α_0, which corresponds to a phase shift of $(\delta\phi_{p,s})_{\alpha 0}=(\phi_p-\phi_s)_{\alpha 0}$; at the end of the taper, the phase shift is $(\delta\phi_{p,s})_{\alpha(L)}=(\phi_p-\phi_s)_{\alpha(L)}$, with $\alpha(L)$ obtained from (2) when $\rho(z)=\rho_L$. Therefore, an approximation for this average phase shift is given by

$$(\delta\phi_{p,s})_{\text{average}} = \frac{(\delta\phi_{p,s})_{\alpha_0}+(\delta\phi_{p,s})_{\alpha(L)}}{2}, \tag{23}$$

and the total phase shift accounting for all reflections inside the sensing region is

$$\Delta\phi_{p,s} \equiv (\phi_p-\phi_s)_{\text{total}} = N_{\text{ref}}(\delta\phi_{p,s})_{\text{average}}. \tag{24}$$

At this stage, it is useful to define the sensitivity of the sensing structure to the refractive index n_s of the surrounding medium, which can be regarded as the basic parameter quantifying the performance of the sensor. A change in the refractive index $n_s=(\varepsilon_s)^{1/2}$ of the surrounding dielectric produces a variation in the propagation constant of the surface plasmon, resulting in the modification of the SPR coupling condition. This can be observed as a change in one or more properties of the light transmiting through the sensing structure. The sensitivity can be defined as

$$S_{n,\xi} = \frac{\delta\xi}{\delta n_s} \tag{25}$$

where $\delta\xi$ is the change in a given property of light (e.g., intensity, resonance wavelength or phase) due to a variation of δn_s in the refractive index of the surrounding medium. Here, we particularly concern with the case of phase interrogation, where the property of light of interest is the phase difference $\Delta\phi_{p,s}$ between the p and s polarization components. In this case, the phase sensitivity to refractive index variations is then defined as $S_{n,\phi}=\delta(\Delta\phi_{p,s})/\delta n_s$.

3. Results and discussion

In this analysis, we consider a multimode silica fiber with the 100-μm core diameter with the numerical aperture of 0.24, tip end diameter of 60μm, and sensing region of L = 1 cm. Accordingly, the number of reflections estimated using (20) is N_{ref} = 60. The refractive indices for the metal and oxide layers are obtained from experimental data available in literatures [32–35]. Concerning the metallic layer, we have found in a previous study that silver is the most adequate one [36]. Figure 3 shows the spectrum of the light returning from the sensing head for different thicknesses of the silver layer, and when combined with the spectral curves obtained when the oxide layer is included, it turns out the best thickness for the silver layer is 80 nm, so this value is set for the analysis that provides the results delivered in the following.

For the outer oxide layer, three different oxide materials are compared, titanium dioxide (TiO_2), silicon dioxide (SiO_2), and aluminum oxide (Al_2O_3). It is known that the introduction into this layer modifies substantially the characteristics of the SPR resonance, most notably the spectral region where it appears [37]. Figure 4 shows the location and shape evolution of the resonance for different thicknesses of the oxide layer. The SPR sensing head is assumed to be surrounded by a non-dispersive external medium with the refractive index of n_s = 1.33.

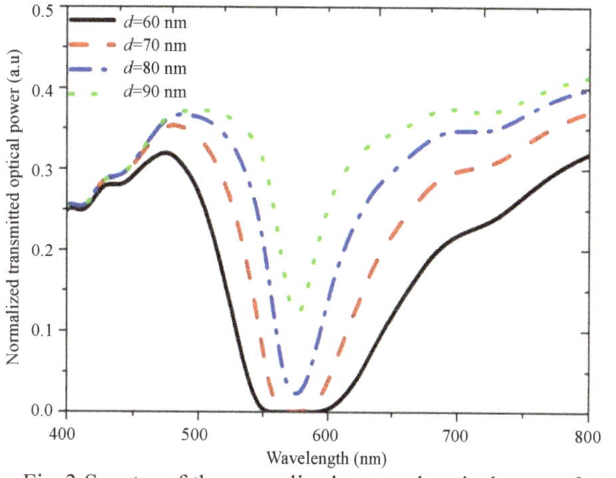

Fig. 3 Spectra of the normalized returned optical power for a fiber optic SPR structure with a single silver layer for several thicknesses (refractive index of the external medium: 1.33).

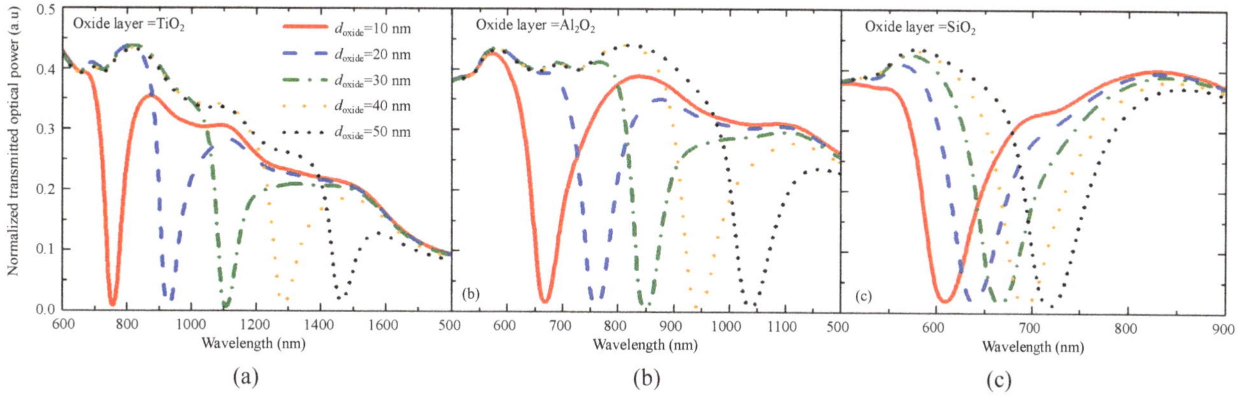

Fig. 4 Spectral behavior of the double-layer fiber optic SPR structure for different thicknesses of the oxide layer (TiO_2, SiO_2, and Al_2O_3 oxide materials are associated with (a), (b), and (c), respectively; refractive index of the external medium: 1.33).

Regarding Fig. 4, we observe that inclusion of a TiO_2 layer permits the largest displacement of the spectral response of the SPR resonance, making it possible to locate it within the 830 nm, 1300 nm, and 1550 nm telecommunication bands by adjusting the layer thickness within an interval which is technically feasible. Also, there are consequences in the width of the resonance, which consistently becomes larger when it moves to longer wavelengths, as shown in Fig. 5 which displays the spectral width of the resonance versus the thickness of the oxide layer.

To perform the study of phase interrogation of fiber optic based SPR sensing structures, a single oscillation angle of the light in the fiber, associated with a specific propagation mode, shall be

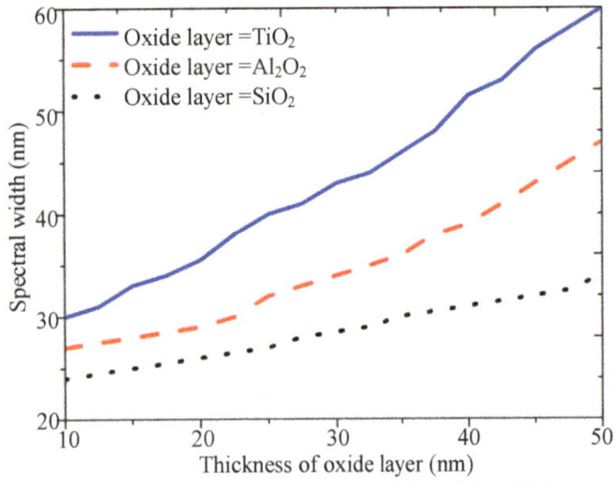

Fig. 5 Evolution of the spectral width of the SPR resonance (full width at the power level 6 dB above the minimum) as a function of the thickness of the oxide layer (the refractive index of the external medium is considered to be 1.33).

considered (oscillation angle defined by the fiber axis and the light ray). The higher the numerical aperture of the fiber is, the larger this angle can be, meaning a smaller angle of incidence in the interface core-cladding of the fiber. In the fiber length where SPR occurs, this condition favors the introduction into a larger phase difference $\Delta\phi_{p,s}$ between the TM and TE polarizations of the light in the reflection process, increasing the sensitivity to variations of the refractive index of the surrounding medium, n_s. As mentioned in the prior section, the phase shift introduced in each reflection is assumed constant, the average of phase shift at the beginning of the taper $(\delta\phi_{p,s})_{\alpha 0} = (\phi_p - \phi_s)_{\alpha 0}$, with $\alpha_0 = 80°$, and at the end of the taper $(\delta\phi_{p,s})_{\alpha(L)} = (\phi_p - \phi_s)_{\alpha(L)}$, with $\alpha_L = 73°$.

One of the topics investigated in this work is the tuning of the characteristics of the thin film layers to obtain sensor operation in the desired wavelength window. Figure 6 shows the normalized transmittance and phase difference $\Delta\phi_{p,s}$ versus n_s for layouts with TiO_2 and Al_2O_3 layers with the thicknesses of 15 nm and 30 nm, respectively, which enables to operate the sensor in the first telecommunications window (the data are shown for λ = 830 nm; from the data given in Fig. 4(c), this wavelength is not accessible by considering a SiO_2 layer). The analysis of the results shows the maximum of the refractive index phase sensitivity, $S_{n,\phi}=\delta(\Delta\phi_{p,s})/\delta n_s$, occurs close to n_s = 1.330, with a value of 5.72×10^4 degrees/RIU for the case of the titanium dioxide layer, while for the situation of aluminum oxide layer a slightly smaller value is obtained (5.57×10^4 degrees/RIU for n_s = 1.327). In both cases, the maximum phase sensitivity is close to the SPR resonance, where the transmitted optical power is rather low. This means almost no light to be detected, therefore decreasing substantially the signal-to-noise ratio. As such, a compromise is needed so that the phase sensitivity is degraded to a certain extent while the signal-noise-ratio in the detection process increases, therefore reaching a situation of the maximum resolution in the measurement of n_s.

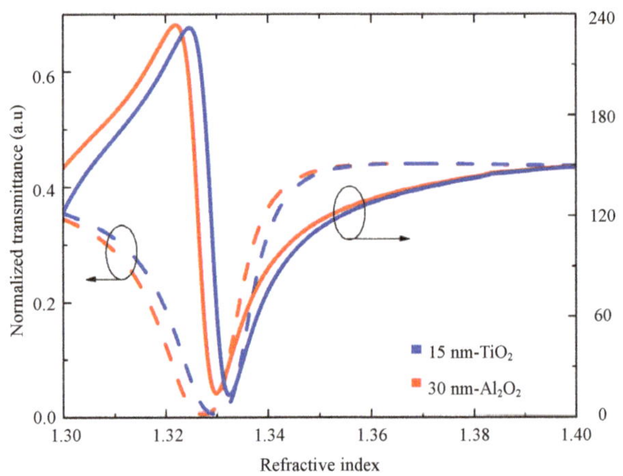

Fig. 6 Normalized transmittance and phase difference (between p and s polarizations) as a function of the refractive index of the surrounding medium for an inner silver layer of 80 nm combined with a TiO_2 layer of 15 nm (solid curves) and a Al_2O_3 layer of 30 nm (dashed curves) (these thicknesses are considered to allow the operation wavelength around 830 nm).

For the sensing head to operate in the second (1300 nm) and third (1550 nm) telecommunications spectral windows, the observation of Fig. 4 indicates the need to consider an external layer of TiO_2 with thicknesses of 40 nm and 55 nm, respectively. The normalized transmittance and $\Delta\phi_{p,s}$ versus n_s are shown in Fig. 7. Regarding these results, it turns out the maximum phase sensitivity has the value of 4.11×10^4 degrees/RIU (at n_s = 1.335) and 3.1×10^4 degrees/RIU (at n_s = 1.332) when the interrogation wavelengths are 1300 nm and 1550 nm, respectively. Due to a wide availability of the high performance optical fiber technology at a relatively low cost, considering SPR sensors supported by optical fiber platforms is advantageous to operate within these telecommunications windows, particularly the one in the 1550-nm region, allowing the design and implementation of sensing structures with rather favorable characteristics.

As mentioned in the above section, since the maximum phase sensitivity occurs in a region close to the dip of the SPR resonance, the available optical power level is rather low, which will degrade the signal-noise ratio to a point that such high sensitivity does not have impact on the final phase reading resolution. Therefore, it shall be considered to operate in a wavelength where there is enough transmitted optical power, which in principle means it can be located in either side of the SPR transmittance curve. To investigate this aspect, the achievable phase sensitivity for different levels of transmittance in both sides of the SPR resonance is analyzed, for the case being selected, which is associated with Fig. 7, i.e., relative to an operating wavelength of 1300 nm. The results obtained are shown in Fig. 8.

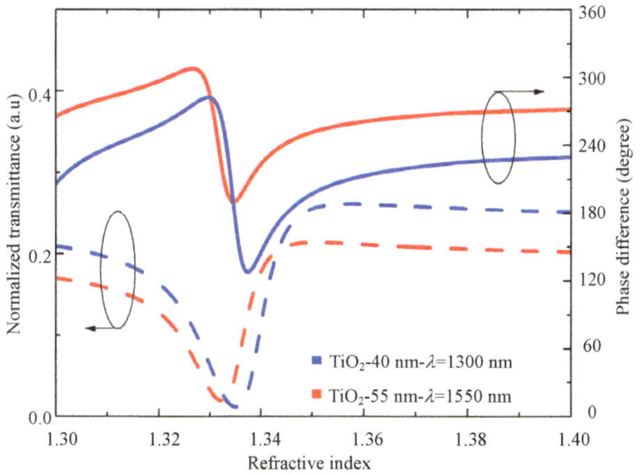

Fig. 7 Normalized transmittance and phase difference (between p and s polarizations) as a function of the refractive index of the surrounding medium for an inner silver layer of 80 nm combined with a TiO_2 layer of 40 nm (solid curves) and a TiO_2 layer of 55 nm (dashed curves), associated with interrogation wavelengths of 1300 nm and 1550 nm, respectively.

With the wavelength fixed, different transmittance levels mean different refractive indices of the surrounding medium. As expected from the observation of Fig. 4(a), due to the asymmetry of the SPR resonance, its left side provides limited freedom to the choice of the operating pair (transmittance and sensitivity), indicating the preference to work out in the right side of the resonance.

Optical sensors based on the SPR phenomenon are inherently highly sensitive to variations in the refractive index of the external medium, permitting to detect very small changes in this measurand, for example variations induced by biochemical processes. Therefore, it is in the context of biochemical sensing that these devices are rather appellative, pointing out the relevance of optimizing their performance in the situation where the external medium is water. Consequently, for each type of the oxide layer, the appropriate thickness is determined in order to attain the maximum sensitivity at the refractive index of water evaluated at the sensor operating wavelength.

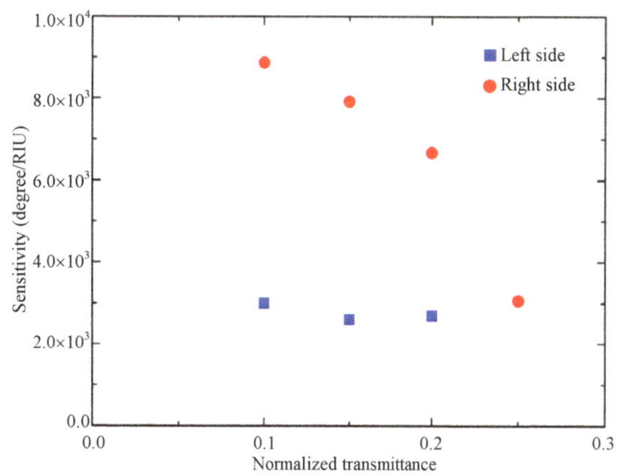

Fig. 8 Phase sensitivity versus normalized transmittance for a fiber optic SPR structure with an inner silver layer of 80 nm combined with a TiO_2 layer of 40 nm (interrogation wavelength of 1300 nm).

Figure 9 shows the maximum phase sensitivity obtained when the operation wavelengths are 630 nm [Fig. 9(a)] and 830 nm [Fig. 9(b)]. The most immediate outcome of the analysis of this data is that for the case of operation at 630 nm it is possible to find the required thickness for the oxide materials considered, TiO_2, SiO_2, and Al_2O_3. However, when moving to 830 nm, the layer of SiO_2 leaves to be an option, as the results shown in Fig. 4(c) already indicated. The same happens when operating at the wavelengths of 1300 nm [Fig. 10(a)] and 1550 nm [Fig. 10(b)], as observed in Fig. 10. This figure also shows the titanium dioxide option for the external layer permits to address operation at both wavelengths, eventually being feasible to consider a layer of aluminum oxide for 1300-nm interrogation, associated with a width of about 86 nm, which is still technically feasible.

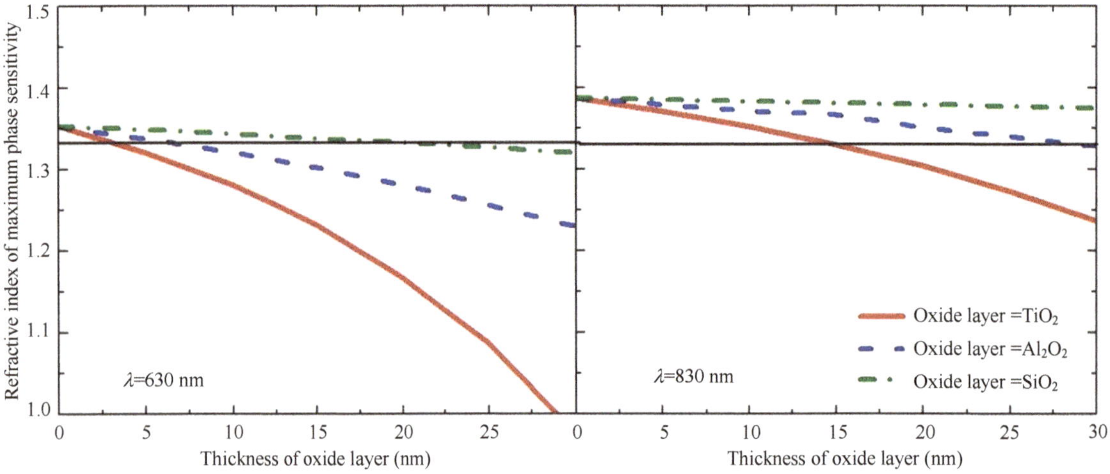

Fig. 9 Thickness of the oxide layer required to operate with the maximum sensitivity to changes in the refractive index of water, determined by the intersection of the horizontal line (that defines in the y-axis the water refractive index at the wavelength of operation) with the curves associated with different materials for the oxide layer [the interrogation wavelengths are 630 nm (a) and 830 nm (b)].

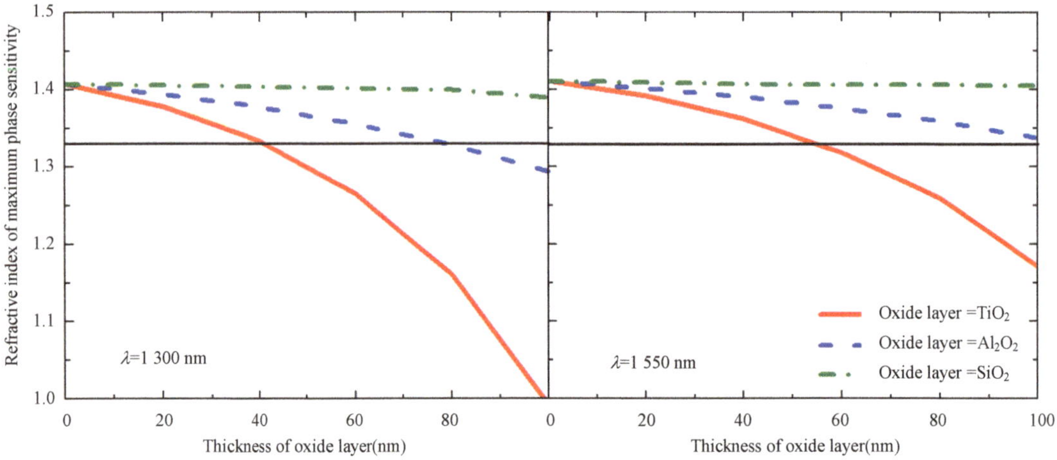

Fig. 10 Same as Fig. 9 but now with interrogation wavelengths of 1300 nm (a) and 1550 nm (b).

4. Final remarks

In this work, an optrode sensing structure based on a linearly tapered fiber tip, covered by a double layer thin film which induces plasmonic behavior at a specific wavelength range, is researched. The characteristics of the plasmonic resonance depend on the refractive index of the surrounding medium, therefore information is obtained on this parameter performing phase interrogation at a specific wavelength. The inner component of the double layer is metallic (silver), while the external one is an oxide dielectric layer. Two main characteristics of this sensing layout are on target: one is the possibility to tune the sensor sensitivity to variations of the refractive index of the surrounding medium, and the other is the amenability to tune the measurement operation to a specific wavelength region, most notably the one around 1550 nm to benefit from the large portfolio of high performance fiber optic components available at these wavelengths. It is found the inclusion of an internal silver layer with the 80-nm-thickness permits the better performance for any of the three different oxide materials considered for the external layer. Also, the results obtained shows the better flexibility is provided by the inclusion of a layer of titanium dioxide, and the approach is compatible with tuning the sensor operation to a large wavelength window that includes operationally important wavelengths (630 nm, 830 nm, 1300 nm, and 1550 nm).

The experimental validation of this study can be done by the application of techniques developed along with the years for the interrogation of polarimetric fiber sensing interferometers illuminated by monochromatic light [38]. Following these approaches, laser light is injected into the multimode fiber at an angle such that the selected core mode is excited. At the input of the sensing head, a fiber polarizer element shall be included in order to have light in the TE and TM polarizations, recombined at the output of the measurement region by means of a second polarizer. This will generate an interferometric signal with a phase proportional to the optical path difference accumulated by the propagation of the TE and TM polarizations, therefore including the phase variations of the light in the TM polarization induced by changes in the surrounding refractive index.

To recover the phase of this signal, a robust and sensitive technique involves the addition to the interferometric phase of a sinewave component, which can be generated by sinewave modulation of the wavelength of the laser light. After photodetection and incorporation of signal processing, which includes a feedback loop to the laser source to tune the direct current (DC) value of the light wavelength, the information on the quasi-static phase of the polarimetric sensing interferometer, therefore on the refractive index of the external medium, can be obtained from the feedback signal [39]. This approach is compatible with a DC phase resolution of the order of 1 mrad, which means from the results presented here refractive index measurement resolutions of the order 10^{-7} shall be possible.

The analysis presented in this work permits to guide the design of fiber optrodes highly sensitive to the surrounding medium refractive index by plasmonic interaction at a certain wavelength window and considering phase reading. These optrodes incorporate an internal metallic layer (Ag) covered with an oxide layer. Different oxide materials (TiO_2, SiO_2, and Al_2O_3) are addressed, and the main characteristics of these structures for refractive index sensing are determined. Paths for the experimental investigation of these sensing devices are also briefly introduced.

Acknowledgment

This research was performed in the framework of Project PTDC/FIS/119027/2010 (Plasmonics Based Fiber Optic Sensing with Enhanced Performance) funded by National Funds through FCT – Fundação para a Ciência e a Tecnologia. L. Coelho acknowledges the support from FCT grant SFRH/BD/78149/2011.

References

[1] J. L. Santos and F. Farahi, *Handbook of optical sensors*. BocaRaton: CRC Press, 2014.

[2] J. Janata, *Principles of chemical sensing*. Berlin: Springer-Verlag, 2009.

[3] F. S. Ligler and C. R. Taitt (Editors), *Optical biosensors: today and tomorrow*. Hungary: Elsevier, 2011.

[4] J. G. Webster and H. Eren (Editors), *Measurement, instrumentation and sensors handbook*. BocaRaton: CRC Press, 2014.

[5] R. W. Wood and P. Mag, "On a remarkable case of

uneven distribution of light in a diffraction grating spectrum," *Proceedings of the Physical Society of London*, 1902, 4: 396–402.
[6] L. Rayleigh, "On the dynamical theory of gratings," *Proceedings of the Royal Society of London A Mathematical Physical & Engineering Sciences*, 2010, 55(532): 399–416.
[7] U. Fano, "The theory of anomalous diffraction gratings and of quasi-stationary waves on metallic surfaces (sommerfeld's waves)," *Journal of the Optical Society of America*, 1941, 31(3): 213–222, 1941.
[8] A. Otto, "Excitation of nonradiative surface plasma waves in silver by method of frustrated total reflection," *Zeitschrift fur Physik A Hadrons & Nuclei*, 1968, 216(4): 398–410.
[9] E. Kretschmann and H. Reather, "Notizen: radiative decay of non radiative surface plasmons excited by light," *Zeitschrift fur Naturforschung A*, 2014, 23(12), 2135–2136.
[10] H. A. Atwater, "The promise of plasmonics," *Scientific American*, 2007, 296(4): 56–62.
[11] R. C. Jorgenson and S. S. Yee, "A fiber-optic chemical sensor based on surface plasmon resonance," *Sensors and Actuators B: Chemical*, 1993, 12(3): 213–220.
[12] S. P. Resurrection, "Surface plasmon resurrection," *Nature Photonics*, 2012, 6(11): 707–794.
[13] H. Raether, *Surface plasmons on smooth and rough surfaces and on gratings*. Berlin: Springer-Verlag, 1988.
[14] J. Homola, S. S. Yee, and G. Gauglitz, "Surface plasmon resonance sensors: review," *Sensors and Actuators B: Chemical*, 1999, 54(1–2): 3–15.
[15] J. Homola, "Present and future of surface plasmon resonance biosensors," *Biosystems*, 2003, 377(3): 528–539.
[16] H. Suzuki, M. Sugimoto, Y. Matsui, and J. Kondoh, "Effects of gold film thickness on spectrum profile and sensitivity of a multimode-optical-fiber SPR sensor," *Sensors and Actuators B: Chemical*, 2008, 132(1): 26–33.
[17] S. K. Ozdemir and G. Turhan-Sayan, "Temperature effects on surface plasmon resonance: design considerations for an optical temperature sensor," *Journal of Lightwave Technology*, 2003, 21(3): 805–814.
[18] H. R. Gwon and S. H. Lee, "Spectral and angular responses of surface plasmon resonance based on the Kretschmann prism configuration," *Materials Transaction*, 2010, 51(5): 1150–1155.
[19] Y. C. Li, Y. F. Chang, L. C. Su, and C. Chou, "Differential-phase surface plasmon resonance biosensor," *Analytical Chemistry*, 2008, 80(14): 5590–5595.
[20] H. P. Chiang, J. L. Lin, and Z. W. Chen, "High sensitivity surface plasmon resonance sensor based on phase interrogation at optimal wavelengths," *Applied Physics Letters*, 2006, 88(14): 141105-1–141105-3.
[21] S. P. Ng, C. M. L. Wu, S. Y. Wu, and H. P. Ho, "White-light spectral interferometry for surface plasmon resonance sensing applications," *Optics Express*, 2011, 19(5): 4521–4527.
[22] A. K. Sharma, R. Jha, and B. D. Gupta, "Fiber-optic sensors based on surface plasmon resonance: a comprehensive review," *IEEE Sensors Journal*, 2007, 7(8): 1118–1129.
[23] A. K. Sharma and B. D. Gupta, "On the performance of different bimetallic combinations in surface plasmon resonance based fiber optic sensors," *Journal Applied Physics*, 2007, 101(9): 093111-1–093111-6.
[24] R. Slavik, J. Homola, and J. Ctyroky, "Miniaturization of fiber optic surface plasmon resonance sensor," *Sensors and Actuators B: Chemical*, 2008, 51(1–3): 311–315.
[25] B. D. Gupta and C. D. Singh, "Evanescent-absorption coefficient for diffuse source illumination: uniform- and tapered-fiber sensors," *Applied Optics*, 1994, 33(13): 2737–2742.
[26] B. Grunewald and G. Holst, "Fiber optic refractive index microsensor based on white-light SPR excitation," *Sensors and Actuators A: Physical*, 2004, 113(2): 174–180.
[27] Y. C. Kim, W. Peng, S. Banerji, and K. S. Booksh, "Tapered fiber optic surface plasmon resonance sensor for analyses of vapor and liquid phases," *Optics Letter*, 2005, 30(17): 2218–2220.
[28] M. Niggemann, A. Katerkamp, M. Pellmann, P. Bolsmann, J. Reinbold, and K. Cam-mann, "Remote sensing of tetrachloroethene with a micro-fibre optical gassensor based on surface plasmon resonance spectroscopy," *Sensors and Actuators B: Chemical*, 1996, 34(1–3): 328–333.
[29] L. A. Obando and K. S. Booksh, "Tuning dynamic range and sensitivity of white-light, multimode, fiber-optic surface plasmon resonance sensors," *Analytical Chemistry*, 1999, 71(22): 5116–5122.
[30] Y. Yinquan, H. Die, H. Li, and L. Min, "Theoretical investigations for surface plasmon resonance based optical fiber tip sensor," *Sensors and Actuators B: Chemical*, 2013, 188(11): 757–760.
[31] M. Born and E. Wolf, *Principles of optics, seventh (expanded) edition*. Cambridge: Cambridge University Press, 1999.
[32] P. Winsemius, F. F. Van Kampen, H. P. Lengkeek, and C. G. Van Went, "Temperature dependence of the optical properties of Au, Ag and Cu," *Surface and Coatings Technology*, 2001, 6(8): 1220–1224.
[33] J. R. DeVore, "Refractive index of rutile and

sphalerite," *Journal of the Optical Society of America*, 1951, 41(6): 416–419.

[34] E. D. Palik, *Handbook of optical constants of solids*. Boston: Academic Press, 1995.

[35] I. H. Malitson, "Interspecimen comparison of the refractive index of fused silica," *Journal of the Optical Society of America*, 1965, 55(10), 1205–1208.

[36] H. Moayyed, I. Leite Teixeira, L. Coelho, J. L. Santos, and D. Viegas, "Analysis of phase interrogated SPR fiber optic sensors with bimetallic layers," *IEEE Sensors Journal*, 2014, 14(10): 3662–3668.

[37] B. Priya and B. D. Gupta, "Surface-plasmon-resonance-based fiber-optic refractive index sensor: sensitivity enhancement," *Applied Optics*, 2011, 50(14): 2032–2036.

[38] B. Culshaw, "Interferometric optical fibre sensors," *IETE Journal of Research*, 1986, 32(4): 311–318.

[39] C. J. Misas, F. M. Moita. Araujo, L. A. Ferreira, J. L. Santos, and J. M. López-Higuera, "Interrogation of low-finesse Fabry-Pérot cavities based on modulation of the transfer function of a wavelength division multiplexer," *Journal Lightwave Technology*, 2001, 19(5): 673–681.

Multi-Wavelength Narrow Linewidth Fiber Laser Based on Distributed Feedback Fiber Lasers

Jingsheng LV[1], Haifeng QI[1*], Zhiqiang SONG[1], Jian GUO[1], Jiasheng NI[1], Chang WANG[1], and Gangding PENG[1,2]

[1]*Shandong Key Laboratory of Optical Fiber Sensing Technologies, Laser Institute of Shandong Academy of Sciences, Jinan, 250014, China*

[2]*School of Electrical Engineering & Telecommunications, The University of New South Wales, NSW, 2052, Australia*

[*]Corresponding author: Haifeng QI E-mail: qihf@sdlaser.cn

Abstract: A narrow linewidth laser configuration based on distributed feedback fiber lasers (DFB-FL) with eight wavelengths in the international telecommunication union (ITU) grid is presented and realized. In this laser configuration, eight phase-shifted gratings in series are bidirectionally pumped by two 980-nm laser diodes (LDs). The final laser output with over 10-mW power for each wavelength can be obtained, and the maximum power difference within eight wavelengths is 1.2 dB. The laser configuration with multiple wavelengths and uniform power outputs can be very useful in large scaled optical fiber hydrophone fields.

Keywords: Multi-wavelength fiber laser; distributed feedback fiber laser; narrow linewidth fiber laser

1. Introduction

Narrow linewidth fiber lasers or semiconductor lasers with fiber output connectors are widely used as laser sources in high sensitive optical fiber sensing fields, such as submarine hydroacoustic detecting and distributed acoustic sensing [1, 2]. Currently, the practical optical fiber hydrophone system is based on the optical fiber interferometer technology. A multi-wavelength narrow linewidth laser or a laser array composed of multiple single-wavelength narrow linewidth lasers, where the multiple wavelengths should be in the international telecommunication union (ITU) grid, is needed for the large scaled hydrophone array [3]. However, the laser array composed of multiple single-wavelength laser modules is always cumbersome and complicated in the configuration. Hence, a laser with the multi-wavelength output is attractive for its compact structure and usability. In recent years, the narrow linewidth fiber laser with the compact structure has been realized in distributed feedback fiber laser (DFB-FL) where a short phase shifted fiber grating in erbium doped fiber as the distributed feedback laser cavity assures the robust single longitudinal mode operation [4−7]. This distributed feedback fiber laser can be used as a sensing element in the high sensitive acoustic detecting field and has been extensively studied previously [8, 9]. Meanwhile, it can be used as a seed laser of the high performance laser source for its excellent single frequency narrow linewidth characteristic and simple structure [10].

In this paper, based on DFB-FLs, a narrow linewidth laser configuration with eight wavelengths in the C band is presented and realized. The eight wavelengths are in the ITU grid with a spectral separation of 1.6 nm (200 GHz). In the laser configuration, eight phase-shifted gratings of eight different wavelengths in the ITU grid are pumped bidirectionally by two 980-nm LDs in series. The final laser output with over 10-mW power for each wavelength can be obtained after the subsequent amplification by an erbium doped fiber amplifier (EDFA) structure. The maximum power difference within the eight wavelengths is 1.2 dB. The laser configuration with multiple wavelengths and uniform power outputs should be a practical laser source in the large scaled optical fiber hydrophone system.

2. Experiments

The schematic diagram of the eight-wavelength narrow linewidth fiber laser configuration based on DFB-FLs is shown in Fig. 1. The eight phase-shifted gratings on the erbium doped fiber have the same structure parameters except the central wavelengths. The eight wavelengths are 1553.292 nm, 1551.788 nm, 1550.010 nm, 1548.475 nm, 1546.883 nm, 1545.301 nm, 1543.658 nm, and 1542.095 nm, respectively, which are just in accord with C30, C32, C34, C36, C38, C40, C42, and C44 ITU channels. The grating length abbreviated as L_g is 40 mm, and a discrete π phase shift is located at $0.4*L_g$ to assure the almost unidirectional output with an over 20-dB backward-to-forward output

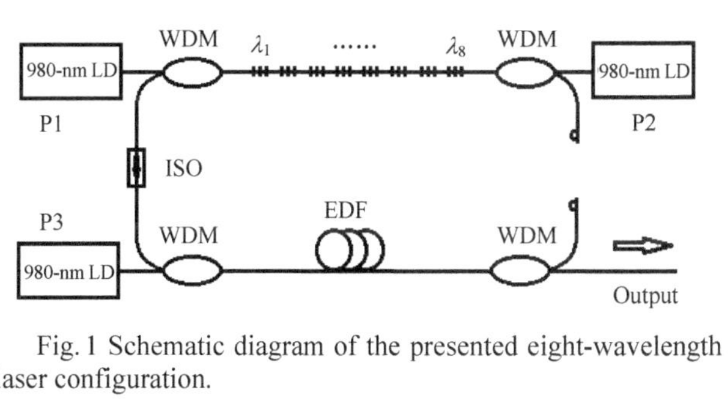

Fig. 1 Schematic diagram of the presented eight-wavelength laser configuration.

power ratio. The detailed fabrication procedure has been reported particularly in previous publications [6, 7]. For each phase-shifted grating, a 4-cm-long phase-shift grating is written on a 5-cm-long photosensitive erbium doped fiber (Nufern, the peak absorption at 1530 nm is 8 dB/m) which is spliced to the passive matching pigtail fiber (Nufern 980Hp) at both ends. The grating is written by scanning 244-nm frequency-doubled harmonic Argon ion continuous wave laser across the phase mask and fiber. A polarization dependent grating is generated by the vertically polarized ultraviolet laser. In the fabrication process, the phase shift is introduced by a simple relative movement of the phase mask to the fiber during the beam scanning. The movement of the phase mask for an accurate π phase-shift is a quarter of the period of the phase mask which is mounted on a piezoelectric transducer (PZT) actuator (PI, P-752.11c) with a nanometer resolution. Different from the shield method and post processing method to get a phase-shifted grating, by using this technique, the generated phase shift is only dependent on the relative displacement of the phase mask without relation to the ultraviolet laser induced index change in the grating. The eight phase-shifted gratings served as eight laser cavities are spliced in series in the same direction. This serial cavity array is pumped from both directions. After adjusting P1 and P2 properly, where P1 is 50 mW and P2 is 260 mw, the output from the isolator (ISO) has a relatively uniform power distribution where the maximum power difference within the eight wavelengths is 1.4 dB (Fig. 2, solid line). The total output power is 1.5 mW measured with EXFO FPM-600 power meter. To get a high and practical power, the laser output from the ISO is subsequently amplified by a forward-pumped EDFA structure. A 500-mW 980-nm LD and a 4.5-m-long erbium doped fiber (EDF) (Er-doped concentration: 4000 ppm) with an absorption of 7 dB/m at 980 nm are applied here. The final output power is 120 mW, and the maximum power difference within the eight wavelengths is 1.2 dB (Fig. 2, dashed line). The laser spectrum in the 50-nm range measured with an optical spectrum analyzer (YOKOGAWA AQ6370) is shown in Fig. 3.

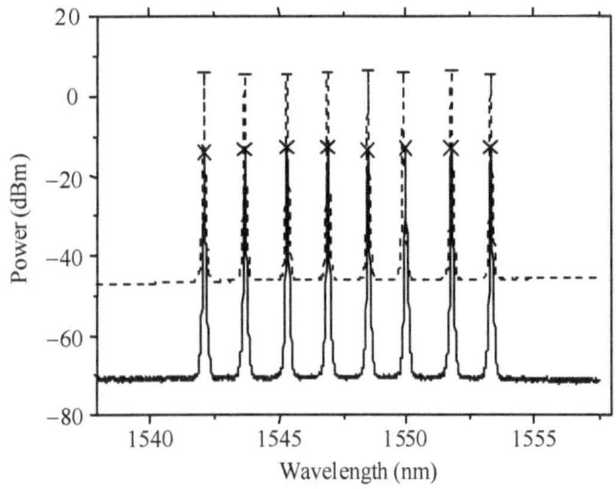

Fig. 2 Final eight-wavelength laser output before (solid line) and after (dashed line) amplification.

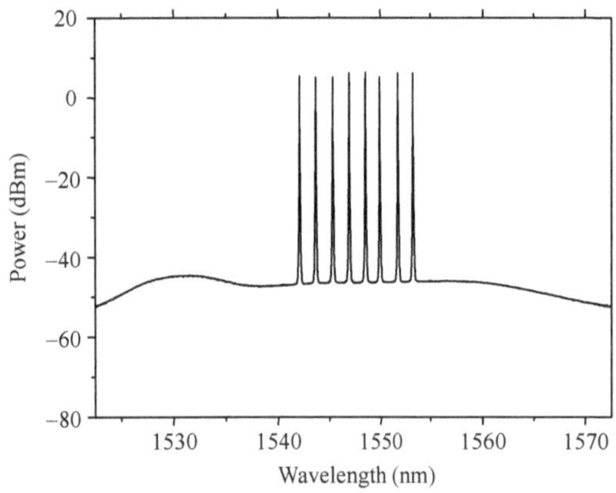

Fig. 3 Final eight-wavelength laser output measured in the 50-nm spectrum range.

For more detailed characteristic analysis, the laser in C34 channel is filtered out by a dense wavelength division multiplexing (DWDM) filter, and its output power is tested to be 14 mW. Its linewidth and relative intensity noise (RIN) are tested to be 1.5 kHz and −100 dB/Hz at the relaxation oscillation frequency (~ 300 kHz), respectively. The schematic diagram for RIN and linewidth test is shown in Fig. 4. For RIN test, the optical fiber interferometer in the soundproof box is removed, and the laser is directly incident to the photodetector (PD). The real algorithm is according to the definition of RIN in the corresponding optical testing handbook and referenced publications [9]. RIN is the ratio of the mean-square optical intensity noise to square of the average optical power:

$$\mathrm{RIN} = \frac{<\Delta P^2>}{P^2} \quad \mathrm{dB/Hz}. \tag{1}$$

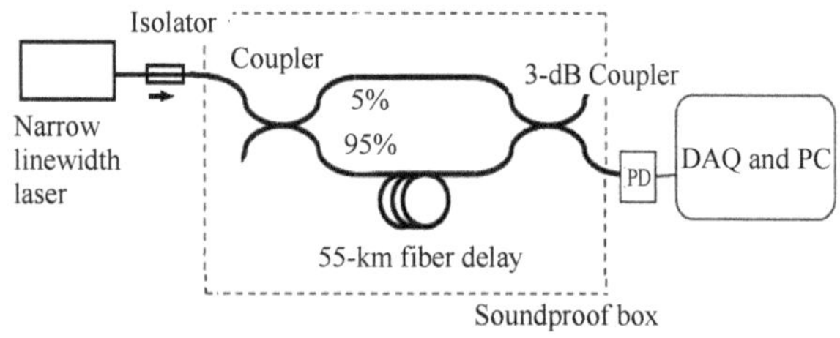

Fig. 4 Schematic diagram for RIN and linewidth test.

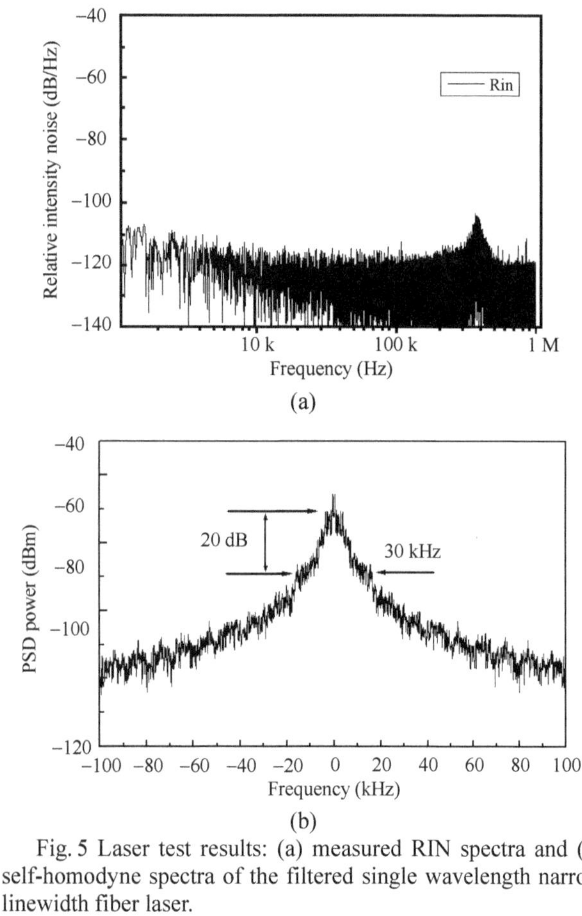

Fig. 5 Laser test results: (a) measured RIN spectra and (b) self-homodyne spectra of the filtered single wavelength narrow linewidth fiber laser.

The linewidth is measured by using the delayed self-homodyne method with a 55-km optical fiber delay line. The beat frequency signal from an arm of the Mach-Zehnder optical fiber interferometer is detected and then processed with a personal computer (PC) to perform the power spectral density analysis. Assuming the linewidth meets the Lorentz linetype, the 20-dB spectrum width is estimated to be 30 kHz, indicating the laser Lorentz linewidth is about 1.5 kHz. The tested results of RIN and linewidth are shown in Fig. 5.

3. Results and discussion

The main advantage of this multi-wavelength narrow linewidth fiber laser configuration is its very compact structure where there are only two pumps for eight cavities and the numbers of the corresponding optical components such as the wavelength division multiplexer (WDM) and ISO are also reduced. However, since the linewidth of the DFB-FL is very sensitive to the ambient noise, the precise temperature control for each cavity, the sound isolation packages for each cavity, and the whole laser configuration are necessary to make it stable and practical.

As for the wavelength number that this configuration can realize, it depends on the pump threshold of each laser cavity, the loss in the array, and the required power uniformity of the laser output. If the wavelength number increases, the power uniformity is more difficult to control. Usually, the pump threshold of the cavity is very low and much smaller than 5 mW, normally ~1 mW. Hence, the laser cavity threshold is not a key problem. The laser cavity, i.e. the phase-shifted grating, is written on a short EDF, and the EDF is spliced to a piece of matched passive fiber at both ends. In the normal laser array configuration, the laser cavities with matched passive fiber sections are spliced to each other in series. Lots of splicing points make the loss in the array large in total. Hence, the real pump power in different cavities may have great differences, and lasers with different wavelengths may have large difference in power duo to the accumulated loss through different laser cavities and splicing points. If the phase-shifted gratings of different wavelengths can be written in a single line EDF without being spliced to each other, the loss in the array can be reduced largely. In the present experiment, the two pumps, P1 and P1, are properly adjusted and optimized to get the uniform power distribution.

4. Conclusions

In this paper, we report a multi-wavelength narrow linewidth fiber laser configuration that has a compact structure and supplies a flat power distribution. It is based on the distributed feedback fiber laser, serial cavity array, and bidirectional pump design. Currently, the eight-wavelength laser with 120-mW total power and 1.2-dB power difference within the wavelengths is obtained. In the next work, more wavelengths should be realized through reducing the loss in the array and adjusting the structural parameters of the laser cavities to meet the requirement in the larger scaled optical fiber sensing. The protype based on this laser configuration is just in progress in the lab.

Acknowledgment

This work is supported by the International Science and Technology Cooperation Program of China (2012DFA10730) and Natural Science Foundation of China (61307101 and 61205083).

References

[1] G. A. Cranch, G. M. H. Flockhart, and C. K. Kirkendall, "Comparative analysis of the DFB fiber laser and fiber-optic interferometric strain sensors," *SPIE*, 2007, 6619: 66192C.

[2] C. Wang, C. Wang, Y. Shang, X. Liu, and G. Peng, "Distributed acoustic mapping based on interferometry of phase optical time-domain reflectometry," *Optics Communications*, 2015, 346: 172−177.

[3] G. A. Cranch, C. K. Kirkendall, K. Daley, S. Motley, A. Bautista, J. Salzano, *et al.*, "Large-scale remotely pumped and interrogated fiber-optic interferometric sensor array," *IEEE Photonics Technology Letter*, 2003, 15(11): 1579−1581.

[4] J. T. Kringlebotn, J. L. Archambault, L. Reekie, and D. N. Payne, "Er3+:Yb3+-codoped fiber distributed

feedback laser," *Optics Letter*, 1994, 19(24): 2102–2103.

[5] Q. Li, F. Yan, W. Peng, T. Feng, S. Feng, S. Tan, *et al*., "DFB laser based on single mode large effective area heavy concentration EDF," *Optics Express*, 2012, 20(21): 23684–23689.

[6] H. Qi, Z. Song, S. Li, C. Wang, and G. Peng, "Short distributed feedback fiber laser with unidirectional output for sensing applications," *Chinese Optics Letter*, 2013, 11(4): 44–46.

[7] H. Qi, Z. Song, J. Guo, C. Wang, and G. Peng, "Apodized distributed feedback fiber laser with asymmetrical outputs for multiplexed sensing applications," *Optics Express*, 2013, 21(9), 11309–11314.

[8] S. Foster, A. Tikhomirov, and M. Englund, "A 16 channel fibre laser sensor array," in *18th Proc. ACOFT/AOS*, Melbourne, Australia, pp. 40–42, 2006.

[9] Y. Léguillon, K. H. Tow, P. Besnard, A. Mugnier, D. Pureur, and M. Doisy, "First demonstration of a 12 DFB fiber laser array on a 100 GHz ITU grid, for underwater acoustic sensing application," *SPIE*, 2012, 8439: 84390J.

[10] H. Qi, Z. Song, J. Guo, J. Ni, C. Wang, and G. Peng, "Narrow-linewidth distributed feedback fiber laser with MOPA," *Chinese Optics Letter*, 2015, 13(S2): S21404.

5

Characteristics of Heat-Annealed Silicon Homojunction Infrared Photodetector Fabricated by Plasma-Assisted Technique

Oday A. HAMMADI*

Department of Physics, College of Education, Al-Iraqia University, Baghdad, IRAQ

*Corresponding author: Oday A. HAMMADI Email: odayata2001@yahoo.com

Abstract: In this work, the effect of thermal annealing on the characteristics of silicon homojunction photodetector was studied. This homojunction photodetector was fabricated by means of plasma-induced etching of p-type silicon substrate and plasma sputtering of n-type silicon target in vacuum. The electrical and spectral characteristics of this photodetector were determined and optimized before and after the annealing process. The maximum surface reflectance of 1.89% and 1.81%, the maximum responsivity of 0.495 A/W and 0.55 A/W, the ideality factor of 1.80 and 1.99, the maximum external quantum efficiency of 76% and 83.5%, and the built-in potential of 0.79 V and 0.72 V were obtained before and after annealing, respectively.

Keywords: Photodetectors; homojunction; spectral responsivity; plasma-induced etching

1. Introduction

Homojunction is essentially a junction between the n- and p-type portions of the same material formed by different impurity doping. The junction is termed as abrupt or graded, depending upon whether the impurity concentration in the material changes abruptly or gradually at the junction region [1, 2]. When two layers of semiconductor materials of opposite carrier types are intimately joined, an exchange of charge carriers takes place, and the Fermi level becomes continuous in both layers [3]. Electrons from n-type portion adjacent to the junction flow into the p-side while holes from p-type to n side, and this flow is due to the density gradient. Then there will be some uncompensated stationary charges forming a dipole array, leading to the barrier formation. This equalizes the Fermi levels and prevents further flow of charges to either side [4, 5]. The energy band diagram of p-n homojunction is shown in Fig. 1.

Silicon photodiodes are responsive to high-energy particles and photons and operated by the absorption of photons or charged particles, generating a flow of current in an external circuit, which is proportional to the incident power. These devices can be used to detect the presence or absence of minute quantities of light with intensities in the range of 10^{-9} mW/cm^2 – 10^2 mW/cm^2 [6]. Silicon photodiodes are efficiently used in many applications, such as spectroscopy, photography, analytical instrumentation, optical position sensors, beam alignment, surface characterization, laser range finders, optical communications, and medical imaging instruments [7, 8].

The photonic and optoelectronic devices based on silicon diodes still have their featured positions in their applications due to several reasons, among which the natural availability, ease fabrication, and open technology are noticeably observed [9]. Despite the reasonably higher efficiencies achieved by similar devices (e.g., solar cells and photodetectors) fabricated from other semiconducting materials, the lower values of silicon devices can be compensated by their lower costs and higher reproducibility in correlation to the revolution in porous and nanocrystalline silicon structures [10, 11].

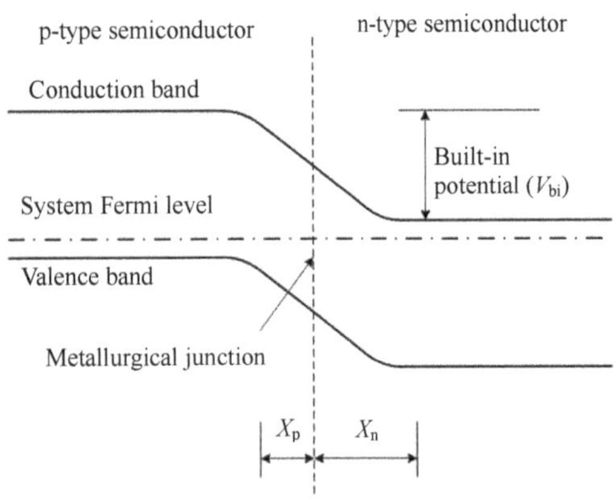

Fig. 1 Energy band diagram of p-n homojunction.

Compared with the avalanche photodiodes, the silicon junction photodetectors have no internal gain but they are very fast (~0.4 ns) and have very large bandwidths (up to 2.5 GHz) [12]. As well, silicon homojunctions are the fundamental structures of bipolar and metal-oxide-semiconductor (MOS) transistors, which represent the foundation of modern semiconductor electronics [13].

In the modern semiconducting materials technology, plasma processing techniques present very good candidates and alternatives for the conventional and expensive ones [14, 15]. Construction and operation of plasma systems for semiconductor processing applications can be performed and optimized to a reasonable degree. Such systems can be easily used for low cost and mass production of diverse structures and devices with very good quality [16, 17].

In this work, a silicon homojunction was fabricated by plasma-assisted technique and characterized for photodetection applications. Also, the effect of thermal annealing on the characteristics of this homojunction was studied.

2. Experiment

Figure 2 shows the experimental steps followed in this work to fabricate the silicon homojunction. A boron-doped p-type silicon wafer with the orientation of <100>, electrical resistivity of 40 cm – 50 cm, diameter of 5 cm, and thickness of 675 m was used for the p-type side of the proposed homojunction. This wafer was etched by argon discharge plasma for 20 min and then heat treated for one hour at 150 ℃ to get homogeneous rough layer of about 20-nm thickness. The sample was then placed inside the sputtering chamber to deposit an n-type silicon layer on its surface. This chamber was evacuated down to 0.001 mbar to prevent any active contaminants may reside inside from the interaction with the silicon samples.

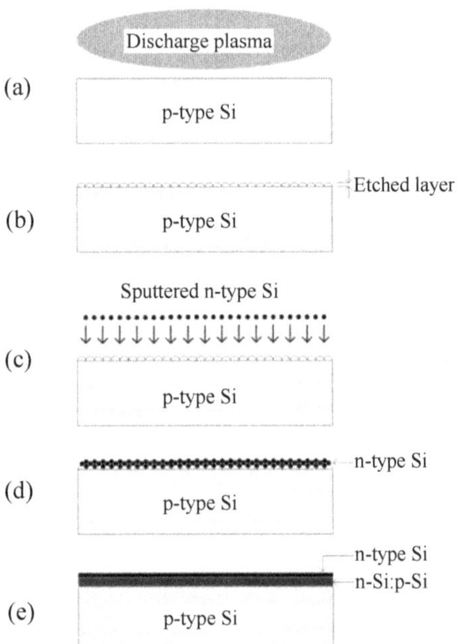

Fig. 2 Fabrication stages of the Si homojunction: (a) plasma-induced etching, (b) formation of etched Si layer, (c) sputtering of n-type Si target, (d) deposition of n-type Si layer, and (e) thermal annealing to form n-Si/p-Si homojunction.

The sputtering target was a phosphorous-doped n-type silicon wafer with the orientation of <111>, electrical resistivity of 25 cm – 30 cm, diameter of 5 cm, and thickness of 675 m. The sputtering target was placed on the cathode while the p-type silicon substrate was placed on the anode and the inter-electrode distance was 4 cm, which was found to be the optimum [18]. The sputtering plasma was generated by the electrical discharge of argon at the pressure of 0.08 mbar, discharge voltage of 3.5 kV, and discharge current of 45 mA. The sputtering time was 2 hours to form a 30-nm-thickness n-type silicon layer over the previously formed 20-nm p-type silicon layer. The final sample formed on the p-type substrate was about 30-nm-thickness n-Si/p-Si layer. In order to introduce the effect of thermal annealing on the characteristics of the fabricated structure, the final sample was gradually annealed for one hour at 20 ℃ – 320 ℃ with a step of 5 ℃/min to induce the formation of the homogeneous structure. Figure 3 shows the scanning electron microscopy (SEM) micrograph of the formed junction.

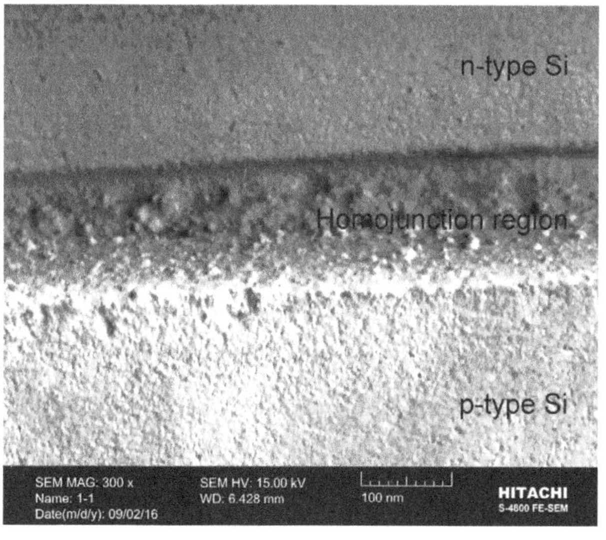

Fig. 3 SEM micrograph of the junction region formed in this work.

The electrical characterization of the fabricated homojunction includes the current-voltage and capacitance-voltage characteristics before and after the thermal annealing process and determination of some parameters, such as ideality factor (n) and built-in potential (V_{bi}). The electrical measurements were carried out using a dc power supply 0 V – 30 V, a Keithley 616 picoammeter, and a Keithley 82 C-V system.

The optical characterization of the fabricated homojunction includes the measurements of the reflectance from the surface, spectral responsivity, and external quantum efficiency before and after the thermal annealing process. The reflectance was measured by the specular hemispherical method, and the spectral responsivity was measured by using the pulsed xenon lamp, monochromator, bandpass filer, charge-coupled device (CCD) array detector, and data acquisition system. All optical measurements were carried out in the spectral range of 350 nm – 1100 nm.

3. Results and discussion

Figure 4 shows the current-voltage (I-V) characteristics of the prepared silicon homojunction before and after the thermal annealing process. It is clearly observed that these characteristics were slightly enhanced by thermal annealing as they were in accordance to the typical behavior of homojunction. As well, the ideality factor of the fabricated homojunction increased from 1.80 to 1.99 after annealing. This enhancement can be attributed to the reduction in the saturation current, which is in turn related to the reduction in the charge carrier recombination. Thermal annealing of the fabricated device presents better properties for charge carrier separation as a consequence of the application of a potential bias. Therefore, the adequate layer characteristics for charge carrier transfer lead to a reduction in charge carrier recombination making up for the lack of charge carrier separation when applying an electric potential bias.

Figure 5 shows the variation of inverse squared capacitance with the applied voltage (C^{-2}-V) for the fabricated homojunction before and after thermal annealing. The capacitance of the depletion layer of this homojunction increased due to thermal annealing. This increase in the capacitance could be attributed to a decrease in the built-in potential from 0.79 to 0.72 as the width of the depletion layer decreased because of the increasing recombination rate on both sides of this layer.

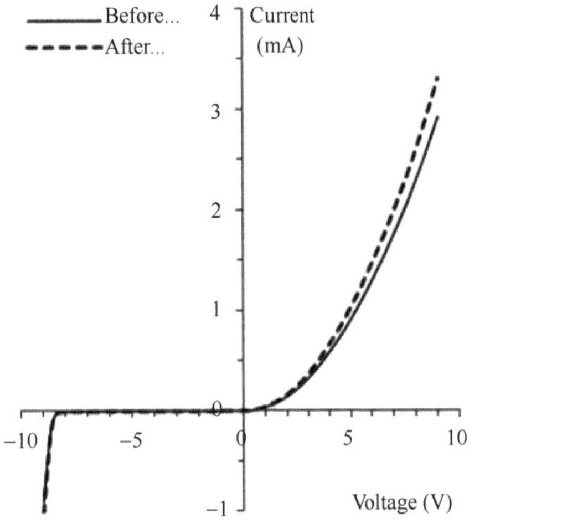

Fig. 4 *I-V* characteristics of the fabricated homojunction before and after thermal annealing.

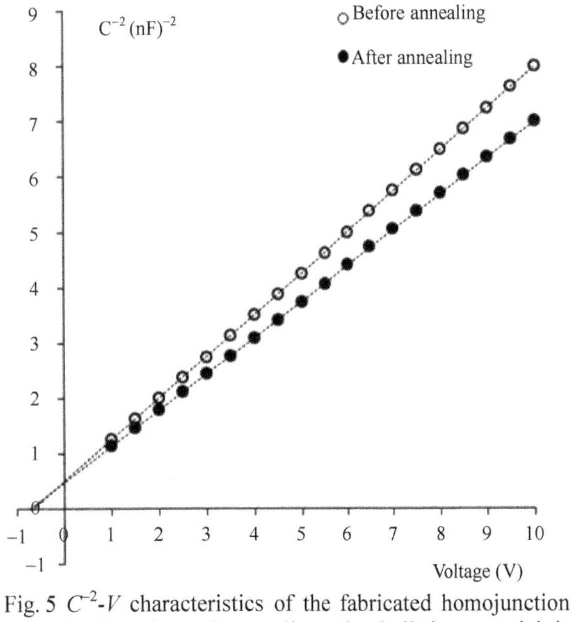

Fig. 5 C^{-2}-V characteristics of the fabricated homojunction before and after thermal annealing (the built-in potential is determined to be 0.72 V before annealing and 0.79 V after annealing).

Figure 6 shows the effect of thermal annealing on the surface reflectance of the fabricated homojunction as it was reduced from 1.89% to 1.81% in the spectral range of 350 nm – 1100 nm. Accordingly, the absorbance of this homojunction is supposed to increase by 0.08% due to this reduction in the surface reflectance, and this enhancement in the absorption may lead to the corresponding enhancement in the spectral response of the homojunction.

The spectral responsivity (R) of the fabricated homojunction was measured in the range of 350 nm – 1100 nm and presented in Fig. 7. The thermal annealing caused an enhancement in the spectral responsivity from 0.495 A/W to 0.55 A/W as the annealing process made the homogeneity of the sensitive layer (n-Si/p-Si) structure better to respond to the incident radiation.

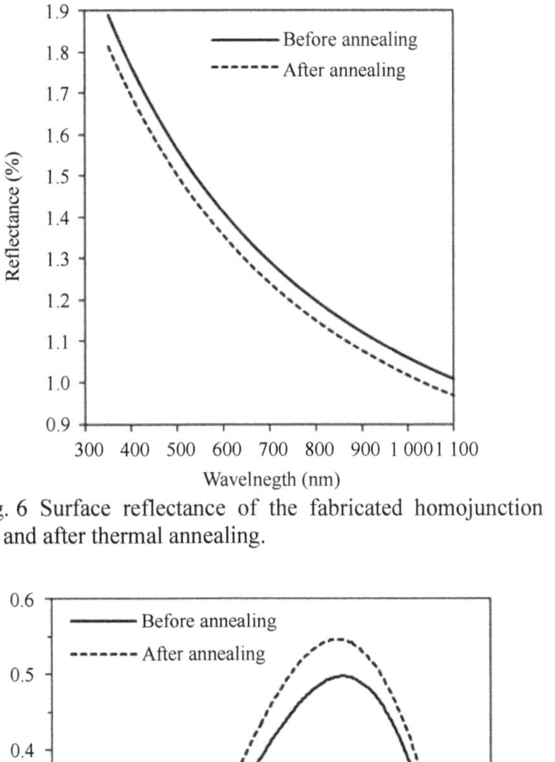

Fig. 6 Surface reflectance of the fabricated homojunction before and after thermal annealing.

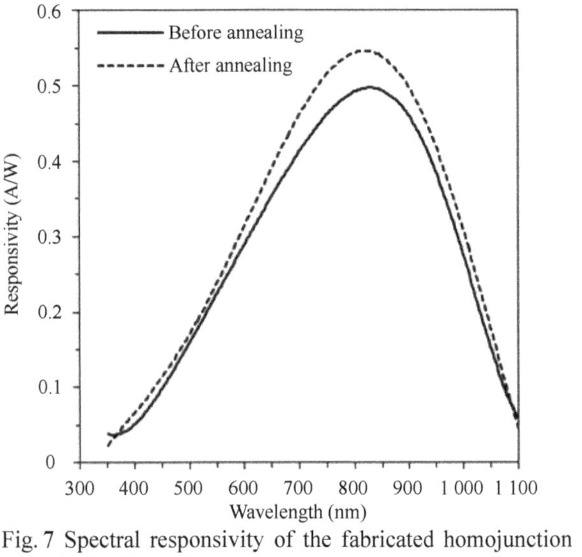

Fig. 7 Spectral responsivity of the fabricated homojunction before and after thermal annealing.

As the fabricated homojunction can be employed

in many applications of photonics and optoelectronics, the performance of this device should be assessed by introducing its efficiency. Therefore, the external quantum efficiency (EQE) of the fabricated homojunction was determined as a function of the wavelength of incident radiation before and after thermal annealing and was enhanced from 76% to 83.5%, respectively, as shown in Fig. 8.

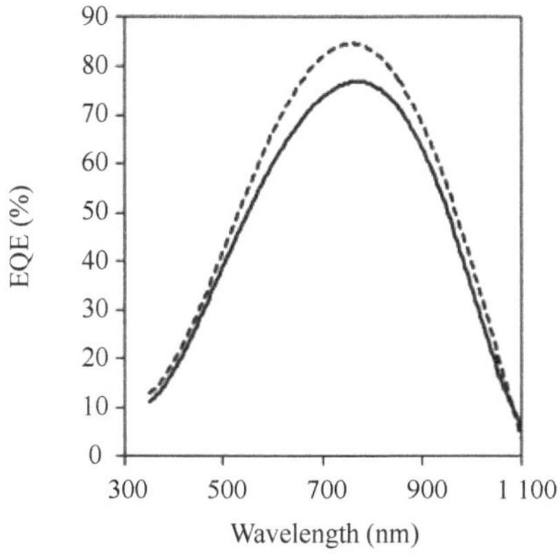

Fig. 8 Quantum efficiency of the fabricated homojucntion photodetector before and after thermal annealing.

4. Conclusions

According to the results obtained from this work, the thermal annealing has reasonable effects on the characteristics of silicon homojunction photodetector fabricated by means of the plasma-induced etching of the p-type silicon substrate and plasma sputtering of the n-type silicon target in vacuum. Due to thermal annealing, the maximum surface reflectance and built-in potential of this photodetector noticeably decreased while the maximum responsivity, the ideality factor, and the maximum external quantum efficiency increased. The fabrication procedure used in this work is reasonably new, low-cost, easily controlled, and reliable.

References

[1] R. A. Ismail, O. A. Abdulrazaq, A. A. Hadi, and O. A. Hamadi, "Full characterization at 904 nm of large area Si p-n junction photodetectors produced by laser-induced diffusion," *International Journal of Modern Physics*, 2007, 19(31): 197−201.

[2] M. Consales, M. Pisco, and A. Cusano, "Lab-on-fiber technology: a new avenue for optical nanosensors," *Photonic Sensors*, 2012, 2(4): 289−314.

[3] C. Chen, H. Wang, Z. Jiang, X. Jin, and J. Luo, "Design, fabrication, and measurement of two silicon-based ultraviolet and blue-extended photodiodes," *Photonic Sensors*, 2014, 4(4): 373−378.

[4] D. Viveiros, J. Ferreira, S. O. Silva, J. Ribeiro, D. Flores, J. L. Santos, *et al.*, "Ammonia sensing system based on wavelength modulation spectroscopy," *Photonic Sensors*, 2015, 5(2): 109−115.

[5] J. Dresner and F. V. Shallcross, "Crystallinity and electronic properties of evaporated CdS films," *Journal of Applied Physics*, 1963, 34(8): 2390−2395.

[6] O. A. Hamadi and K. Z. Yahiya, "Optical and electrical properties of selenium-antimony heterojunction formed on silicon substrate," *Journal of Pure and Applied Sciences*, 2007, 4(2): 1−11.

[7] O. A. Hamadi, B. A. M. Bader, and A. K. Yousif, "Electrical characteristics of silicon p-n junction solar cells produced by plasma-assisted matrix etching technique," *Engineering and Technics*, 2008, 26(8): 995−1001.

[8] R. A. Ismail, O. A. Abdulrazaq, A. A. Hadi, and O. A. Hamadi, "Full characterization at 904 nm of Si p-n junction photodetectors produced by LID technique," *International Journal of Modern Physics B*, 2007, 19(31): 197−201.

[9] A. K. Yousif and O. A. Hamadi, "Plasma-induced etching of silicon surfaces," *Bulgarian Journal Physics*, 2008, 35(3): 191−197.

[10] O. A. Hamadi, "Characteristics of CdO-Si heterostructure produced by plasma-induced bonding technique," *Proceedings of the Institution of Mechanical Engineers Part L Journal of Materials Design and Applications*, 2008, 222: 65−71.

[11] O. A. Hamadi, "Effect of annealing on the electrical characteristics of CdO-Si heterostructure produced by plasma-induced bonding technique," *Iraqi Journal of Applied Physics*, 2008, 4(3): 34−37.

[12] A. A. K. Hadi and O. A. Hamadi, "Optoelectronic characteristics of As-doped silicon photodetectors produced by LID technique," *Iraqi Journal of Applied Physics*, 2008, 1(2): 23−26.

[13] O. A. Hammadi, "Photovoltaic properties of thermally-grown selenium-doped silicon photodiodes for infrared detection applications," *Photonic Sensors*, 2015, 5(2): 152−158.

[14] O. A. Hammadi, M. K. Khalaf, and F. J. Kadhim, "Silicon nitride nanostructures prepared by reactive sputtering using closed-field unbalanced dual magnetrons," *Proceedings of the Institution of Mechanical Engineers Part L Journal of Materials

Design and Applications, 2015, 229(5).
[15] O. A. Hammadi, M. K. Khalaf, and F. J. Kadhim, "Fabrication of UV photodetector from nickel oxide nanoparticles deposited on silicon substrate by closed-field unbalanced dual magnetron sputtering techniques," *Optical and Quantum Electronics*, 2015, 47(2): 1−9.
[16] O. A. Hammadi, M. K. Khalaf, and F. J. Kadhim, "Fabrication and characterization of UV photodetectors based on silicon nitride nanostructures prepared by magnetron sputtering," *Proceedings of the Institution of Mechanical Engineers Part N Journal of Nanoengineering and Nanosystems*, 2015, 230(1): 32−36.
[17] O. A. Hammadi and N. E. Naji, "Electrical and spectral characterization of CdS/Si heterojunction prepared by plasma-induced bonding," *Optical and Quantum Electronics*, 2016, 48(8): 1−7.
[18] O. A. Hammadi, M. K. Khalaf, F. J. Kadhim, and B. T. Chiad, "Operation characteristics of a closed-field unbalanced dual-magnetrons plasma sputtering system," *Bulgarian Journal Physics*, 2014, 41(1): 24−33.

6

Optimal Design of 850 nm 2×2 Multimode Interference Polymer Waveguide Coupler by Imprint Technique

Yuchen SHAO[1], Xiuyou HAN[1*], Xiaonan HAN[1], Zhili LU[1], Zhenlin WU[1], Jie TENG[2], Jinyan WANG[3], Geert MORTHIER[4], and Mingshan ZHAO[1]

[1]*School of Physics and Optoelectronic Engineering, Dalian University of Technology, Dalian, 116024, China*

[2] *No. 38 Research Institute, China Electronics Technology Group Corporation, Hefei, 230088, China*

[3] *School of Chemical Engineering, Dalian University of Technology, Dalian, 116024, China*

[4]*Photonics Research Group, Department of Information Technology, Ghent University-IMEC, 9000 Ghent, Belgium*

[*]Corresponding author: Xiuyou HAN E-mail: xyhan@dlut.edu.cn

Abstract: A 2×2 optical waveguide coupler at 850 nm based on the multimode interference (MMI) structure with the polysilsesquioxanes liquid series (PSQ-Ls) polymer material and the imprint technique is presented. The influence of the structural parameters, such as the single mode condition, the waveguide spacing of input/output ports, and the width and length of the multimode waveguide, on the optical splitting performance including the excess loss and the uniformity is simulated by the beam propagation method. By inserting a taper section of isosceles trapezoid between the single mode and multimode waveguides, the optimized structural parameters for low excess loss and high uniformity are obtained with the excess loss of –0.040 dB and the uniformity of –0.007 dB. The effect of the structure deviations induced during the imprint process on the optical splitting performance at different residual layer thicknesses is also investigated. The analysis results provide useful instructions for the waveguide device fabrication.

Keywords: Polymer waveguide; coupler; multimode interference

1. Introduction

The optical waveguide coupler, as a basic unit, implements the function of splitting or combining lightwave in photonic integrated circuits. Compared with the conventional directional coupler, the coupler based on the multimode interference (MMI) structure has many advantages including compact structure, good fabrication tolerance, low optical loss, and polarization insensitivity [1–3]. The MMI based coupler can constitute the high speed optical switch [4, 5], large extinct ratio modulator [6], high Q waveguide microring filter [7], multi-wavelength laser array [8], and compact optical 90° hybrid for a balanced detection receiver [9].

Among the optical materials for integrated waveguide devices, the polymer has the features of low cost, flexible tuning of refractive index, and easy forming of film by spin-coating [10]. The polymer based integrated waveguide devices can be fabricated by the simple imprint technique [11], which avoids the complicated processes of growth and etching for fabrication of inorganic optical waveguide devices with silicon, silica or semiconductor materials. Also, polymer waveguide devices can be integrated with a light source and a photo detector on a common board, which promote the reduction in cost and minimization of the optical module and subsystem [12, 13]. The research on polymer based optical waveguide devices has been attracting more and more attention in the field of optical interconnection on board [14], high speed coherent receiver [15], and optical biosensor [16, 17]. Especially for the optical biosensor, the polymer materials have several unique advantages. Among these, polymer surfaces have good biocompatibility and can be easily modified to immobilize a wide range of biomolecules, which is beneficial to achieve the label-free detection [18]. Additionally, most analytes are aqueous with a large optical absorption loss at the lightwave band of 1550 nm, which greatly deteriorates the performance of silicon on insulator based biosensors [19]. The polymer waveguides can work at the lightwave bands of 650 nm or 850 nm [20–22] where the aqueous analytes have a very low optical absorption loss [23].

Many researches on the polymer waveguide MMI coupler have been carried out [23–28], which are generally around 1550 nm. In this paper, the MMI based 2×2 optical waveguide coupler at 850 nm with the polysilsesquioxanes liquid series (PSQ-Ls) polymer material [29] and the imprint technique [30] is optimally designed. The influence of the structural parameters such as the single mode conditions, the waveguide spacing of input/output waveguides, and the width and length of the multimode waveguide, on the device performance including the excess loss and the uniformity, is simulated and analyzed. The optimized structural parameters for low excess loss and high uniformity are obtained. The influence of the structural parameter deviations induced during the imprint process on the device performance under different residual layer thicknesses is also studied. The analysis results provide useful instructions for the following waveguide device fabrication.

2. Structure design

Figure 1 shows the 2×2 optical waveguide coupler based on the MMI structure, which has two input single mode waveguides: (1) the multimode waveguide and (2) output single mode waveguides. The single mode conditions of input/output waveguides, the waveguide spacing D, and the width W_{MMI} and length L_{MMI} of the multimode waveguide are the main structural parameters to be designed.

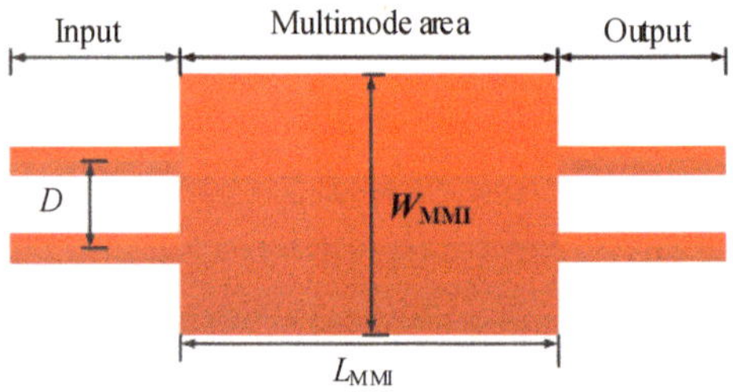

Fig. 1 2×2 optical waveguide coupler based on the MMI structure.

The excess loss and the uniformity are the two important performance indices of the coupler, with the unit of dB. The excess loss is defined as [31]

$$EL = 10\lg\left(\frac{\sum P_{out}}{\sum P_{in}}\right) \quad (1)$$

where $\sum P_{out}$ is the sum of output optical power, and $\sum P_{in}$ is the sum of input optical power.

The uniformity is defined as [31]

$$UF = 10\lg\left(\frac{P_{min}}{P_{max}}\right) \quad (2)$$

where P_{min} and P_{max} are the minimum and maximum optical powers of the two output ports, respectively.

The structure model of the 2×2 coupler based on MMI is built and simulated by the beam propagation method (BPM). The wavelength of the input optical field is set as 850 nm, and the input power is normalized as 1. The power of two output ports is monitored when the waveguide structural parameters, such as the waveguide spacing D, the width W_{MMI}, and length L_{MMI} of the multimode waveguide, are adjusted. Then the excess loss and uniformity are calculated with (1) and (2). The structural parameters are optimized by simulating the influence of the different structural parameters on the excess loss and the uniformity.

2.1 Waveguide cross section

Figure 2 shows the cross section of the input/output waveguide of the MMI based coupler. The material of the lower cladding is the PSQ-Ls with the low index of 1.458 named as PSQ-LL, and the core is the PSQ-Ls with the high index of 1.529 named as PSQ-LH [29]. The upper cladding is set as air with the index of 1. As well known, the height H and width W of the core should fulfill the single mode condition. In addition, the residual layer with the thickness of Δh will be generated when the core is fabricated by the imprint technique [30], and the influence of the residual layer on the performance of the coupler could not be ignored.

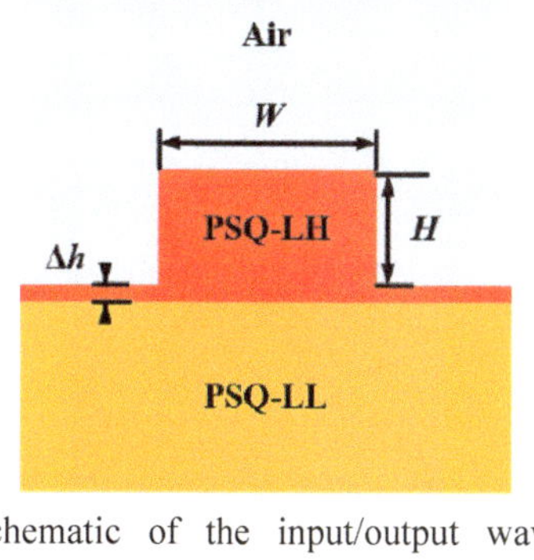

Fig. 2 Schematic of the input/output waveguide cross section.

Figure 3 shows the effective refractive index versus different widths W of the core with the height of the core H=1 μm and he thickness of the residual layer Δh=0.2 μm, respectively. It can be seen from Fig. 3 that the higher-order mode will appear when the width is equal to or larger than 2 μm. Therefore, the width of the core is chosen as 1.8 μm. The inset of Fig. 3 shows the calculated mode field with the core width of 1.8 μm. It can be seen clearly that only the fundamental mode transmits in the waveguide.

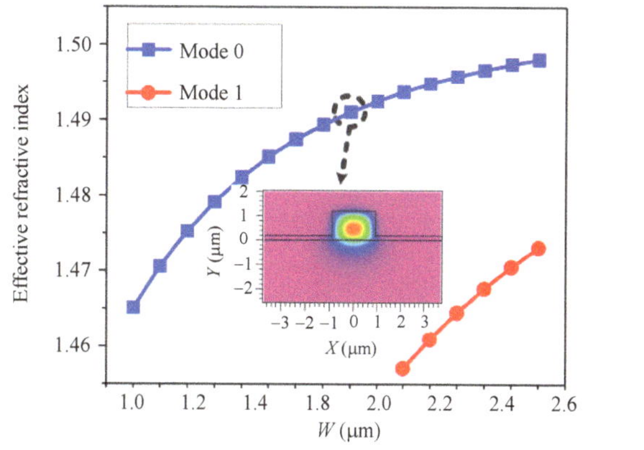

Fig. 3 Effective refractive index versus different waveguide core widths.

2.2 Waveguide spacing

The mode coupling occurs when the two waveguides are close to each other, which impacts the optical splitting property of the MMI based coupler. The model of two parallel straight waveguides with the length of 500 μm is set up as the inset in Fig. 4. The lightwave field with the normalized power of 1 is input to Waveguide 1, and the output power from Waveguide 2 is monitored when the waveguide spacing D changes.

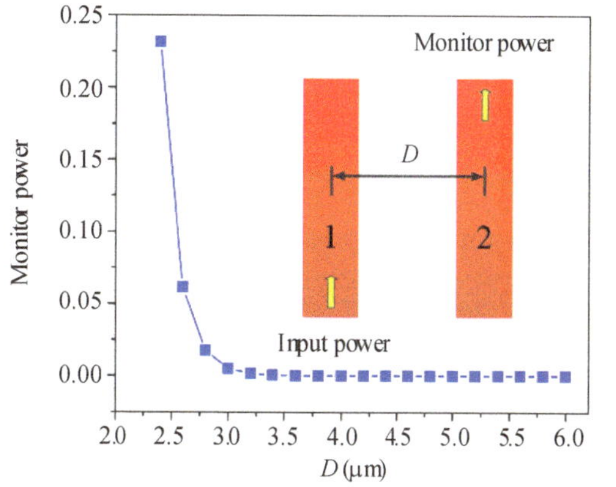

Fig. 4 Power output from Waveguide 2 versus different waveguide spacings D.

Figure 4 shows the output power from Waveguide 2 versus different waveguide spacings D. The lightwave power coupled from Waveguide 1 into Waveguide 2 gradually decreases with an increase in the waveguide spacing D, and the power approaches zero when the waveguide spacing increases to 3.4 μm. The power does not change as the waveguide spacing further increases. The waveguide spacing D of the input/output waveguides is chosen as 5 μm in consideration of no coupling between the two parallel waveguides and the feasibility of the waveguide fabrication with the imprint technique.

2.3 Multimode waveguide

For the multimode waveguide design, the width W_{MMI} and length L_{MMI} of the waveguide are the main structural parameters. The two input/output waveguides of the 2×2 MMI based coupler with 3-dB splitting ratio are located at $W_e/3$ and $2W_e/3$ of the multimode waveguide, respectively, according to the self-image principle [31]. The effective width of the waveguide is defined as W_e and is approximately equal to the practical width of the waveguide:

$$W_{MMI} \approx W_e = 3D. \quad (3)$$

The length of the multimode waveguide meets the following relationship:

$$L_{MMI} = \frac{1}{2} L_\pi = \frac{1}{2} \frac{4n_{core} W_e^2}{3\lambda_0} \approx \frac{2n_{core} W_{MMI}^2}{3\lambda_0} \quad (4)$$

where L_π is the beat length of the two lowest-order modes, n_{core} is the index of the waveguide core, λ_0 is the wavelength of the input lightwave. With the above chosen waveguide spacing D of 5 μm, the width of the multimode waveguide is about 15 μm, and the length of the multimode waveguide is about 270 μm according to (3) and (4), respectively.

Then the model of the 2×2 MMI based coupler is established according to above parameters. The excess loss and uniformity are simulated and calculated when the input/out waveguide spacing D is tuned with a step of 0.1 μm. Figure 5(a) shows the excess loss and uniformity with different spacings D. Obviously, the change in the waveguide spacing has the primary influence on the excess loss which approaches the minimum as D is 5.1 μm. Figure 5(b) shows the excess loss and uniformity with different the multimode waveguide lengths L_{MMI}. It can be seen that the excess loss approaches the minimum as L_{MMI} is 273 μm. Therefore, the width W_{MMI} and length L_{MMI} of the multimode waveguide are chosen as 5.1 μm and 273 μm, respectively.

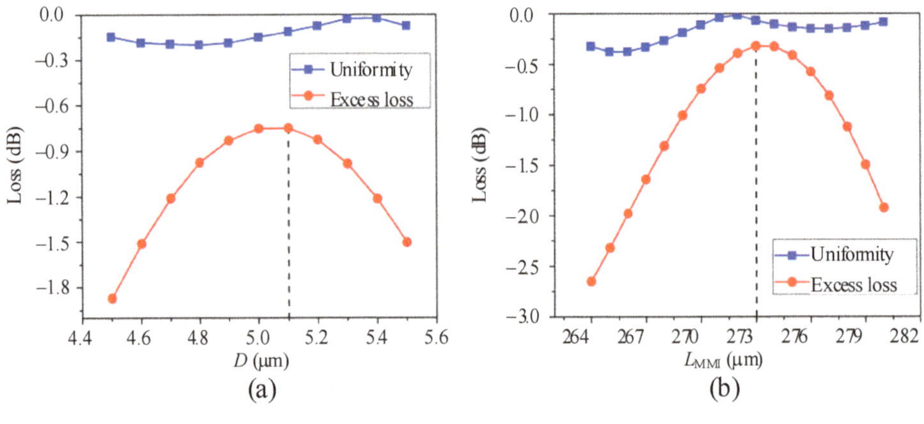

Fig. 5 Excess loss and uniformity: (a) different input/output waveguide spacings D and (b) different lengths of the multimode waveguide L_{MMI}.

The optical splitting property of the 2×2 MMI based coupler is simulated with above structural parameters, and the result is shown in Fig. 6. From Fig. 6, it can be observed that the normalized power of two output ports is 0.4712 and 0.4647, respectively. Then the excess loss is –0.325 dB, and the uniformity is –0.071 dB calculated by (1) and (2).

The structural parameters and performance indices of the 2 × 2 MMI based coupler are listed in Table 1.

Table 1 Structural parameters and performance indices of the 2×2 MMI optical waveguide coupler.

W	H	D	W_{MMI}	L_{MMI}	EL	UF
1.8μm	1μm	5.1μm	15μm	273μm	−0.325dB	−0.071dB

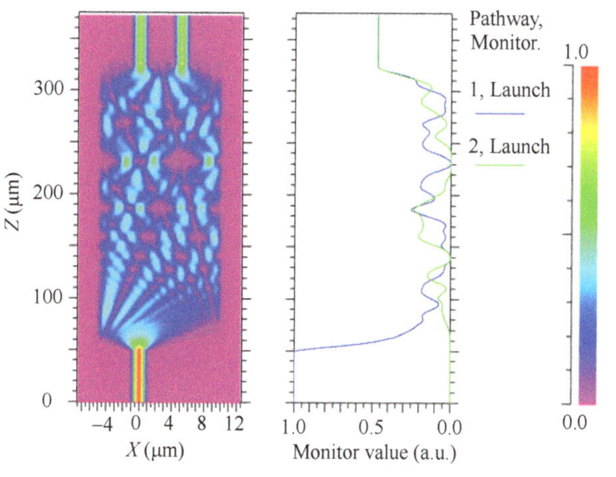

Fig. 6 BPM simulation result of the 2×2 MMI based coupler.

3. Structure optimization

From Table 1, it can be seen that the excess loss of the 2×2 MMI based coupler is relatively high. It is due to the mismatch loss between the single mode and multimode waveguides. A taper section of isosceles trapezoid with the length L_{taper} and the width W_{taper}, as shown in Fig. 7, is inserted between the single mode and multimode waveguides to suppress the mismatch loss [32, 33].

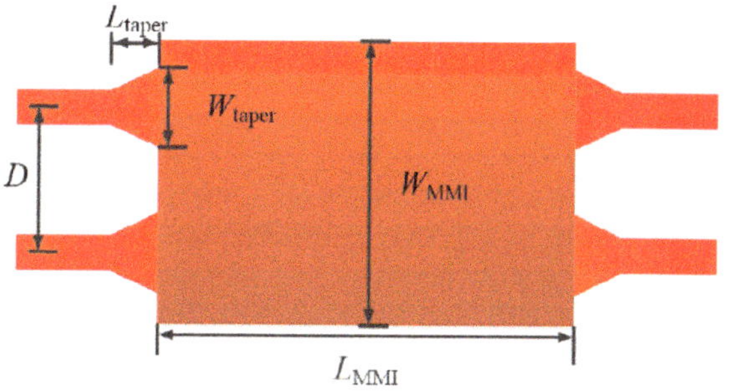

Fig. 7 2×2 MMI based coupler with the taper section.

The structural parameters of the taper section are optimized. At first, the excess loss with different widths W_{taper} of the taper section is simulated with the length L_{taper} of 100 μm, and the result is shown in Fig. 8(a). The excess loss decreases gradually with an increase in the width W_{taper}. It tends to be the minimum when W_{taper} is 3.1 μm, and it doesn't change basically when W_{taper} increases further. Then the width W_{taper} of the taper section is selected as 3.1 μm to analyze the excess loss with different lengths L_{taper} of the taper section, and the result is shown in Fig. 8(b). L_{taper} is selected as 100 μm to ensure the minimum excess loss with the possible short length of the taper section.

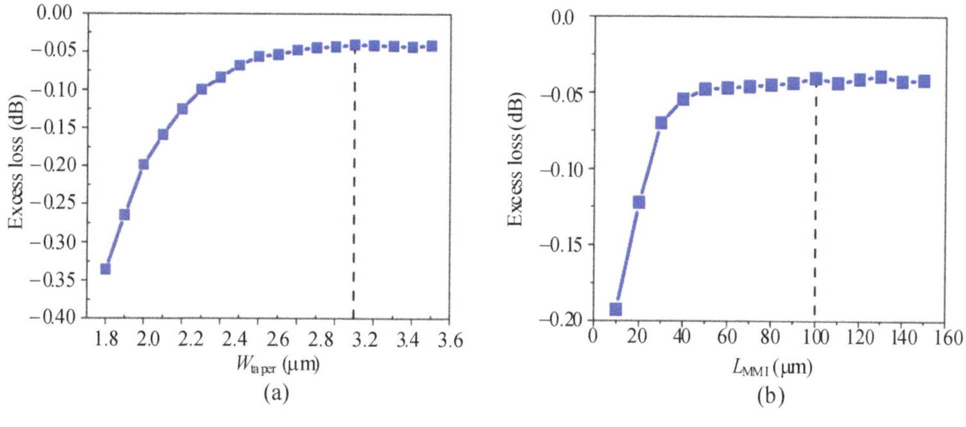

Fig. 8 Excess loss with different widths W_{taper} (a) and lengths L_{taper} (b) of the taper section.

The optical splitting performance of the 2×2 MMI based coupler is simulated with the optimized structural parameters, and the result is shown in Fig. 9. The normalized powers of two output ports are 0.4949 and 0.4958, respectively. Then the excess loss is –0.040 dB, and the uniformity is –0.007 dB calculated by (1) and (2). The structure and property parameters of the optimized coupler are listed in Table 2. Comparing the parameters between Tables 1 and 2, the excess loss is reduced obviously after introducing the taper section between single mode and multimode waveguides. At the same time, the uniformity is also improved significantly.

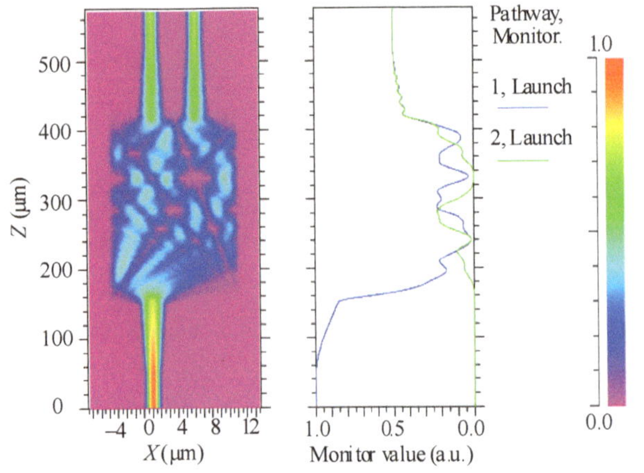

Fig. 9 Simulated result of the 2×2 MMI based coupler with the taper section.

Table 2 Structural parameters and performance indices of the 2×2 MMI based coupler with the taper section.

W	H	D	W_{MMI}	L_{MMI}	W_{taper}	L_{taper}	EL	UF
1.8μm	1μm	5.1μm	15μm	273μm	3.1μm	100μm	–0.040dB	–0.007dB

4. Discussion of fabrication tolerance

The optimized 2×2 MMI based coupler has the good characteristics of optical splitting, but the structural parameters such as the waveguide spacing D, length L_{MMI}, and width W_{MMI} of the multimode waveguide may deviate from the optimal values due to the horizontal offset of the mold during the fabrication process by the imprint technique. At the same time, the thickness of the residual layer may vary in a small range due to the tiny change in pressure, which also exerts negative effect on the optical splitting performance of the coupler. Therefore, it is necessary to analyze the structural parameter tolerance of the coupler to give instructions for the device fabrication and performance improvement.

At first, the influence of deviation of the multimode waveguide width W_{MMI} from the optimized value on the excess loss and the uniformity is analyzed. The input/output waveguide spacing D and the length L_{MMI} of the multimode waveguide are set as the optimized values of D = 5.1 μm and L_{MMI} = 237 μm. The thickness of the residual layer is set as 0.1 μm, 0.2 μm, and 0.3 μm, respectively. The power from the two output ports of the coupler are obtained by simulation with the BPM, with which the excess loss and uniformity are calculated by (1) and (2). The analysis results are shown in Fig. 10. It can be seen that the deviations of the multimode waveguide width W_{MMI} have a larger influence on the excess loss than on the uniformity. For the excess loss, as shown in Fig. 10(a), it increases gradually as W_{MMI} deviates from the optimized value. The changes in the residual layer thickness Δh also affect the excess loss a bit. However, it would be acceptable if the deviation of W_{MMI} is in a certain range. The maximum values of the excess loss and uniformity are set as –0.3 dB and –0.1 dB, respectively. The tolerable deviation of W_{MMI} is between –0.1 μm and 0.12 μm (0.67%) from the optimized value of 15 μm.

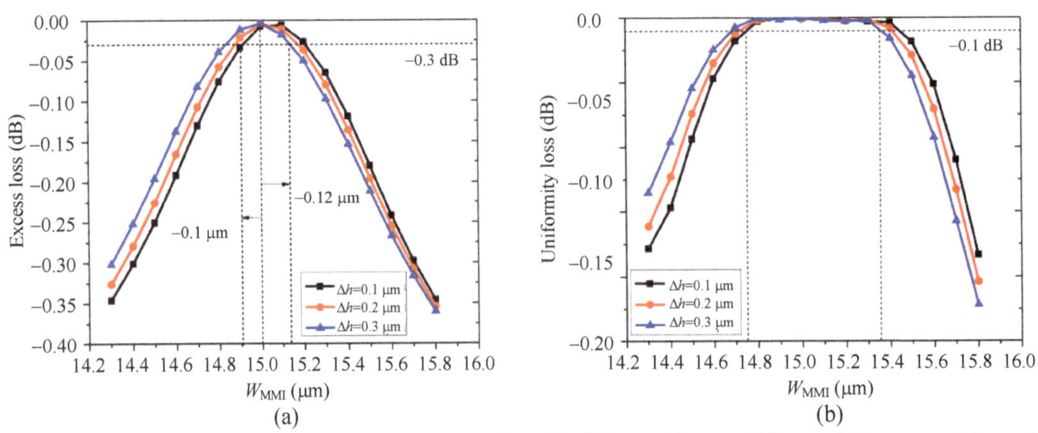

Fig. 10 Splitting performance: (a) excess loss and (b) uniformity with different W_{MMI} at different thicknesses of the residual layer.

Then the influence of deviations of the multimode waveguide length L_{MMI} and the waveguide spacing D of the input/output port from the optimized values on the excess loss and uniformity is analyzed by the similar process. The analysis results are shown in Figs. 11 and 12, respectively. For the maximum values of excess loss and uniformity of –0.3 dB and –0.1 dB, the tolerable deviation of L_{MMI} is between –5 μm and 3 μm (1.10%) from the optimized value of 273 μm, and the tolerable deviation of D is between –0.44 μm and 0.35 μm (6.86%) from the optimized value of 5.1 μm.

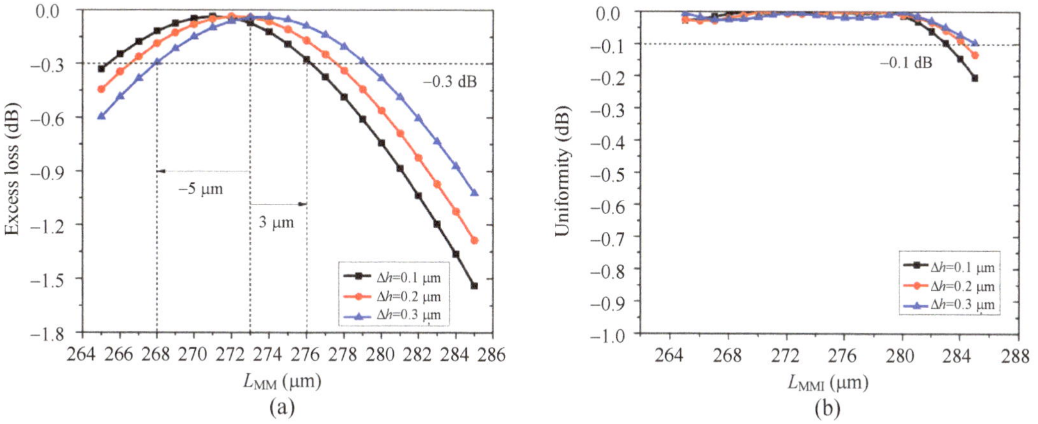

Fig. 11 Splitting performance: (a) excess loss and (b) uniformity with different L_{MMI} at different thicknesses of the residual layer.

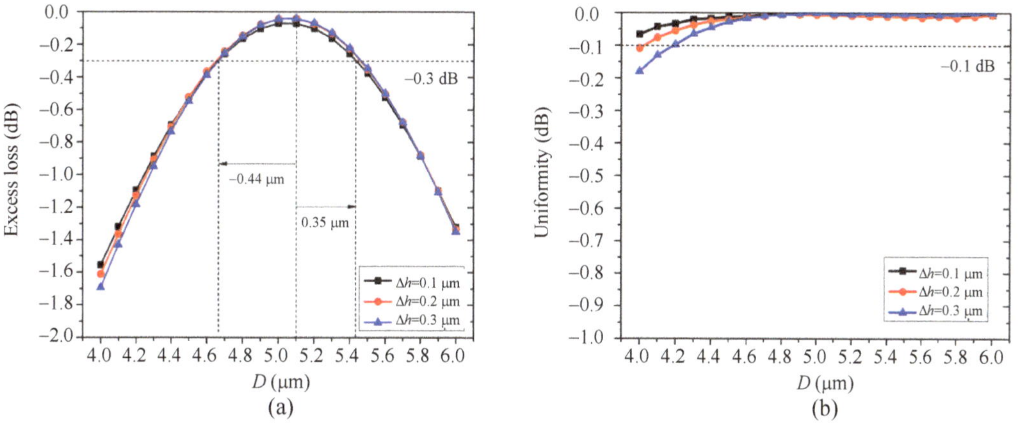

Fig. 12 Splitting performance: (a) excess loss and (b) uniformity with different D at different thicknesses of the residual layer.

According to the above analysis, the length L_{MMI} and the width W_{MMI} of the multimode waveguide have less tolerance than the waveguide spacing D of the input/output port. Therefore, they should be controlled precisely during the waveguide fabrication process. On the other hand, the thickness of the residual layer Δh could be a favorable measure to improve the performance of the coupler. For example, Δh could increase properly in the fabrication process if the practical L_{MMI} is larger than the designed value. Overall, the structural parameters of the 2×2 MMI based coupler, such as the waveguide spacing D, the length L_{MMI}, the width W_{MMI} of the multimode waveguide, and the residual layer thickness Δh should be considered comprehensively in order to improve the performance of the coupler.

5. Conclusions

A 2×2 MMI based optical waveguide coupler at 850 nm with the polymer of PSQ-Ls and imprint technique has been optimally designed by the BPM. The height and width of the waveguide core for single mode transmission is designed as 1 μm and

1.8 μm, respectively. The optimized structural parameters including the waveguide spacing D of the input/output port, the length L_{MMI}, and the width W_{MMI} of the multimode waveguide are obtained by minimizing the excess loss and uniformity and inserting the taper section of isosceles trapezoid between the single mode and multimode waveguides. The excess loss and the uniformity of the MMI based coupler are –0.040 dB and –0.007 dB, respectively, with the optimized structural parameters. The fabrication tolerance of structural parameters has also been investigated in order to achieve an MMI based coupler with good performance by the imprint technique. The analysis results provide useful instructions for the further waveguide device fabrication.

Acknowledgment

This work was supported in part by the International Science & Technology Cooperation Program of China (No. 2014DFG32590), National Natural Science Foundation of China (No.61077015, 61307040), National R&D Program (No.2012AA040406), Natural Science Foundation of Liaoning Province (No. 2014020002), Opening Project of Shanghai Key Laboratory of All Solid-state Laser and Applied Techniques (No. 2013ADL04), and Fundamental Research Funds for the Central Universities (DUT13JB01, DUT15ZD231, and DUT2015TD47).

Reference

[1] L. W. Cahill and T. T. Le, "Optical signal processing using MMI elements," in *10th Anniversary International Conference on Transparent Optical Networks*, Greece, vol. 4, pp. 114–117, 2008.

[2] H. F. Zhou, J. F. Song, C. Li, H. J. Zhang, and P. G. Lo, "A library of ultra-compact multimode interference optical couplers on SOI," *IEEE Photonics Technology Letter*, 2013, 25(12): 1149–1152.

[3] L. Lu, L. Zhou, S. Li, Z. Li, X. Li, and J. Chen, "4 × 4 nonblocking silicon thermo-optic switches based on multimode interferometers," *Journal of Lightwave Technology*, 2015, 33(4): 857–864.

[4] N. Xie, T. Hashimoto, and K. Utaka, "Very low-power, polarization-independent, and high-speed polymer thermooptic switch," *IEEE Photonics Technology Letters*, 2009, 21(24): 1861–1863.

[5] A. M. Al-Hetar, A. B. Mohammad, A. S. M. Supa'at, and Z. A. Shamsan, "MMI-MZI polymer thermo-optic switch with a high refractive index contrast," *Journal of Lightwave Technology*, 2011, 29(2): 171–178.

[6] R. W. Chuang, M. T. Hsu, Y. C. Chang, Y. J. Lee, and S. H. Chou, "Integrated multimode interference coupler-based Mach-Zehnder interferometric modulator fabrication a silicon-on-insulator substrate," *IET Optoelectronics*, 2012, 6(3): 147–152.

[7] L. Jin, J. W. Wang, X. Fu, B. Yang, Y. C. Shi, and D. X. Dai, "High-Q microring resonators with 2 × 2 angled multimode interference couplers," *IEEE Photonics Technology Letters*, 2013, 25(6): 612–614.

[8] J. P. Leidner and J. R. Marciante, "Tapered multi-mode interference waveguide for high-power self-organizing single-mode semiconductor laser arrays," *IEEE Journal Quantum Electronics*, 2011, 47(7): 965–971.

[9] S. H. Jeong and K. Morito, "Compact optical 90° hybrid employing a tapered 2×4 MMI coupler serially connected by a 2×2 MMI coupler," *Optics Express*, 2010, 18(5): 4275–4288.

[10] H. Ma, A. K. Y. Jen, and L. R. Dalton, "Polymer-based optical waveguides: materials, processing, and devices," *Advanced Materials*, 2002, 14(19): 1339–1365.

[11] L. J. Guo, "Recent progress in nanoimprint technology and its applications," *Journal of Physics D: Applied Physics*, 2004, 37(11): R123–R141.

[12] J. Wang, C. Zawadzki, N. Mettbach, W. Brinker, Z. Y. Zhang, D. Schmidt, *et al.*, "Polarization insensitive 25-Gbaud direct D(Q)PSK receiver based on polymer planar lightwave hybrid integration platform," *Optics Express*, 2011, 19(13): 12197–12207.

[13] Z. Zhang, A. Maese-Nove, A. Polatynski, T. Mueller, G. Irmscher, D. Felipe, *et al.*, "Colorless, dual-polarization 90° hybrid with integrated VOAs and local oscillator on polymer platform," in *Optical Fiber Communications Conference and Exhibition*, California, vol. TH1F3, pp.1–3, 2015.

[14] N. Bamiedakis, A. Hashim, R. V. Penty, and I. H. White, "A 40 Gb/s optical bus for optical backplane interconnections," *Journal of Lightwave Technology*, 2014, 32(8): 1526–1537.

[15] J. Wang, M. Kroh, A. Theurer, C. Zawadzki, D. Schmidt, R. Ludwig, *et al.*, "Dual-quadrature

coherent receiver for 100G Ethernet applications based on polymer planar lightwave circuit," *Optics Express*, 2011, 19(26): B166–B172.

[16] L. Wang, J. Ren, X. Han, T. Class, X. Jian, P. Bienstman, *et al.*, "A label-free optical biosensor built on a low cost polymer platform," *IEEE Photonics Journal*, 2012, 4(3): 920–930.

[17] C. Delezoide, M. Salsac, J. Lautru, H. Leh, C. Nogues, J. Zyss, *et al.*, "Vertically coupled polymer micro-racetrack resonators for label-free biochemical sensors," *IEEE Photonics Technology Letters*, 2012, 24(4): 270–272.

[18] J. Ren, L. Wang, X. Han, J. F. Cheng, H. L. Lv, J. Y. Wang, *et al.*, "Organic silicone sol-gel polymer as non-covalent carrier of receptor proteins for label-free optical biosensor application," *ACS Applied Materials & Interfaces*, 2013, 5(2): 386–894.

[19] S. Schmidt, J. Flueckiger, W. X. Wu, S. M. Grist, S. T. Fard, V. Donzella, *et al.*, "Improving the performance of silicon photonic rings, disks, and Bragg gratings for use in label-free biosensing," *SPIE*, 2014, 9166(91660M): 1–38.

[20] M. H. Salleh, A. Glidle, M. Sorel, J. Reboud, and J. M. Cooper, "Polymer dual ring resonators for label-free optical biosensing using microfluidics," *Chemical Communications*, 2013, 49(30): 3095–3097.

[21] J. Halldorsson, N. B. Arnfinnsdottir, A. B. Arnfinnsdottir, B. Agnarsson, and K. Leosson, "High index contrast polymer waveguide platform for integrated biophotonics," *Optics Express*, 2010, 18(15): 16217–16226.

[22] X. B. Wang, J. Sun, Y. F. Liu, J. W. Sun, C. M. Chen, X. Q. Sun, *et al.*, "650-nm 1×2 polymeric thermo-optic switch with low power consumption," *Optics Express*, 2014, 22(9): 11119–11128.

[23] L. Kou, D. Labrie, and P. Chylek, "Refractive indices of water and ice in the 0.65- to 2.5-µm spectral range," *Applied Optics*, 1993, 32(19): 3531–3540.

[24] F. Wang, J. Y. Yang, L. M. Chen, X. Q. Jiang, and M. H. Wang, "Optical switch based on multimode interference coupler," *IEEE Photonics Technology Letters*, 2006, 18(2):421–423.

[25] L. Yang, B. Yang, Z. Sheng, J. W. Wang, D. X. Dai, and S. L. He, "Compact 2×2 tapered multimode interference couplers based on SU-8 polymer rectangular waveguides," *Applied Physics Letters*, 2008, 93(20): 203304-1–203304-3.

[26] L. Jin, J. W. Wang, X. Fu, B. Yang, Y. C. Shi, and D. X. Dai, "High-Q microring resonators with 2 × 2 angled multimode interference couplers," *IEEE Photonics Technology Letter*, 2013, 25(6): 612–614.

[27] N. Xie, T. Hashimoto, and K. Utaka, "Design and performance of low-power, high-speed, polarization-independent and wideband polymer buried-channel waveguide thermo-optic switches," *Journal of Lightwave Technology*, 2014, 32(17): 3067–3072.

[28] J. L. Dong, K. S. Chiang, and W. Jin, "Compact three-dimensional polymer waveguide mode multiplexer," *Journal of Lightwave Technology*, 2015, 33(22): 4580–4588.

[29] H. B. Zhang, J. Y. Wang, L. K. Li, Y. Song, M. S. Zhao, and X. G. Jian, "A study on liquid hybrid material for waveguides-synthesis and property of PSQ-Ls for waveguides," *Journal of Macromolecular Science Part A*, 2008, 45(3): 232–237.

[30] X. Y. Han, L. H. Wang, Y. Wang, P. Zou, Y. Y. Gu, J. Teng, *et al.*, "UV-soft imprinted tunable polymer waveguide ring resonator for microwave photonic filtering," *Journal of Lightwave Technology*, 2014, 32(20): 3924–3932.

[31] L. B. Soldano and E. C. M. Pennings, "Optical multi-mode interference devices based on self-imaging: principles and applications," *Journal of Lightwave Technology*, 1995, 13(4): 615–627.

[32] J. Wang, M. H. Qi, Y. Xuan, H. Y. Huang, Y. Li, M. Li, *et al.*, "Proposal for fabrication-tolerant SOI polarization splitter-rotator based on cascaded MMI couplers and an assisted bi-level taper," *Optics Express*, 2014, 22(23): 27869–27879.

[33] K. Debnath, R. Moore, A. Liles, and L. O'Faolain, "Toolkit for photonic integrated circuits based on inverted rib waveguides," *Journal of Lightwave Technology*, 2015, 33(19): 4145–4150.

7

A New Detection Method Based on CFAR and DE for OFPS

Zezheng QIU[1], Tong ZHENG[2], Hongquan QU[2*], and Liping PANG[1]

[1]*School of Aeronautic Science and Engineering, Beijing University of Aeronautics and Astronautics, Beijing, 100191, China*

[2]*College of Electrical & Information Engineering, North China University of Technology, Beijing, 100144, China*

*Corresponding author: Hongquan QU E-mail: qhqphd@ncut.edu.cn

Abstract: Optical fiber pre-warning system (OFPS) is widely utilized in pipeline transport fields. The intrusions of OFPS need to be located. In this system, the original signals consist of noises, interferences, and intrusion signals. Here, noises are background and harmless interferences possessing with high power, and the intrusion signals are the main target of detection in this system. Hence, the study stresses on extracting the intrusion signals from the total ones. The proposed method can be divided into two parts, constant false alarm rate (CFAR) and dilation and erosion (DE). The former is applied to eliminate noises, and the latter is to remove interferences. According to some researches, the feature of noise background accords with the CFAR spatial detection. Furthermore, the detection results after CFAR can be presented as a binary image of time and space. Besides, interferences are relatively disconnected. Consequently, they can be eliminated by DE which is introduced from the image processing. To sum up, this novel method is based on CFAR and DE which can eliminate noises and interferences effectively. Moreover, it performs a brilliant detection performance. A series of tests were developed in Men Tou Gou of Beijing, China, and the reliability of proposed method can be verified by these tests.

Keywords: Optical fiber pre-warning system; constant false alarm rate; dilation and erosion

1. Introduction

Optical fiber pre-warning system (OFPS) is employed in monitoring the transport of petroleum and gas as well as forecasting the leakage to pipelines [1−5]. Moreover, great losses have been avoided in the field of pipeline transport. Actually, the optical fibers are laid parallel with a pipeline in the same ditch. They are used not only to transmit data but also to extract vibration signals [6−7]. In other words, intrusions on the ground are supposed to be forecasted by analyzing the extracted signals. Consequently, OFPS is of certain value to be studied [8−9].

The vibration signals in OFPS consist of noises, interferences, and intrusion signals. Here, noises are background for OFPS. Interferences with high power enable may lead to fibers vibrate. However, the vibration does no harm to the fibers and pipelines. In other words, interferences are harmless. In addition, harmful signals from digging, hitting, and mechanical vibrations are harmful. In OFPS, a number of scholars have researched on the signal processing for OFPS. According to the previous study, signals can be extracted by refraction or reflection system [10–14]. Here, Zengguang Qin, from Ottawa University, Canada, closely studied the distribution of optical fiber vibration sensors which laid theoretical basis for the signal extraction in OFPS [14]. Meanwhile, most scholars are inclined to recognize harmful intrusions based on analyzing vibration signals directly. The team, from Tian Jin University, China, put forward a method to classify various vibration signals through support vector machine (SVM) [7]. However, with an increase in background complexity degree, interference cases are variable. Therefore, it is of great necessity to eliminate interferences and noises before recognition. Meantime, the detection accuracy can be improved by this proposal.

In this paper, we propose a new detection method to remove noises and interferences. From our early studies, we found that the vibration signals are non-stationary in the time dimension and its background distribution can be divided into two orthogonal Gaussian noises. At the same time, the envelope of synthesis signal obeys Rayleigh distribution. These features are identical to the requirements of constant false alarm rate (CFAR). Consequently, we are advised conducting the spatial detection by applying CFAR firstly. The traditional CFAR can be divided into three parts which are CA-CFAR [15], OS-CFAR [16], and A-CFAR [17]. In our system, the CA-CFAR is applied because it has an outstanding detection performance in the homogeneous background and relatively low computation cost. The corresponding threshold is derived by the given constant false alarm rate. Then the noised can be removed. Yet powerful interference exists on practical surrounding which is supposed to be detected by CFAR [18–20]. We can draw the conclusion that the interference needs further dispose. On the other hand, the result of detection can be presented as a binary image which means alarming and non-alarming, respectively. Therefore, dilation and erosion (DE), borrowed from image processing, is considered in OFPS. Here, dilation is used to bridge little disconnect cases within continuous alarm points. Meantime, erosion is applied to remove some discrete alarm points which are some false alarm points with high probability. In brief, interferences are supposed to be wiped off effectively.

In addition, we conducted practical tests for three typical time and space in Men Tou Gou district, Beijing. Various data were dealt by the new method which is provided in this paper. The results show that the detection method is very valid to improve the detection performance of harmful intrusions.

2 Detection method

2.1 Detection method outline

OFPS mainly comprises hardware and software modular. Its general structure is shown in Fig. 1. Its working principle can be listed as follows.

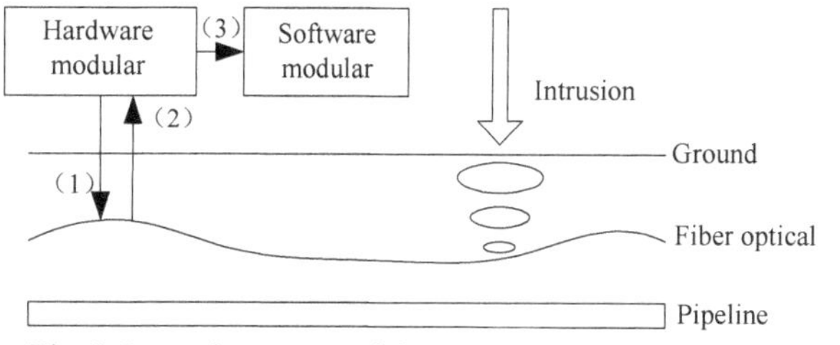

Fig. 1 General structure of OFPS.

(1) The light impulses are produced by the hardware modular and transmitted in the optical fibers;

(2) The changing refractive index of fiber, which is caused by elastic-optic effect, results in the change in light intensity. Finally, this kind of change returns to the hardware modular.

(3) The detection results of intrusions can be obtained by the signal processing and shown in the host computer.

The flow charts of the detection method which is presented in this paper are shown in Fig. 2. In detecting, noises eliminated by CFAR and interferences removed by DE are performed successively. After processing, the types of data vary distinctly. Firstly, the original data consist of noises, interferences, and intrusions. After the CFAR detection, massive noises are eliminated. Then, the interferences are removed by DE. In other words, data after detection only include intrusions.

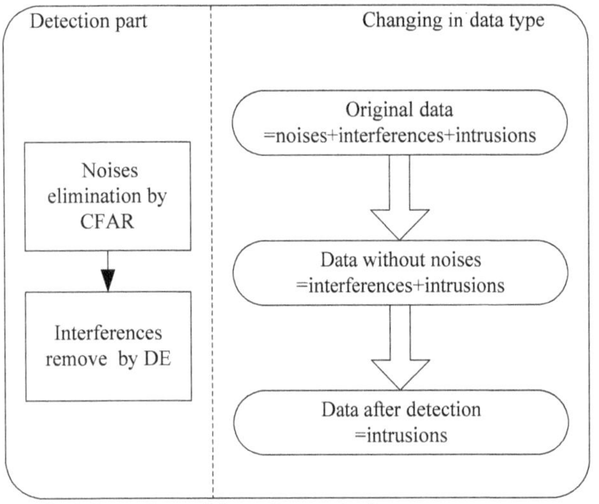

Fig. 2 Procedure of detection method.

2.2 Principle of provided method

The principle of CFAR and DE will be declared in this section. Meantime, we illustrate some workable rules in the view of OFPS.

2.2.1 CFAR

CFAR is a detection method which can derive threshold adaptively in accordance with the defined constant false alarm rate P_{f0}. The structure of CA-CFAR can be got by [15]. The test cell is surrounded by protection cells which are also surrounded by N reference cells. The reference cells are considered as the detection background. Firstly, the noise level can be obtained for different CFAR methods. Then, the threshold can be confirmed. Finally, the test cell is compared with corresponding threshold. If the test cell is great, the alternative hypothesis H_1 is fulfilled, giving an alarm to the test cell. Otherwise, null hypothesis, H_0, is achieved, and test cell is identified as noise. The derivations of specific formulas can be seen as follows.

The background of OFPS is Gaussian noise, of which the envelope follows Rayleigh distribution and its square obeys exponential distribution. Hence, its probability distribution function (PDF) is modeled as

$$f(x)=\frac{1}{2\mu}\exp\left(-\frac{x}{2\mu}\right),\ x>0 \tag{1}$$

where x denotes vibration signals. If vibration signals only consist of noises, then μ equals the variance of noises, σ^2. Otherwise, $\mu=\sigma^2(1+\text{SNR})$ and SNR are the signal-to-noise rate (SNR). Moreover, the false alarm probability is

$$\begin{aligned} P_f &= \int_0^{\infty} P(X>TZ\mid H_0)f_Z(z)dz \\ &= E_Z\left[P(X>TZ\mid H_0)\right] \end{aligned} \tag{2}$$

where T is multiplier of threshold, Z denotes noise level, $f_z(z)$ is PDF of Z, $E_z(\cdot)$ represents a mean of Z, and H_0 is null hypothesis.

Substitute (1) into (3), then

$$\begin{aligned} P_f &= E_Z\left[\int_{TZ}^{\infty} f(x\mid H_0)dx\right] \\ &= E_Z\left[\int_{TZ}^{\infty}\frac{1}{2\sigma^2}\exp\left(-\frac{x}{2\sigma^2}\right)dx\right] \\ &= E_Z\left[\exp\left(-\frac{TZ}{2\sigma^2}\right)\right]=M_Z\left(\frac{T}{2\sigma^2}\right) \end{aligned} \tag{3}$$

where $M_Z(\cdot)$ represents moment generating function (MGF) of Z.

Hence, the false alarm probability, P_f, is only decided by T when the distribution of Z is confirmed. T can be obtained by designing. P_{f0} is a designed false alarm probability before the detection.

CA-CFAR is typical of ML-CFAR, and we usually write Z as

$$Z=\sum_{i=1}^{N}x_i \tag{4}$$

where N is the data quantity of reference cell.

If the signals of reference cell are independent and identically distributed (IID), an excellent detection performance will be guaranteed. Moreover, exponential distribution is a special case of Gamma when $\alpha=1$.

$$\begin{aligned} p(v) &= \frac{\beta^{-\alpha}v^{\alpha-1}}{\Gamma(\alpha)}\exp\left(-\frac{v}{\beta}\right) \\ & v\ge 0,\alpha\ge 0,\beta\ge 0 \end{aligned} \tag{5}$$

where $\Gamma(\cdot)$ is a Gamma function, and α and β are two parameters, $v\sim G(\alpha, \beta)$.

Furthermore, the accumulation of random variables obeying the same Gamma distribution also follows a Gamma distribution. Here, α needs to be accumulated.

Hence, Z obeys the following Gamma distribution:

$$Z \sim G\left(N, 2\sigma^2\right). \quad (6)$$

We can obtain $M_v(u)=(1+\beta u)^{-\alpha}$ with $v \sim G(\alpha, \beta)$. According to (3), the false alarm probability of CA-CFAR is

$$P_f = M_Z\left(\frac{T}{2\sigma^2}\right) = \left(1+2\sigma^2 \frac{T}{2\sigma^2}\right)^{-N} = \left(1+T\right)^{-N}. \quad (7)$$

Hence, threshold multiplier can be identified as

$$T = P_{f0}^{-1/N} - 1. \quad (8)$$

In addition, the detection threshold S is

$$S = Z \cdot T. \quad (9)$$

Finally, the test cell is compared with the corresponding threshold S. If the test cell is great, it is thought as an alarm point. Otherwise, the test cell is identified as noise.

2.2.2 DE

(1) Dilation

Translating structural element A for X results in A_x. If the intersection of A_x and X is not null, give a tag for x. Assemble of X, which satisfies the above feature, is the so called dilation of X by A. We write it as

$$X \oplus A = \left\{x \mid A_X \cup x \neq \phi\right\}. \quad (10)$$

The method of dilation is comparing points in A with those in X one by one. If a point in A is in X, then the corresponding point in X is the image. Therefore, we apply it to bridge the gap of alarms in OFPS.

(2) Erosion

Translating structural element A of x results in A_x. If A_x is included in X, give a tag to x. Assemble of x, which satisfies the above-mentioned feature, is the so called erosion of X by A. We can express it as

$$X \Theta A = \left\{x \mid A_X \cap x \subset A_X\right\}. \quad (11)$$

The rule of erosion is doing the same steps like dilation. If the points in A are all included in X, then the points in X should be maintained. Otherwise, they do not belong to the erosion results of X by A. Hence, the disconnected points which have been dilated are finally eliminated by erosion in OFPS.

After vast simulation experiments, we found that $(1, 1, \cdots, 1)_{100\times1}$ and $(1, 1, \cdots, 1)_{1000\times1}$ can be considered as structural elements in DE of OFPS, respectively.

3. Experiment and analysis

The practical objections of OFPS are shown in Fig.3, which are optical fibers, optical fiber fusion splicer, signal processing bar, and software display in Figs. 3(a)–(d), respectively.

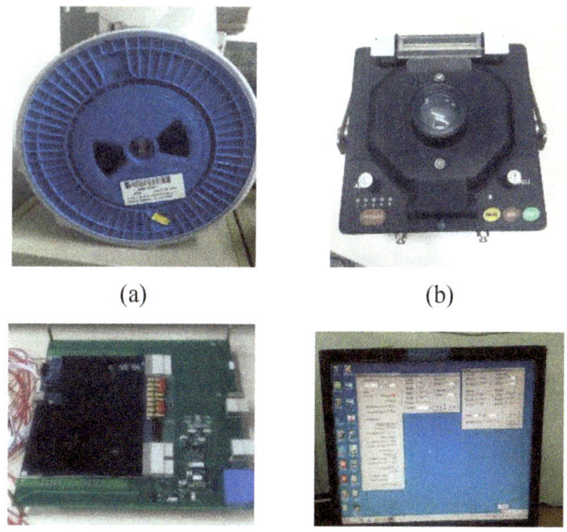

(a) (b) (c) (d)

Fig. 3 Practical objections of OFPS: (a) optical fibers, (b) fusion splicer, (c) signal processing board, and (d) software display.

To verify the reliability and efficiency of algorithm proposed in this paper, we conducted practical tests in Men Tou Gou district of Beijing, China. Three kinds of intrusion results are listed in this paper, which are digging, hitting, and electric breaker. Three cases of experiment are shown in Table 1.

Table 1 Experiment background.

Cases	Vibration types	Situation (km)	Time (ms)
Case 1	Digging	41	1500–6000
		50	500–7500
		51	0–7500
Case 2	Hitting	47	0–7500
		48	0–7500
		58	0–7500
Case 3	Electric breaker	41	0–7500
		47	0–7500
		51	2000–6000

The cases of experiments are shown as follows:

• Case 1: The event of digging occurs at 41 km, 50 km, and 51 km. Their durations are 1500 ms–6000 ms, 500 ms–7500 ms, and 0 ms–7500 ms, respectively.

• Case 2: The event of hitting occurs at 47 km, 48 km, and 58 km. Their durations are all the same in 0 ms–7500 ms.

• Case 3: The event which is caused by electric breaker occurs at 41 km, 47 km, and 51 km. Their durations are 0 ms–7500 ms, 0 ms–7500 ms, and 2000 ms–6000 ms, respectively.

3.1 CFAR results

The consequences of processing the three sets of data by CA-CFAR are shown in Fig. 4.

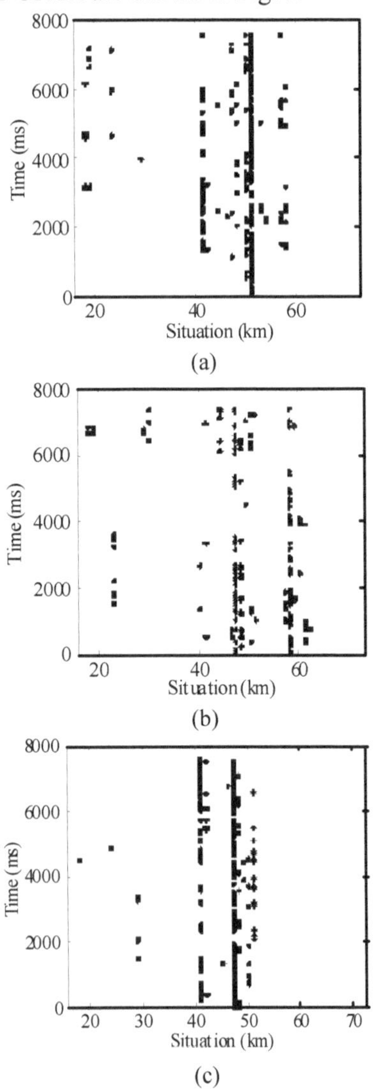

Fig. 4 Results of CFAR: (a) Case 1, (b) Case 2, and (c) Case 3.

According to the analysis of Fig. 4, the conclusions that we can obtain are listed as follows:

• Case 1: In Fig. 4(a), digging signals at 41 km, 50 km, and 51 km can be detected. Meanwhile, the vibration time is corresponded to the practical one. However, the interference at 16 km–28 km and 42 km–49 km can be detected through CFAR as well.

• Case 2: In Fig. 4(b), signals caused by hitting at 47 km, 48 km, and 58 km can be detected. And the vibration time is the same with the one in reality. Yet the interference at 18 km–41 km are also detected by CFAR.

• Case 3: In Fig. 4(c), signals of electric breaker at 47 km, 48 km, and 51 km are detected. Meantime, the vibration time is consistent with the actual time. But interference at 18 km, 24 km, and 29 km are detected by CFAR likewise.

According to above experiments, we can draw the conclusion that intrusion signals can be detected efficiently. However, interferences with high power are detected by CFAR. So they need to do further processing to eliminate interferences.

3.2 DE results

The data after CFAR are disposed by DE and the results are shown in Fig. 5.

Some conclusions can be obtained as follows:

• Case 1: In Fig. 5(a), digging at 41 km, 50 km, and 51 km can be detected. Meanwhile, the vibration time is corresponded to the practical one. Interferences detected by CFAR are eliminated effectively as well.

• Case 2: In Fig. 5(b), signals caused by hitting at 47 km, 48 km, and 58 km are detected. The vibration time is the same with the one in reality. Interferences detected by CFAR are also eliminated effective.

• Case 3: In Fig. 5(c), signals of electric breaker at 47 km, 48 km, and 51 km are detected. Meantime, the vibration time is consistent with the actual time. Interferences detected by CFAR are eliminated effectively likewise.

Through the processing of DE, we can find that

interferences are supposed to be validly eliminated. Meanwhile, real intrusions are detected effectively.

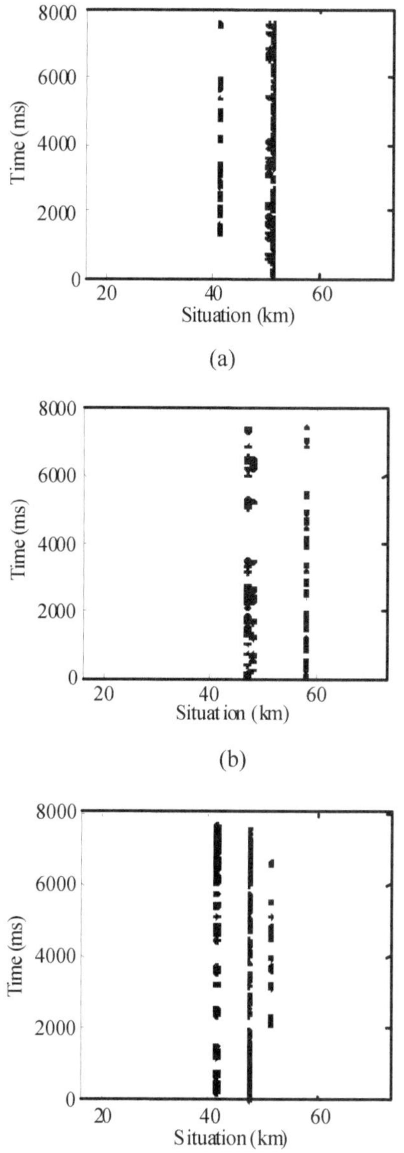

(c)

Fig. 5 Results of DE: (a) Case 1, (b) Case 2, and (c) Case 3.

4. Discussion and conclusions

According to the features of vibration signals, a brand-new method based on CFAR and DE is proposed in this paper to detect harmful intrusion signals. The method consists of two parts, noise elimination by CFAR and interference removal by DE. In the former part, CA-CFAR is adopted. It is the simplest one in a class of spatial CFARs and performs brilliantly in the homogeneous background. In addition, intrusion signals and interference of high power can be detected efficiently after the space-dimensional detection. Yet interference still exists after CFAR which cannot remove interferences. In the latter part, DE, which is introduced from the image processing, is adopted to do the further procession. From the results, the interferences are eliminated effectively. Field test data are used to verify the reliability of the proposed detection method. The measured data were collected in Men Tou Gou district of Beijing, China. After processing, the time and situations of intrusion signals can be located precisely and clearly. In other words, pre-warning function is achieved.

Acknowledgement

The authors are grateful to the anonymous reviewers for their critical and constructive review of the manuscript. This work was funded by the National Natural Science Foundation of China (No. 61571014).

References

[1] W. Liang, L. Lu, and L. Zhang, "Coupling relations and early-warning for 'equipment chain' in long-distance pipeline," *Mechanical Systems and Signal Processing*, 2013, 41(1): 335–347.

[2] J. Kang and Z. H Zou, "Time prediction model for pipeline leakage based on grey relational analysis," *Physics Procedia*, 2010, 25(2): 2019–2024.

[3] T. Zhang, Y. Tan, H. Yang, J. Zhao, and X. Zhang, "Locating gas pipeline leakage based on stimulus-response method," *Energy Procedia*, 2014, 61: 207–210.

[4] Q. Fu, H. J. Wang, and F. Qiu, "Pipeline leak detection based on fiber optic early-warning system," *Procedia Engineering*, 2010, 7: 88–93.

[5] Z. Qu, S. Jin, and Y. Zhou, "Study on the distributed optical fiber pipeline leakage pre-warning system and the method of signal analysis," in *Proceedings of the 6th International Pipeline Conference*, Calgary, Canada, pp. 677–681, 2006.

[6] W. Liang, L. Zhang, Q. Xu, and C. Yan, "Gas pipeline leakage detection based on acoustic technology," *Engineering Failure Analysis*, 2013, 31(6): 1–7.

[7] Z. Qu, H. Feng, Z. Zeng, J. Zhuge, and S. Jin, "A SVM-based pipeline leakage detection and pre-warning system," *Measurement*, 2010, 43(4): 513–519.

[8] Q. Lv, L. Li, H. Wang, Q. Li, and X. Zhong, "Influences of laser on fiber-optical distributed disturbance sensor based on Φ-OTDR," *Infrared and Laser Engineering*, 2014,12(43): 3919–3923.

[9] H. F. Martins, S. Martin-Lopez, P. Corredera, and M. L. Filograno, "Coherent noise reduction in high visibility phase-Sensitive optical time domain reflectometer for distributed sensing of ultrasonic waves," *Journal of Lightwave Technology*, 2013, 31(23): 3631–3637.

[10] Q. Li, C. Zhang, L. Li, and X. Zhong, "Localization mechanisms and location methods of the disturbance sensor based on phase-sensitive OTDR," *Optik-Interntional Journal for Light and Electron Optics*, 2014, 125(9): 2099–2103.

[11] Q. Lin, C. Zhang, and C. Li, "Fiber-optic distributed sensor based on phase- sensitive OTDR and wavelet packet transform for multiple disturbances location," *Optik-Interntional Journal for Light and Electron Optics*, 2014, 125(24): 7235–7238.

[12] A. R. Bahrampour and F. Maaoumi, "Resolution enhancement in long pulse OTDR for application in structural health monitoring," *Optical Fiber Technology*, 2010, 16(24): 240–249.

[13] L. Lu, Y. Song, X. Zhang, and F. Zhu, "Frequency division multiplexing OTDR with fast signal processing," *Optics and Laser Technology*, 2012, 44(7): 2206–2209.

[14] Z. Qin, "Distributed optical fiber vibration sensor based on Rayleigh backscattering," Ph.D. dissertation, University of Ottawa, Ottawa, 2013.

[15] H. Rohling, "Radar CFAR thresholding in clutter and multiple target situations," *IEEE Transactions on Aerospace and Electronic Systems*, 1983, AES-19(4): 608–621.

[16] M. E. Smith and P. K. Varshney, "Intelligent CFAR processor based on data variability," *IEEE Transactions on Aerospace and Electronic Systems*, 2000, 36(3): 837–847.

[17] S. Ward, M. Bélanger, D. Donovan, A. Horsman, and N. Carrier, "Automatic censored CFAR detection for nonhomogeneous environments," *IEEE Transactions on Aerospace and Electronic Systems*, 1992, 28(1): 286–304.

[18] R. Zhang, W. Sheng, and X Ma, "Improved switching CFAR detector for non-homogeneous environments," *Signal Processing*, 2013, 93(1): 35–48.

[19] G. V. Weinberg, "Management of interference in Pareto CFAR processes using adaptive test cell analysis," *Signal Processing*, 2014, 104: 264–273.

[20] B. Shi, C. Hao, C. Hou, X. Ma, and C Peng, "Parametric Rao test for multichannel adaptive detection of range-spread target in partially homogeneous environments," *Signal Processing* 2015, 108: 421–429.

8

Research on One-Piece Structure Target Flow Sensing Technology Based on Fiber Bragg Grating

Chuntong LIU, Zhengyi ZHANG*, Hongcai LI, Zhenxin HE, Xiaofeng ZHAO, and Rongjing WU

Department Two, Rocket Force University of Engineering, Xi'an 710025, China

*Corresponding author: Zhengyi ZHANG E-mail: 13022719922@163.com

Abstract: In view of problems existing in the detection of the traditional hydraulic system, such as the large volume of sensor and the low measurement accuracy, a new one-piece target type flow sensor is designed and researched based on fiber Bragg grating (FBG). A compact structure is designed, which is convenient to be dismantled, processed, and installed, based on the analysis of the principle of FBG and the structure of target type flow sensor. The force of target put in fluid flow is turned into the FBG wavelength drift, with a corresponding relationship. The problem on the cross sensitivities of the temperature and strain is solved effectively by using double FBG symmetrically pasted on the both surfaces of the cantilever. The impact on the fluid state is analyzed through simulation in the software FLUENT, and the results show that the impact was smaller than that of the traditional structure. The results of experiments in the hydraulic system show that there is a good linear relationship between the change in the dual FBG central wavelength and mass loading on the target sheet has a good linear relationship, and the sensitivity is twice that of a single FBG sensitivity.

Keywords: Fiber Bragg grating; flow sensor; target type structure; equal intensity cantilever beam

1. Introduction

In the industrial production process and automation control field, the flow is one of the parameters being detected and controlled. Flow meter is also used in the detection of widely used varied and mechanical rotor flow sensor [1]. But due to the complexity of the mechanical structure, its measurement accuracy is low. Ultrasonic flowmeter [2], the electromagnetic flowmeter [3], and acoustics Doppler current meter [4] have a high measurement accuracy, and the operation is simple, but their cost is high, and they are interferenced easily by electromagnetic. Fiber Bragg grating (FBG) flow sensor is a new type of sensing technology developed in recent 20 years [5, 6], and its principle is that the central Bragg reflected wavelength is sensitive to physical quantities of the external physics, through the special package and structure design to realize measurement of various parameters [7, 8]. Compared with the traditional electrical sensors, FBG has lots of advantages, such as small volume, light weight, high sensitivity, the absolute measurement, and strong ability of resistance against electromagnetic interference. It combines sensing and transmission, and it is easy to be reused and constitutes a sensor network [9, 10]. Therefore, it's very suitable for a hydraulic system in detection of flow parameters, such as high precision and multiple point detection requirements.

In recent years, Yu *et al.* [11], Wang *et al.* [12], Wang *et al.* [13], and others have applied FBG in the traditional target flowmeter and have realized the accurate and quick measurement of flow by FBG based on the circular target and a cantilever beam structure, but its structure is complex and not easy to be installed and sealed, and it fails to give full play to the advantages of FBG sensor. According to the above background, this paper proposes a set of choke target slices and sensitive elements in one of equal strength cantilever structures, on the basis of analyzing the measuring principle, and after the structure design and material processing, the feasibility of this structure is verified by experiment, and the sensor is proved sensitive and reliable.

2. Basic principle

2.1 Basic principle of FBG sensing

According to coupled-mode theory, for broadband incident light, the refractive index of FBG cycle disturbances affects only a very narrow range of wavelengths of spectral period of impact, that is, only to meet the Bragg condition

$$\lambda_B = 2n_{\text{eff}}\Lambda \qquad (1)$$

where λ_B is the center of Bragg reflection wavelength, n_{eff} is the effective refractive index, and Λ is the grating period.

Light waves can be reflected by grating, the transmission spectra of the rest are not affected, and FBG will play the role of reflector or filter. Equation (1) shows that λ_B changes with Λ and n_{eff}. Temperature and strain are the physical parameters which FBG is directly sensitive to, and they can cause Λ and n_{eff} to change. FBG is directly sensitive to the strain and temperature, and they will also change Λ and n_{eff}. λ_B is influenced by temperature through elastic-optic effect and thermal expansion effect, and λ_B is influenced by strain through elastic-optic effect and the grating period. The drifting of λ_B caused by strain and temperature can be deduced as

$$\Delta\lambda_B = (\alpha+\xi)\lambda_B\Delta T + (1-P_e)\lambda_B\varepsilon = K_T\Delta T + K_\varepsilon\varepsilon \qquad (2)$$

where α and ξ are the thermal expansion coefficient and thermal optical coefficient of the fiber respectively; P_e is the effective elastic-optic coefficient; ε is the axial strain; K_T and K_ε are the sensitivity coefficients of temperature and strain. Equation (2) is the theoretical basis of analyzing FBG sensing process. When there is outside temperature and stress field on FBG, the appropriate information can be accurately obtained accurately by measuring the change in the reflection wavelength with the wavelength demodulation, which is the basic principle of FBG sensing.

2.2 Target type flow sensing principle of FBG

According to the basic principle of FBG sensing, FBG is sensitive to temperature and strain directly, so to realize the precise measurement to flow, the special structure should be designed to turn flow into the physical quantity to which FBG is sensitive.

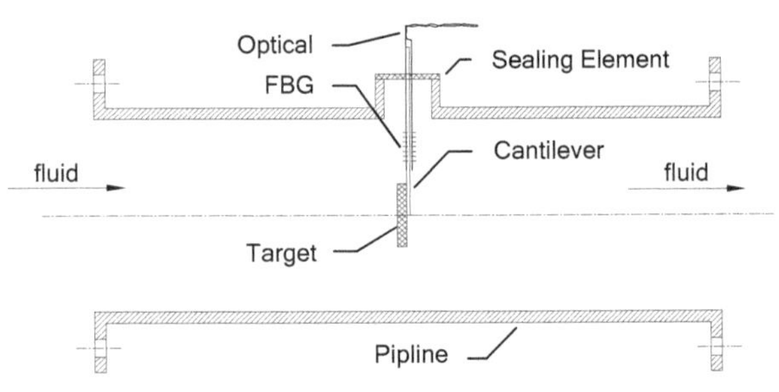

Fig. 1 Target type flow sensing principle diagram.

In the paper, the structure of the target type flow measurement is used, and the structure principle is shown in Fig. 1. At the axis of the pipeline which the fluid flows through, a disc-shaped target sheet is placed, and an annular space is formed. Target piece is fixed on the triangle cantilever beam, and the force from the flow on the target piece is converted into the strain of the triangle cantilever beam which is measured by the sensitive element.

When the target is impacted by fluid, it is subjected to a resultant force which is composed of three forces: the fluid dynamic pressure, fluid static pressure (differential pressure), and viscous friction, and the first two forces play a leading role when the flow is large. By the Bernoulli equation, we can derive [15]

$$A_1(p_0-p)+A_1\left(\frac{pv_0^2}{2}-\frac{pv^2}{2}\right)=\frac{1}{2}\rho\xi v^2A_1 \quad (3)$$

where ρ is the fluid density; A_1 is the target area; ζ is the local resistance coefficient; v_0 and p_0 are the average flow velocity and pressure of fluid in front of the target, respectively; v and p are the average flow velocity and pressure of fluid in behind of the target, respectively; $A_1(p_0-p)$ is the static pressure on the target and referred to as F_1; $A_1(\rho v_0^2/2-\rho v^2/2)$ is the dynamic pressure on the target and referred to as F_2; F_1+F_2 is the resultant force on the target and referred to as F, and F within v is given as

$$F=\frac{1}{2}\rho\zeta v^2A_1. \quad (4)$$

The FBG attached to both sides of the central axis of the triangle cantilever beam is deformed by the force on the triangle cantilever beam caused by the flowing fluid, and the central wavelength of reflection wave is changed. The central wavelength of the grating in the direction of flow drifts in the direction of increasing, the central wavelength of the grating against the direction of flow drifts in the direction of decreasing, and the amount of drift is the same. Because they are influenced by the temperature at the same time, the change in the two FBG wavelength is given as [16]

$$\Delta\lambda_{B1}/\lambda_B=(1-P_e)\varepsilon+(\xi+\alpha)\Delta T \quad (5)$$

$$\Delta\lambda_{B2}/\lambda_B=-(1-P_e)\varepsilon+(\xi+\alpha)\Delta T. \quad (6)$$

The change in double FBG wavelength $\Delta\lambda$ within the grating axial strain ε is given as

$$\Delta\lambda=\Delta\lambda_{B1}-\Delta\lambda_{B2}=2\varepsilon(1-P_\varepsilon)\lambda_B. \quad (7)$$

Equation (7) shows that the sensitivity is improved by the differential structure of double FBG, and at the same time, the problem on the cross sensitivities of the temperature and strain is solved effectively. According to the mechanics of materials, the strain of the triangle cantilever beam ε within the loading force in the free end F is given as [17]

$$\varepsilon=\frac{6FL}{Ebh^2} \quad (8)$$

where b is the bottom width of beam; h is the thickness; E is the young's modulus of cantilever beam material; L is the length of the cantilever beam fixed end.

We can discern from (7) and (8) that $\Delta\lambda$ within F is given as

$$\Delta\lambda=\frac{12FL}{Ebh^2}(1-P_e)\lambda_B. \quad (9)$$

Supposing oil in the pipeline is incompressible fluid, according to the continuity equation of the total fluid flow, we can discern from (4), (8), and (9) that the total flow Q in the pipeline can be deduced as

$$Q=A_2v=(D^2-d^2)\sqrt{\frac{\pi bh^2E}{24(1-P_\varepsilon)L\zeta\rho d^2}}\cdot\sqrt{\Delta\lambda}=K\cdot\sqrt{\Delta\lambda} \quad (10)$$

where D is the pipe diameter; d is the target path; A_2 is the annulus area between D and d; Q is the flow through A_2.

P_e of the typical quartz fiber is equal to 0.22, b, h, L, and E are constants if cantilever beam size and target size are determined, and ρ and ζ are fixed values for a particular fluid. So from (10) we can get that the fluid characteristic parameters under the given conditions can be accurately measured as long as measuring the amount of wavelength change of FBG.

3. Structure design

3.1 One-piece structure design of the target type flow sensor

The principle of the traditional target flowmeter is that the resistance strain gauges is sensitive to deformation of the target caused by the fluid flow, and the sensitive part of the electrical devices must be isolated from measured media, so the flowmeter is very difficult to be processed, handled, and sealed. And the impact force on the target sheet must be achieved by the force transmission rod. However,

FBG is non-corrodible and non-electrical component, so in the paper, the force transmission mechanism is omitted, and the target sheet and the triangle cantilever beam are made organically integrate to simplify design.

In Fig. 2, the sensor is mainly composed of four parts of valve body, convex platform, end cover, and one-piece target. Body, convex platform, and end cover are processed with hard aluminum alloy, and one-piece target piece are processed entirely by the method of linear cutting. The center of target sheet must be coincided with the axis of pipeline in the whole installation process, and target sheet is perpendicular to the direction of fluid flow, which relies mainly on the convex part of the convex platform to achieve. Type O sealing ring is used to face seal between the valve body and convex platform, end cover and convex platform, and the sealing cap is fastened by screws. Two FBGs are pasted on both sides of the central axis of the triangle cantilever beam symmetrical, and the differential configuration is formed. The impact on the target piece from the fluid causes the deformation of the cantilever and FBG posted on both sides, so the size of the flow can be measured according to the offset of the central reflection wavelength.

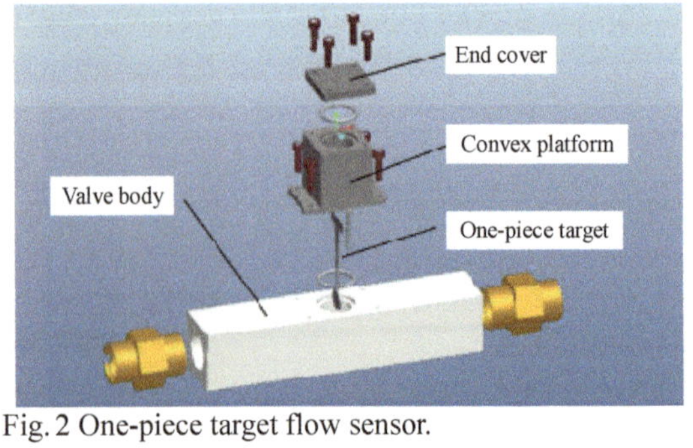

Fig. 2 One-piece target flow sensor.

3.2 One-piece structure design of the target piece

The sensitivity and performance of the target type flow sensor mainly depends on the rationality of the structural design of the one-piece target piece. The design principles is as follows:

(1) The size of the ratio of pipe-diameter should be moderate, and the sensitivity and pressure loss should be taken into account.

(2) The measuring range of sensor should be considered when determining the size of the triangle cantilever beam, and the accuracy should meet the requirements within the range.

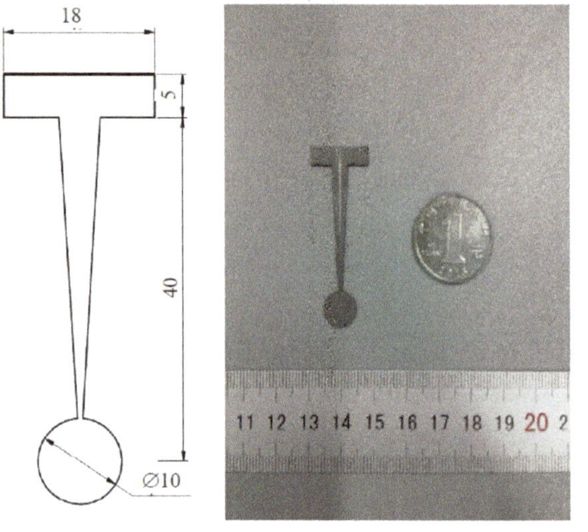

Fig. 3 One-piece target: (a) processing drawings and (b) material object.

When the pipe diameter is determined, differential pressure gradually increases with an increase in diameter of target piece, and that is to say, the pressure loss through the one-piece target type sensor is larger. Combing with the practical application, we select the body diameter D equal to 16 mm and the target diameter equal to 10 mm. The range of the flow sensor is designed between 0 L/s and 2 L/s, so the speed of the fluid flowing through the gap between the pipeline and target sheet is at a range of 0 to 16.3 m/s, and then you can calculate the range of the impact on the one-piece target. The measurement accuracy should be increased as far as possible in the process of designing the one-piece target on the premise that the strength requirements are met. The one-piece target is designed as shown in Fig. 3. The processing drawing is shown in Fig. 3(a), and the material object is shown in Fig. 3(b). The material of the one-piece target is 304# stainless steel, the thickness is equal to 1mm, the length of the cantilever l is equal to 40mm, the bottom width b is equal to 5mm, the diameter of circular target sheet is equal to 10mm, and the center of the circular target sheet is coincided with the vertex of the triangle cantilever beam. The length of the rectangle area above the target piece is equal to 18 mm, and the width is equal to 5 mm. In the process of installation, it is placed in the grooves of convex platform to fasten the one-piece target and ensure that the target sheet plane is perpendicular to the direction of the fluid flow.

4. FLUENT simulation analysis

We use FLUENT software for one-piece target type flow sensor fluid analyzing. Through the simulation experiment, we observe the status of fluid and compare it with that of traditional target type flow sensors, to determine whether the design of one-piece target structure is feasible.

The 3D geometric model is set up in the GAMBIT software, and flow sensor body part is designed to a tee model, of which the pipe diameter is 16mm, the horizontal line is 150mm, and vertical pipeline is 40mm. Oil in a horizontal pipe flows along the X-axis direction. Solution 1 is for the structure of one-piece target type, and Solution 2 is for the traditional target type structure, except for the target piece of different structures, other conditions of the two schemes are consistent. Meshing with Tet/Hybrid (the form of tetrahedral mesh is given priority to the appropriate position including hexahedral grid, cone, and wedge) and Tgrid (it will be divided into tetrahedron grid cell body, and in the proper position it can also contain the cone, hexahedron, and wedge unit) type, the grid size is set to 0.5. Target around the grid uses refinement uprocessing by wall partition boundary grid. After meshing, the model diagrams as shown in Figs. 4(a) and 4(b) are one-piece target type structure and the traditional model of the target structure, respectively.

After the model is established in this paper, we export mesh file from GAMBIT, open the FLUENT software, choose three dimensional single precision calculating body, import the mesh file, and check the meshing condition. According to the actual situation, we set the parameters of the segregated solver and implicit solver. The k-epsilon model is chosen as the viscosity model, and the oil is chosen as the fluid. The boundary condition is that the inlet velocity is 10 m/s, and other conditions are initial settings of FLUENT software.

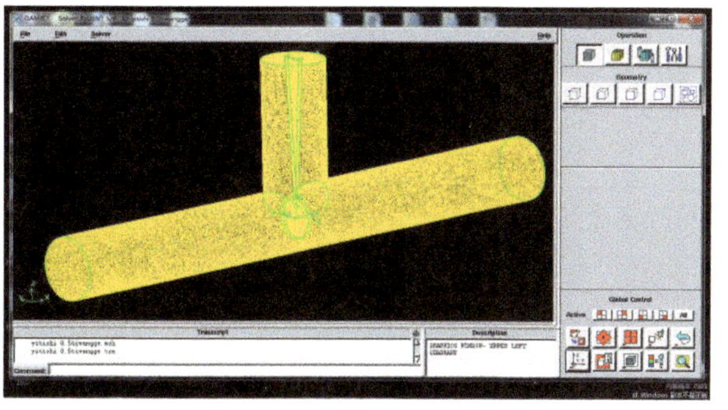

(a)

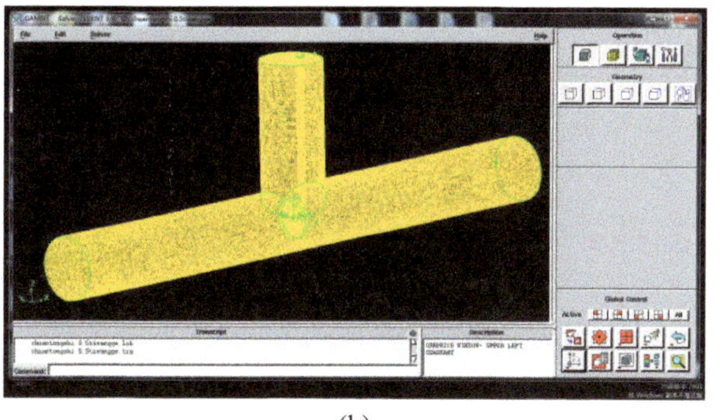

(b)

Fig. 4 One-piece target type flow sensor structure model diagram: (a) one-piece target type structure and (b) traditional model of the target structure.

When solving initialization, the inlet should be selected. The convergence precision k and epsilon are both set to 0.001. Then start to solve, after solving, set up three sections as XOZ (main view), XOY (top view), and YOZ view (left view).

As shown in Fig. 5, there are two kinds of scheme of the top view of the dynamic pressure contours (plane XOY). Figure 6 shows the main view of the velocity contours (plane XOZ). Figure 7 shows the top view of the flow chart (plane XOY).

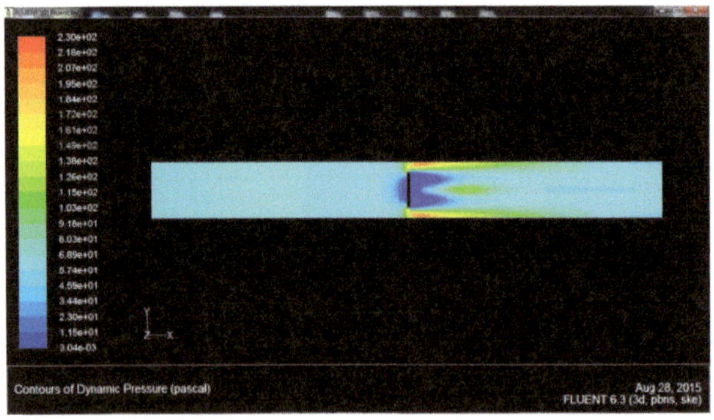

(a)

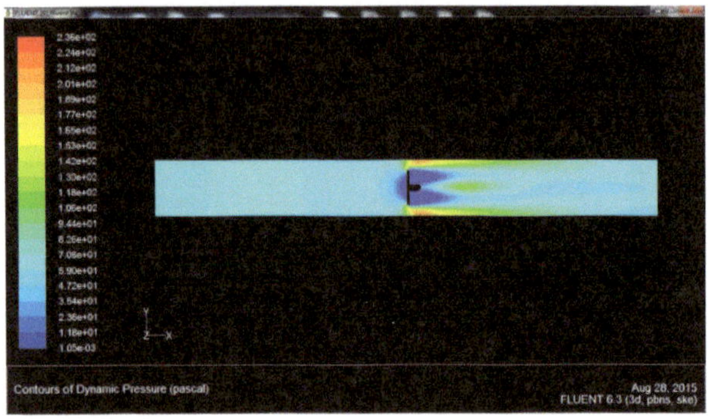

(b)

Fig. 5 Dynamic stress nephogram: (a) one-piece target type structure and (b) traditional model of the target structure.

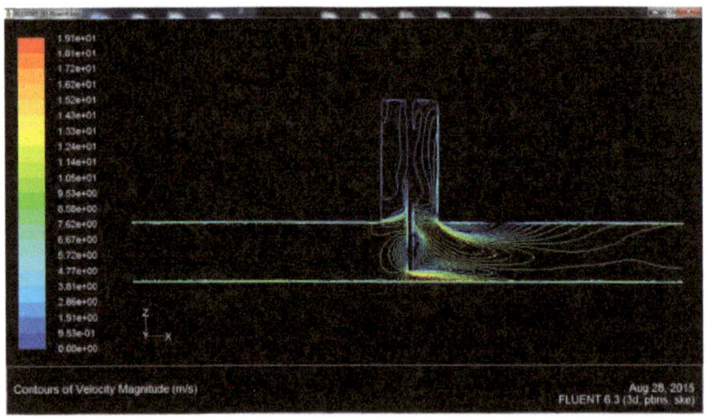

(a)

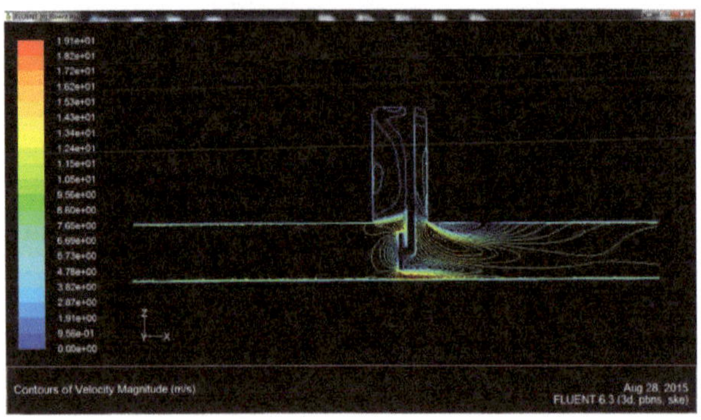

(b)

Fig. 6 Velocity isoline: (a) one-piece target type structure and (b) traditional model of the target structure.

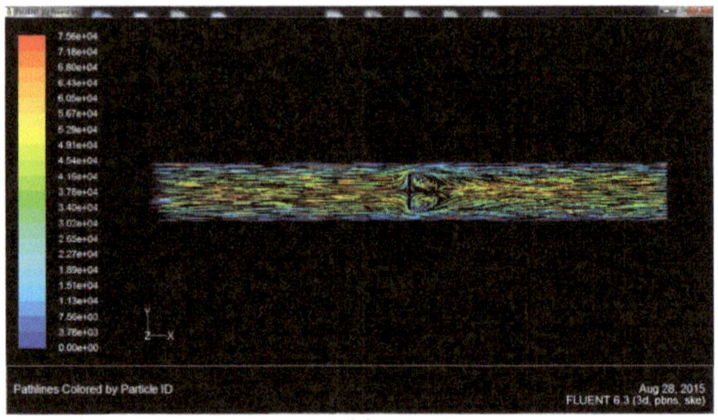

(a)

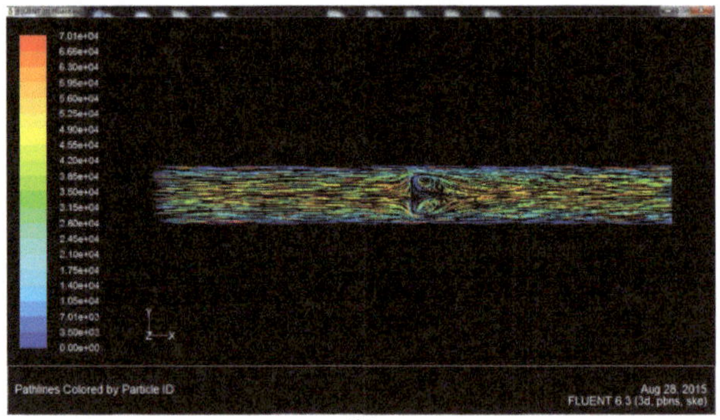

(b)

Fig. 7 Motion pattern: (a) one-piece target type structure and (b) traditional model of the target structure.

Through the above simulation results, it can be seen that with the same flow, one-piece flow sensor target type's influence on the fluid is smaller than that of the traditional target type flow sensor. After the fluid flows through the target piece, state of fluid can restore equilibrium within a short distance. Thus it can be seen that design of the structure of one-piece target type flow sensor is reasonable and feasible.

5. FBG flow sensing experiment

The hydraulic loop is built as shown in Fig.8 in order to further verify the sensing properties of one-piece target type flow sensor based on FBG. Mobile oil source device is mainly composed of hydraulic pumps and fuel tanks and supplies oil for the system. The one-piece target type flow sensor is in the hydraulic circuit. A smart turbine flowmeter is in series as the standard of flow measurement and the flow of the circuit is adjusted by the throttle. The measuring range is between 0.5 m^3/h and 10 m^3/h that between 0.14 L/s and 2.78 L/s, the measurement accuracy is 1.0, the nominal diameter is 10 mm, and the pressure limit is 25 MPa. In the connection of the system, there should be two pipes with the diameter of 8D and 5D on both sides of the target piece according to the requirements of the target flowmeter and two pipes with the diameter of 20D and 5D according to the requirements of the turbine flowmeter. The liquid can only flow through the turbine flowmeter in one way.

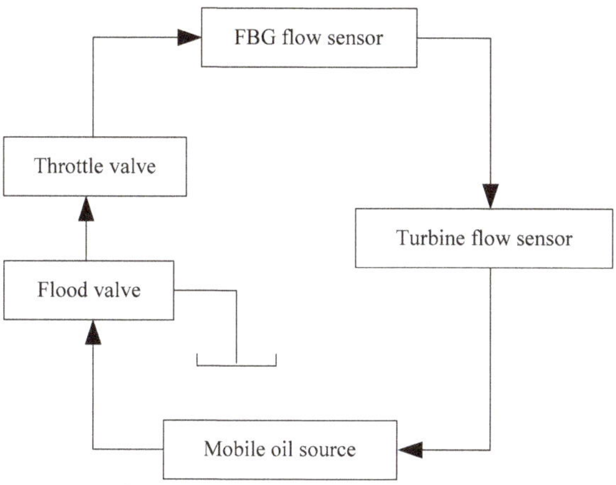

Fig. 8 Flow test system circuit diagram.

Figure 9 shows the assembled one-piece target type flow sensor, and two FBGs with the central wavelength λ_B =1550 nm are pasted symmetrically on both sides of the central axis of the triangle cantilever beam by using the modified epoxy resin adhesives, and they are fused in series by using the optical fiber fusion splicer (OFFS). The optical fiber is between the plywood and pressure diaphragm, and FC plug is fused using the OFFS.

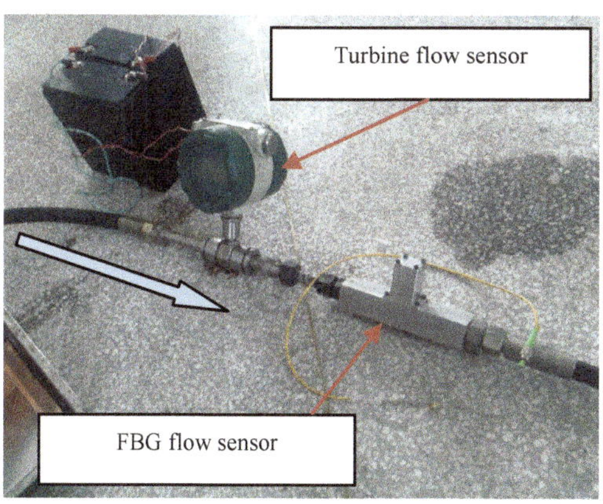

Fig. 9 Flow sensor test system experimental photos.

Fig. 10 shows the experimental photographs. In the hydraulic circuit of flow test, the oil is pumped out from the pump into the overflow valve, throttle valve, the flow testing equipment, the turbine flowmeter, the one-piece flow target type optical FBG sensor, and the tank at last.

Fig. 10 Flow sensor test system experimental photos.

In the experiment, we connect the test oil and the sensing light path, start the oil source, read the central wavelength of a pair of FBG in the demodulation equipment as the experimental data when the flow is 0 L/s and then start the oil source, adjust the opening size of the throttle valve to increase the flow gradually in the test loop, and record the central wavelength values of FBG1 and FBG2 target type flow sensor respectively every 0.2 L/s after the system is stable. The test data is shown in Table 1.

Table 1 Flow sensing experiment.

Flow (L/s)	Central wavelength (nm)				
	FBG1	$\Delta\lambda_{B1}$	FBG2	$\Delta\lambda_{B2}$	$\Delta\lambda_B$
0	1550.083	0.083	1549.944	0.056	0.139
0.2	1550.163	0.163	1549.862	0.138	0.301
0.4	1550.227	0.227	1549.796	0.204	0.431
0.6	1550.355	0.355	1549.672	0.328	0.683
0.8	1550.474	0.474	1549.551	0.449	0.923
1.0	1550.58	0.58	1549.447	0.553	1.133
1.2	1550.761	0.761	1549.266	0.734	1.495
1.4	1550.923	0.923	1549.101	0.899	1.822
1.6	1551.105	1.105	1548.922	1.078	2.183
1.8	1551.332	1.332	1548.688	1.312	2.644
2.0	1551.507	1.507	1548.52	1.48	2.987
2.2	1551.792	1.792	1548.227	1.773	3.565

Figure 11 shows the offset of the central wavelength of FBG1, FBG2, and double FBG after fitting the experimental data.

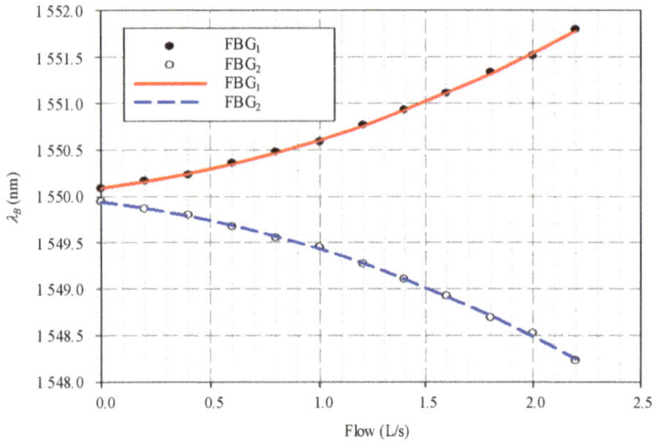

(a)

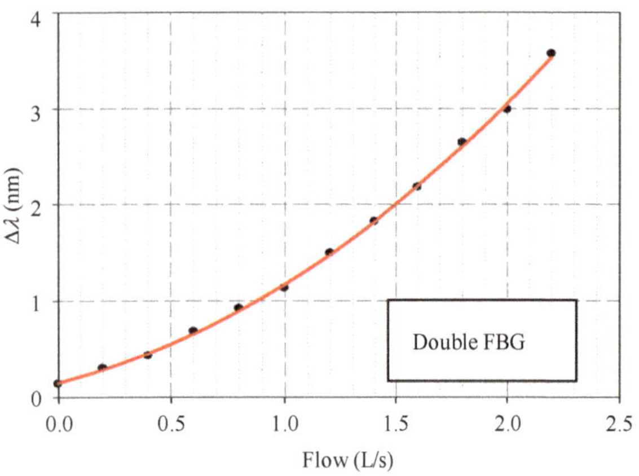

(b)

Fig. 11 Flow sensor experimental data fitting: (a) FBG_1 and FBG_2 central wavelength offset fitting processing and (b) double FBG wavelength offset fitting processing.

According to (10), the offset of the FBG central wavelength in the one-piece target type flow sensor is proportional to the square of the flow, and the test data after fitting are obviously quadratic, so the fitting curve equation is given as

$$\begin{cases} \lambda_{B1} = 1550.087 + 0.213Q^2 \\ \lambda_{B2} = 1549.938 + 0.216Q^2 \\ \Delta\lambda_B = 0.1419 + 0.423Q^2. \end{cases} \tag{11}$$

Fitting curves prove that the offset of the central wavelength of the sensor is proportional to the square of the flow. The offset of the central wavelength of double FBG is twice as large as that of one FBG, and the flow sensor sensitivity of the differential configuration of the double FBG is doubled. And we can get the resolution of the one-piece target type flow sensor is equal to 0.049 L/s from that the quadratic term coefficient is equal to 0.423 nm/(L/s)^2, and the resolution of the demodulation instrument is equal to 1 pm.

6. Conclusions

In the paper, we design a construction which combines target sheet and sensitive element into one product based on the sensing principle of the FBG target type flowmeter. The power transmission structure is omitted, and the triangle cantilever beam is selected as the fastener and the force sensor of the target sheet. Two FBGs, of which the central wavelength of which are equal, are affixed on the central axis of both sides of the cantilever beam because of its sensitivity to the deformation of the cantilever beam caused by the forces on the target sheet. The FLUENT software is employed for the simulation analysis of the influence that one-piece target piece has on the fluid state and the conclusion is reached that the influence of the one-piece target piece structure on the fluid state is less than that of traditional structures. The feasibility and the reliability of the sensor design are verified through flow experiments, the sensor sensitivity is improved effectively, and the measurement repeatability of the sensor is excellent. The one-piece structure target flow sensor based on FBG is compact and convenient to be sealed, dismantled, and installed. What is more, it can adapt to the inflammables, explosion-prone objects, strong electromagnetic radiation, the long-distance test, and the distributed test, so it has broad application prospects.

Acknowledgment

This work was supported by the National Natural Science Foundation of China (No. 41404022) and the Shanxi National Science Foundation (No.2015JM4128).

References

[1] L. Z. Liang, Z. J. Chen, and Y. B. Wang, "Signal detector of propeller flow velocity meter," *Process Automation Instrumentation*, 2003, 24(5): 41–43.

[2] D. D. Zheng, P. Y. Zhang, and T. Zhang, "A method based on a novel flow pattern model for the flow adaptability study of ultrasonic flowmeter," *Flow Measurement and Instrumentation,* 2013, 29(1): 25–31.

[3] T. Leeungculsatien and G. P. Lucas, "Measurement of velocity profiles in multiphase flow using a multi-electrode electromagnetic fow meter," *Flow Measurement and Instrumentation*, 2013, 31(6): 86–95.

[4] C. L. Hu, Z. S. Lin, and Y. Y. Chen, "Portable laser Doppler flowmeter for microcirculation detection," *Biomedical Engineering Letters*, 2013, 3(2): 109–114.

[5] D. S. Wang, The study of flow measurement based on the optical fiber gating system, Ph.D. dissertation, Yanshan University, 2013.

[6] X. Z. Zhang and X. Y. Li, "Optical fiber flow sensor," *Transducer Technology*, 1999, 18(1): 57–59.

[7] M. Majiimder, T. K. Gangopadhyay, and A. K. Chakraborty, "Fiber Bragg gratings in structural health monitoring-present status and applications," *Sensors and Actuators*: *A Physical*, 2008, 147(1): 150–164.

[8] Y. B. Wang and H. J Lan, "Study of fiber Bragg grating sensor system based on wavelength-division multiplexing/time-devision mltiplexing," *Acta Photonica Sinica*, 2010, 30(8): 2196–2201.

[9] D. C. Robertson, P. Niewczas, and J. R. Mcdonald, "Interrogation of a dual-fiber-Bragg-grating sensor using an arrayed waveguide grating," *IEEE Transactions on Instrumentation and Measurement,* 2007, 56(6): 2641–2645.

[10] M. S. Jiang, Q. M. Sui, and L. Jia, "FBG-based ultrasonic wave detection and acoustic emission linear location system," *Optoelectronics Letters*, 2012, 8(3): 220–223.

[11] D. K. Yu, Z. A. Jia, and X. G. Qiao, "FBG sensors combining target with cantilever beam for simultaneous measurement of flow-rate/ temperature," *Journal of Optoelectronics Laser*, 2010, 21(5): 710–713.

[12] Z. F. Wang, J. Wang, and Q. M. Sui, "A target-type fiber Bragg grating flow velocity sensor," *Journal of Optoelectronics Laser*, 2012, 23(1): 1–8.

[13] D. S. Wang, Y. T. Wang, and W. W. Pan, "A circular target flow measurement system," *Applied Mechanics and Materials*, 2013, 427–429: 1233–1236.

[14] H. C. Li, C. T. Liu, and Y. B. Feng, "An embedded nozzle differential pressure style flow sensor based on FBG," *Journal of Optoelectronics Laser*, 2014, 25(10): 1886–1891.

[15] S. Takashima, H. Asanuma, and H. Niitsuma, "A water flowmeter using dual fiber Bragg grating sensors and cross-correlation technique," *Sensors and Actuarors A: Physical*, 2004, 116(1): 66–74.

[16] Y. Zhao, K. Chen, and J. Yang, "Novel target type flowmeter based on a differential fiber Bragg grating sensor," *Measurement*, 2005, 38(3): 230–235.

[17] Y. R. Hu, Y. H. Tian, and C. Li, "Fiber Bragg grating flow sensor," *Chinese Journal of Sensors and Actuators*, 2010, 23(4): 471–474.

9

All-Optical Flip-Flop Based on Coupled SOA-PSW

Lina WANG, Yongjun WANG*, Chen WU, and Fu WANG

School of Electronic Engineering, Beijing University of Posts and Telecommunications, Beijing, 100876, China

*Corresponding author: Yongjun WANG E-mail: wangyj@bupt.edu.cn

Abstract: The semiconductor optical amplifier (SOA) has obvious advantages in all-optical signal processing, because of the simple structure, strong non-linearity, and easy integration. A variety of all-optical signal processing functions, such as all-optical wavelength conversion, all-optical logic gates and all-optical sampling, can be completed by SOA. So the SOA has been widespread concerned in the field of all-optical signal processing. Recently, the polarization rotation effect of SOA is receiving considerable interest, and many researchers have launched numerous research work utilizing this effect. In this paper, a new all-optical flip-flop structure using polarization switch (PSW) based on polarization rotation effect of SOA is presented.

Keywords: Polarization switch; all-optical flip-flop; nonlinear polarization effect; semiconductor optical amplifier

1. Introduction

All-optical signal processing techniques mainly rely on the devices' nonlinear effects to achieve. The intensity, frequency, phase and polarization state of the incoming optical signals can be modulated by using nonlinear effects, which can achieve wavelength conversion, logic gates [1, 3], signal regeneration, and other all-optical signal processing functions [2]. The implementation of the flip-flop [3] in this paper is based on semiconductor optical amplifiers (SOAs). The nonlinear effects of SOA can be summarized into the following four forms: cross-gain modulation (XGM) [3], cross-phase modulation (XPM), four-wave mixing (FWM), and nonlinear polarization rotation [1, 4] (NPR) effects. Compared with several other nonlinear effects, NPR effect can consider the SOA's polarization state of polarized sensitive effectively.

All-optical flip-flop [5–8] is the basic processing unit in the optical signal processing devices. Through a combination between the various flip-flops, it can achieve a variety of high-speed optical signal processing functions. The applications of flip-flops based on SOAs are widely used nowadays. In [3], all-optical flip-flop based on coupled SOA fiber ring lasers was demonstrated. It described four different types of flip-flops. In [9], a set-reset all-optical flip-flop based on Mach-Zehnder interferometer (MZI) [5–7] of SOA was proposed. All-optical flip-flop memory based on two coupled polarization switches was proposed in [10]. Each flip-flop has a different performance and structure. In this paper, a type of clocked all-optical flip-flop structure utilizing NPR effect and polarization switch is presented. We use the bit rates with 1 Gbit/s and 2.5 Gbit/s to demonstrate the feasibility of this structure.

2. SOA-based optical logic gates

2.1 Basic equations of SOA

All-optical logic gate including optical switches, optical computing, and optical regeneration is one of the key components in the field of all-optical signal processing. In this paper, the implementations of all-optical logic gates are based on the NPR effect of SOA. The basic equations describing the working characteristics of SOA are carrier concentration rate equation and optical power transmission equation.

The rate equation of the carrier concentration change is

$$\frac{\partial N(z,t)}{\partial t} = \frac{I}{eV} - R(N) - \frac{\Gamma g(N,\lambda)P(z,t)}{h\nu A} \quad (1)$$

where $N(z,t)$ is the carrier density of the SOA active region at position z in time t, P is the optical signal power of the SOA active region, and $g(N,\lambda)$ is the material gain coefficient of the optical signal.

The equation of the power change is

$$\frac{\partial P_k(z,t)}{\partial z} = \left[\Gamma g(N,\lambda_k) - \alpha_{\text{int}}\right] P_k(z,t). \quad (2)$$

The SOA adopts the segmentation model to simulate

$$\Delta z = L / M \quad (3)$$

where L is the length of the active region and M is the number of segments.

The decomposition of the polarized optical field divides into a transverse electric (TE) and transverse magnetic (TM) component. The output of light power P is given by

$$P_{\text{out}} = P^{\text{TE}} + P^{\text{TM}} \pm 2\sqrt{P^{\text{TE}} P^{\text{TM}}} \cos \Delta\phi \quad (4)$$

where the phase difference is $\Delta\phi = \phi^{\text{TE}} - \phi^{\text{TM}}$. Equations (1)–(4) are the basic equations we use in the numerical calculations.

2.2 All-optical logic gates

In order to achieve all-optical SR and JK flip-flops, it needs two types of logic gates including "AND" and "$A \cap B \cap \overline{C}$". The paper utilizes a simple logical structure to implement logic gates. The schematic setup of all-optical logic gates based on SOA-NPR effect is shown in Fig. 1. The signal A and the signal B simultaneously inject into the SOA. The wavelength of signal A is λ_1, and the wavelength of signal B is λ_2. After two beams are coupled into the SOA, as a result of NPR effect, it implements two types of logic gates. Port 2 outputs "AND" logic gate while Port 1 outputs the "$A \cap \overline{B}$" logic gate. It needs two structures to implement the $A \cap B \cap \overline{C}$ logic gate.

The simulation results of the logic gate are shown in Figs. 2 and 3. The two results utilize different bit rates with 1 Gbit/s and 2.5 Gbit/s. The clock pulse widths are 0.3 ns and 0.15 ns, respectively. The signal A is probe light, signal B is pump light, and the extinction ratio of logic gates reaches more than 30 dB.

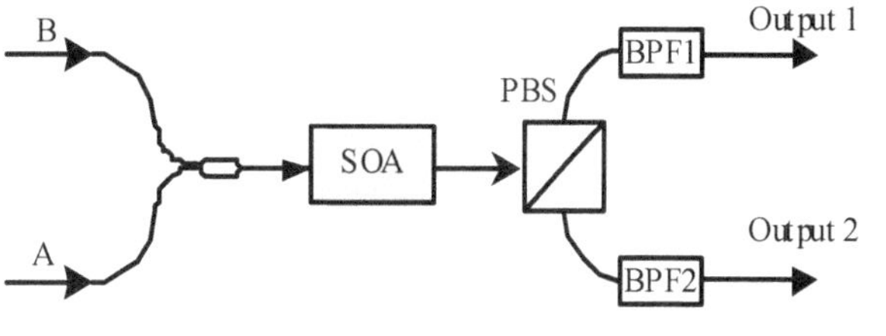

Fig. 1 SOA-based optical logic gates structure.

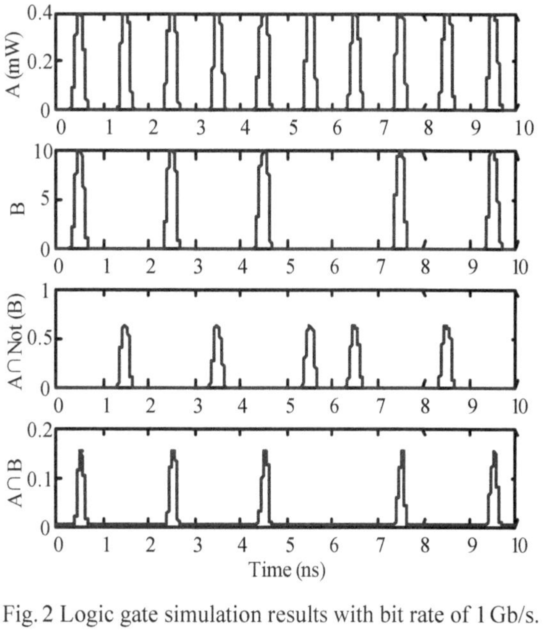

Fig. 2 Logic gate simulation results with bit rate of 1 Gb/s.

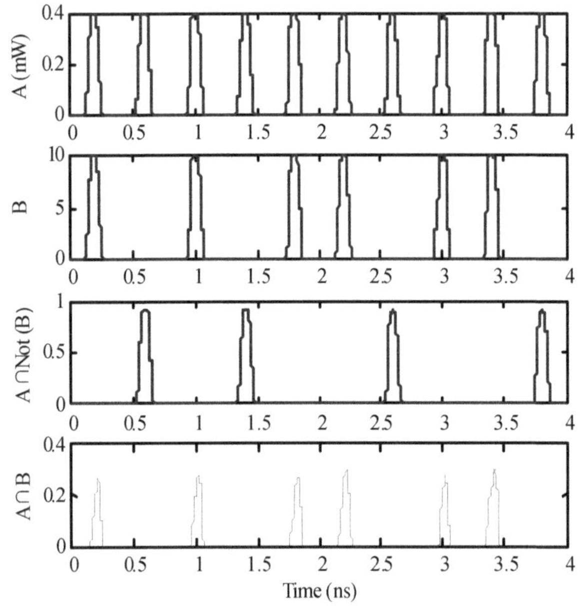

Fig. 3 Logic gate simulation results with bit rate of 2.5 Gb/s.

3. Optical SR latch based on SOA-PSW

As shown in Fig. 4, the optical SR latch has two different states. In State 1, Laser1 emits continuous wave (CW) light at wavelength λ_1 that is amplified by SOA_1, and then the output light is sent into polarizing beam splitter (PBS). The refractive index of the TE mode and TM mode is changed by the light that is fed into the SOA_2 through PBS. It makes the Laser2's output light with wavelength λ_2 after through SOA_2 can not enter PSW_1 (PSW: polarization switch). At this time, the PSW_1 is dominant and PSW_2 is subordinate. The port Output1 emits the light at wavelength λ_1 while the port Output2 has no output. Similarly, in State 2, the port Output2 emits the light at wavelength λ_2 while the port Output1 has no output. In order to change the state of the flip-flop, the light pulse of set or reset is injected into the dominant PSW, which changes the refractive index of the TE and TM modes, so that the PSW's light polarization state fails to pass PBS. Thus the SOA of the original slave PSW will not contain the injected high power light, so the light in the original slave PSW can go through PBS and thereby enter the dominant PSW, thus two PSWs swap positions and complete the state switching.

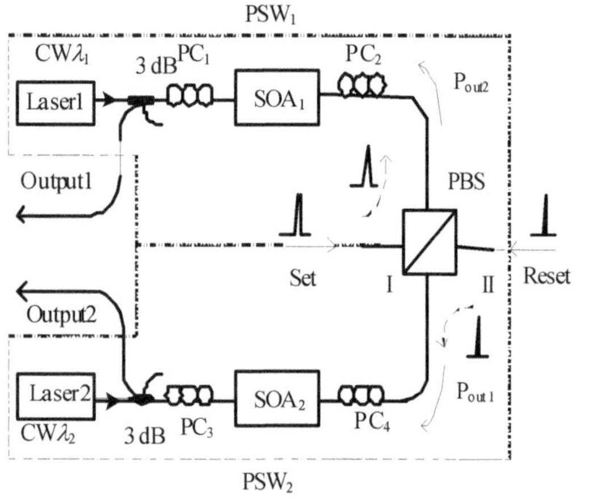

Fig. 4 Experimental setup of all-optical SR latch based on two coupled SOA-PSW.

The advantage of this structure is using two symmetrical polarization switches and an interference structure. In addition, the set and reset operations are symmetrical, the state switching speed is fast, and it only depends on the length of fiber between the two SOAs.

4. Two types of all-optical clocked flip-flops

In order to have a more detailed description of the proposed structure, this paper describes and illustrates two types of all-optical clocked flip-flops including SR and JK flip-flops. The all-optical clocked SR and JK flip-flops work with a clocked synchronization signal.

4.1 SR flip-flop

The SR flip-flop setup is shown in Fig. 5. This logic structure includes two "AND" logic gates and one optical SR latch. The clock (CLK) signal and S signal compose one "AND" logic gate, and the CLK signal and R signal form the other "AND" logic gate. Then the two outputs are separately sent into the set and reset latch ports.

In the stable operation time, the flip-flop is in a steady state. Only when a clock pulse comes in, depending on the S and R values at that time, the state transition occurs, that is so-called the state switching time. In addition to the above cases, the flip-flop no longer changes its state. That is called flip-flop's clocked synchronization.

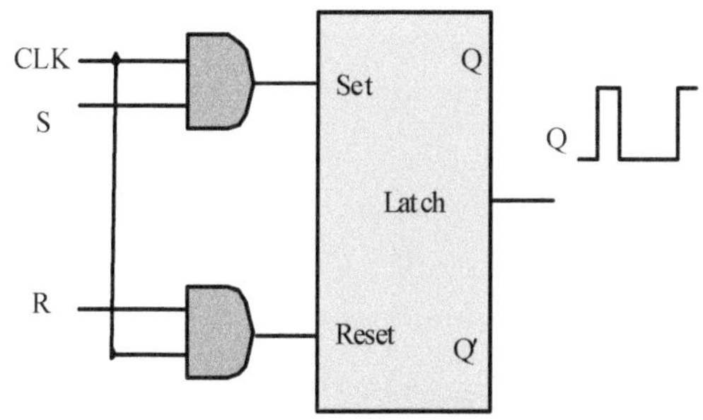

Fig. 5 SR flip-flop logic structure.

The clocked SR flip-flop experimental results are shown in Figs. 6 and 7. In Fig. 6, the clock pulse has a repetition rate of 1 GHz with a pulse width of 0.3 ns. In Fig. 7, the clock pulse has a repetition rate of 2.5 GHz with a pulse width of 0.15 ns. The figures show that the flip-flop's falling edge time is shorter than the rising edge time, because the gain increases slowly to its recovery time after a decrease in the SOA gain.

4.2 JK flip-flop

The JK flip-flop setup is shown in Fig. 8. This logical structure includes two $A \cap B \cap \overline{C}$ logic gates and one SR latch. The CLK signal, J signal, and Q signal compose one $A \cap B \cap \overline{C}$ logic gate, and the CLK signal, K signal, and inverted Q signal form the other $A \cap B \cap \overline{C}$ logic gate. Then the two outputs are separately sent into SR latch ports.

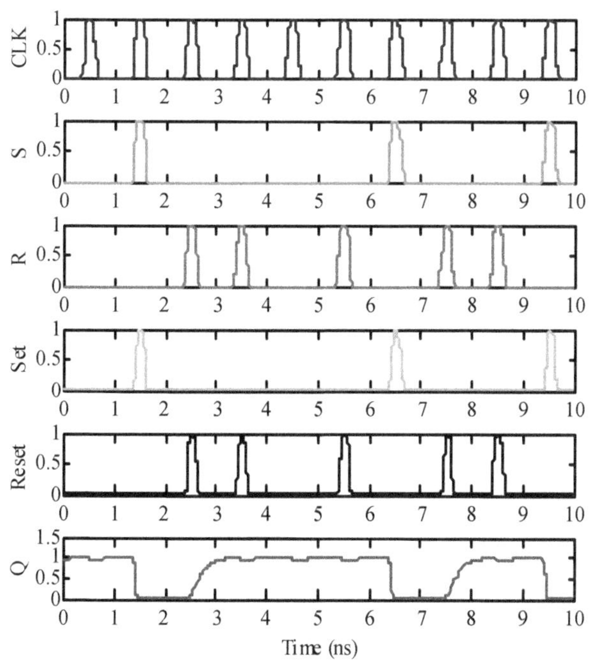

Fig. 6 SR flip-flop simulation result with bite rate of 1 Gb/s.

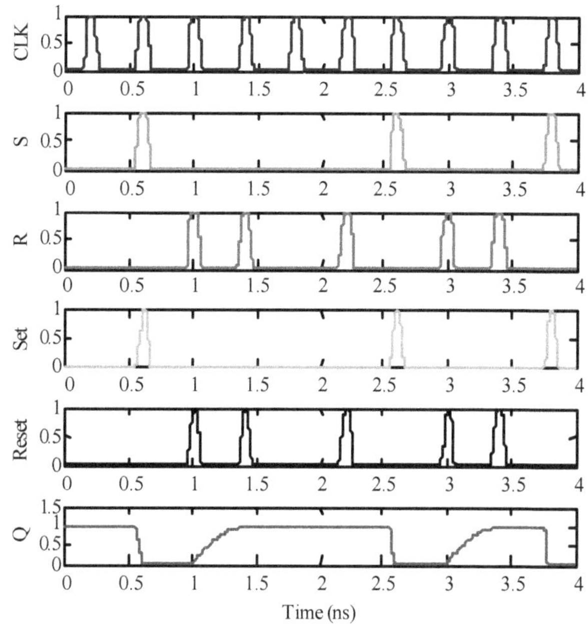

Fig. 7 SR flip-flop simulation result with bit rate of 2.5 Gb/s.

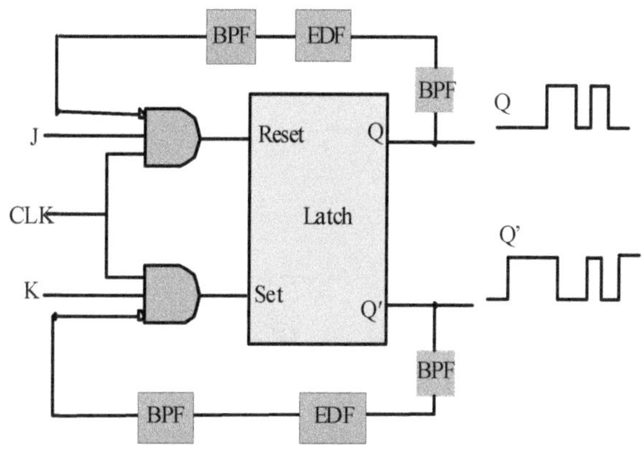

Fig. 8 JK flip-flop logic structure.

The JK flip-flop state depends on the injection of external input signals and two previous opposite states. The difference with the SR flip-flop is that the JK flip-flop requires two feedback signals as input.

In the stable operation time, the flip-flop is in a steady state. Only when a clock pulse comes in, depending on the J, K values and two feedback signals at that time, the state transition occurs. The same with the SR flip-flop, the clock pulse has a repetition rate of 1 GHz and 2.5 GHz. The external input signals' pulse width are 0.3 ns and 0.15 ns, respectively. The transition time of 2.5 Gb/s bit rate is slightly better than 1 Gb/s. In Fig. 10, the falling edge time is about 30 ps while the falling time is about 40 ps in Fig. 9. Figs. 9 and 10 show that the flip-flop's falling edge time is shorter than the rising edge time, because the gain increases slowly to its recovery time after a decrease in SOA gain. Two different results indicate that this structure can handle different bit rates.

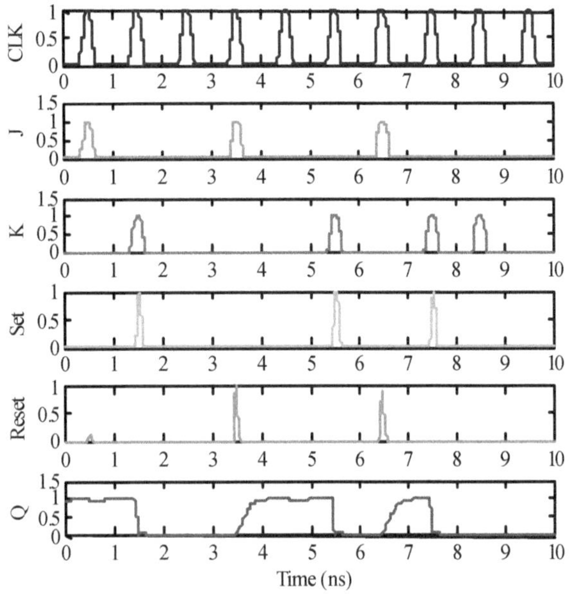

Fig. 9 JK flip-flop simulation results with bit rate of 1 Gb/s.

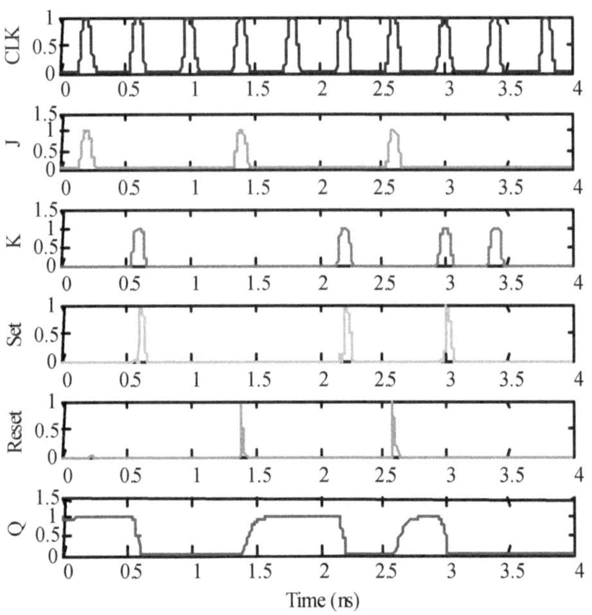

Fig. 10 JK flip-flop simulation results with bit rate of 2.5 Gb/s.

5. Conclusions

In this paper, we have discussed two aspects including the logic gates based on SOA-NPR effect and the flip-flops based on two coupled nonlinear polarization switch of SOA. According to the working principle of SOA, we achieve all-optical SR and JK flip-flops with the clock signal pulse repetition frequency of 1 GHz and 2.5 GHz, and the clock pulse widths are 0.3 ns and 0.15 ns, respectively. In the simulation, the SOA parameters such as carrier density, gain factor, the carrier lifetime, and optical gain are important to ensure the correct of the flip-flop operation. The final results show that the work in this paper has a certain significance for the study of all-optical flip-flop.

Acknowledgment

Thanks to the financial supports from the National Natural Science Foundations of China (61378061 and 61575026).

References

[1] L. Yi, W. Hu, and H. He, "Reconfigurable multi-logic gates based on SOA nonlinear polarization rotation effect," in *Optoelectronics and Communications Conference*, Sapporo, pp. 194–195, 2010.

[2] B. Dagens, A. Labrousse, R. Brenot, B. Lavigne, and M. Renaud, "SOA-based devices for all-optical signal processing," in *Optical Fiber Communications Conference*, Atlanta, vol. 2, pp. 582–583, 2003.

[3] J. Wang, G. Meloni, G. Berrettini, L. Poti, and A. Bogoni, "All-optical clocked flip-flops and binary counting operation using SOA-based SR latch and logic gates," *IEEE Journal of Selected Topics in Quantum Electronics*, 2010, 16(5): 1486–1494.

[4] H. J. S. Dorren, D. Lenstra, Y. Liu, M. T. Hill, and G. D. Khone, "Nonlinear polarization rotation in semiconductor optical amplifiers: theory and application to all-optical flip-flop memories," *IEEE Journal of Quantum Electronics*, 2003, 39(1): 141–148.

[5] P. Bakopoulos, K. Vyrsokinos, D. Fitsios, T. Alexoudi, D. Apostolopoulos, H. Avramopoulos, *et al.*, "All-optical t-flip-flop using a single SOA-MZI-based latching element," *IEEE Photonics Technology Letters*, 2012, 24(9): 748–750.

[6] T. Chattopadhyay, C. Reis, P. André, and A. Teixeira, "Theoretical analysis of all-optical clocked D flip-flop using a single SOA assisted symmetric MZI," *Optics Communications*, 2012, 285(9): 2266–2275.

[7] H. Uenohara, S. Shimizu, T. Kato, Y. Tatara, K. Goto, N. Fukui, *et al.*, "All-optical flip-flop circuit based on SOA-MZI," in *Asia Communications and Photonics Conference and Exhibition*, Shanghai, pp. 691–692, 2010.

[8] D. K. Gayen, T. Chattopadhyay, and K. E. Zoiros, "All-optical D flip-flop using single quantum-dot semiconductor optical amplifier assisted Mach-Zehnder interferometer," *Journal of Computational Electronics*, 2015, 14(1): 129–138.

[9] J. Sohrabtash, A. Zarifkar, and M. H. Sheikhi, "A high performance all-optical set-reset flip-flop based on SOA-MZI," *Optoelectronics Letters*, 2014, 10(6): 430–433.

[10] Y. Liu, M. T. Hill, H. D. Waardt, G. D. Khone, D. Lenstra, and H. J. S. Dorren, "All-optical flip-flop memory based on two coupled polarization switches," *Electronics Letters*, 2002, 38(16): 904–906.

10

A Low-Cost, Portable Optical Sensing System With Wireless Communication Compatible of Real-Time and Remote Detection of Dissolved Ammonia

Shijie DENG[1*], William DOHERTY[2], Michael AP MCAULIFFE[1], Urszula SALAJ-KOSLA[3], Liam LEWIS[1], and Guillaume HUYET[1,2]

[1]*Centre for Advanced Photonics & Process Analysis, Cork Institute of Technology and Tyndall National Institute, Cork, Ireland*

[2]*Department of Physical Science, Cork Institute of Technology, Cork, Ireland*

[3]*Department of Chemical and Environmental Science, University of Limerick, Limerick, Ireland*

[*]Corresponding author: Shijie DENG E-mail: shijie.deng@cit.ie

Abstract: A low-cost and portable optical chemical sensor based ammonia sensing system that is capable of detecting dissolved ammonia up to 5 ppm is presented. In the system, an optical chemical sensor is designed and fabricated for sensing dissolved ammonia concentrations. The sensor uses eosin as the fluorescence dye which is immobilized on the glass substrate by a gas-permeable protection layer. A compact module is developed to hold the optical components, and a battery powered micro-controller system is designed to read out and process the data measured. The system operates without the requirement of laboratory instruments that makes it cost effective and highly portable. Moreover, the calculated results in the system can be transmitted to a PC wirelessly, which allows the remote and real-time monitoring of dissolved ammonia.

Keywords: Ammonia sensing; optical chemical sensor; portable optical sensing system; remote sensing

1. Introduction

Dissolved ammonia is an environmental toxicant that is especially problematic for aquatic organisms. Ammonia (NH_3) accumulates easily in aquatic systems because it is a natural byproduct of fish metabolism. All animals excrete some waste in the process of metabolizing food into the energy, nutrients, and proteins they use for survival and growth. For fish, the principal metabolic waste product is ammonia. Because it is continuously excreted and potentially lethal, successful aquaculture operations must therefore incorporate methods to detect and eliminate ammonia before it can accumulate and harm the aquatic life [1, 2]. As a result, there is a real requirement to develop a cost effective and low maintenance system to allow for continuous monitoring of dissolved ammonia concentrations in aquatic environments.

Advances in optical chemical sensors including sensitive, highly selective, easy to miniaturize, electrical, and magnetic interference-free, reference electrode independent and operational in varied environments (e.g. air or liquid) have allowed them to be used in a wide range of applications such as medical application, environmental and pharmaceutical analysis, and process control [3–6]. In addition, the wide availability of the miniature photo-detectors and light sources and the broad usage of optical fibers make the optical chemical sensors very attractive for applications requiring portable and compact sensing solutions. In this work, to develop a portable ammonia sensing system, it is decided to design an optical chemical sensor for the detection of ammonia in the liquid.

A small number of optical chemical sensors based ammonia sensing systems have been developed for the detection of dissolved ammonia [7–10]. However, those systems require laboratory instruments to operate (e.g. laser source for incident light and a spectrometer or charge-coupled device (CCD)/color camera for sensor results readout). This makes them bulky and expensive, which limits the portability of the device.

In this work, an optical chemical sensor based ammonia sensing system that operates independently of laboratory instruments is developed. It is able to detect dissolved ammonia up to 5 ppm and a lower detection limit of 0.1 ppm is achieved with longer response time. In the system, an optical chemical sensor is developed and used for sensing dissolved ammonia. Eosin is used as the fluorescence dye which is immobilized on a glass substrate by a gas-permeable protection layer. A compact module is designed for housing the optical components, and a battery powered micro-controller based circuit is designed to control the incident light signal from the light-emitting diode (LED) and read out the fluorescence signal from the sensor. This makes the system highly portable and enhances its usefulness for compact applications. Moreover, the system allows users to transmit the measurement data to a PC through the wireless transceivers that makes it possible for real-time and remote monitoring.

2. System description

Figure 1 shows the block diagram of the ammonia sensing system developed, which consists of two main parts: A. the optical components and sensor module, which includes the optical ammonia sensor, an LED, a photo-detector, optical filters, and the holder for the components; B. the battery powered micro-controller based printed circuit board (PCB), which is used to modulate the light signal from the LED, process the fluorescence signal collected by the detector and transmit the data to a PC through the wireless transceivers.

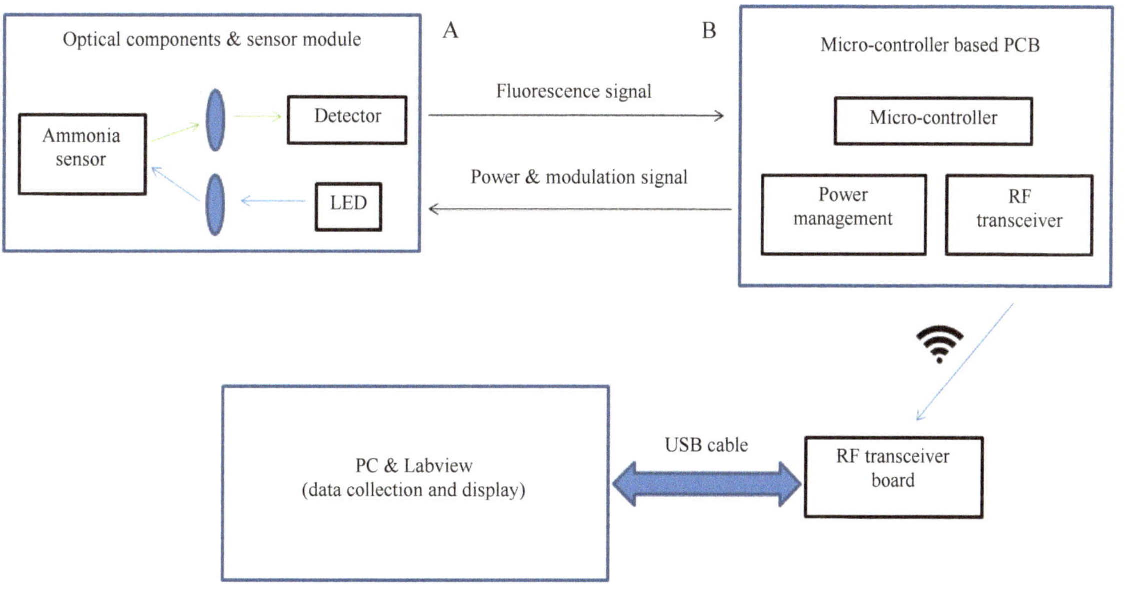

Fig. 1 Block diagram of the ammonia detection system. A and B highlight the optical and sensor module and the micro-controller based PCB, respectively.

2.1 Development of optical ammonia sensor

Ammonia (NH_3) is an essential nutrient for plant and is not accessible in its molecular form. Bacteria can convert the dissolved ammonia to nitrite (NO_2^-) and nitrates (NO_3^-) which can then be used by plants, a process known as "fixing". On interacting with water, ammonia is in equilibrium with ammonium ion and hydroxide ion as shown in (1):

$$NH_3 + H_2O \leftrightharpoons NH_4^+ + OH^- . \quad (1)$$

The equilibrium lies to the right, meaning the vast majority of nitrogen exists in the ammonium ion form, rather than the ammonia form. In the ammonium form, it is quite harmless, and larger concentrations can be supported.

However, this equilibrium position is dependent on the pH of the system. Changes in pH can have a detrimental effect on the system. Given that the dissociation constant K_A of a system is by definition constant and applying Le Châtelier's principle to the following equation, the effect of pH change can be seen. The dissociation constant of ammonium-ammonia system can be seen in (2) and (3):

$$NH_4^+ \leftrightharpoons NH_3 + H^+ \quad (2)$$

$$K_A = \frac{[NH_3][H^+]}{[NH_4^+]} . \quad (3)$$

If the pH increases, effectively there are less ions. The system, which can be considered originally in equilibrium, has been disturbed, and will rearrange the distribution of concentration of components of the reaction so as to minimize the effects of the disturbance, i.e. K_A must be kept constant. This is achieved by producing more ions, which also produces more NH_3. Therefore, increasing the pH releases more NH_3 into the system.

The ammonia sensor designed in this work is based on the change in optical properties associated with a change in pH. To achieve this, a three-layer sensor is developed which consists of a solid inert support, a pH sensitive matrix, and a gas-permeable protection layer (see Fig. 2). Supports used are glass microscope slides. The pH dye chosen is eosin which is a fluorescent dye with the advantage of having a low fluorescent state in low pH conditions and higher fluorescence in higher pH conditions [pH>3.0, greater than the pKa (acid dissociation constant at logarithmic scale) of eosin]. It has an absorption maximum at 517 nm and an emission maximum at 538 nm [11].

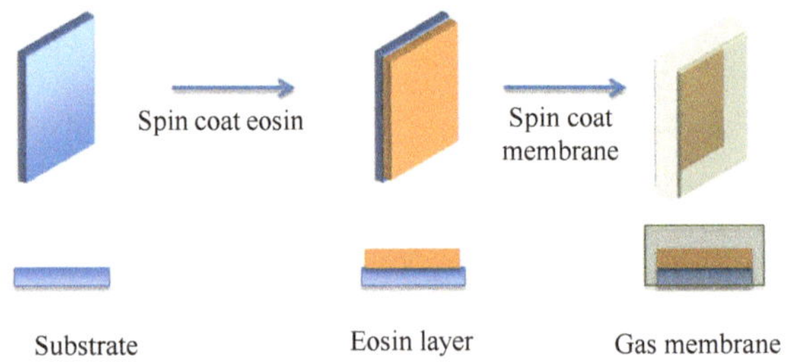

Fig. 2 Fabrication of the ammonia sensor.

The sensing film is comprised of eosin (1 mM), microcrystalline cellulose (MCC) (5% w/w), and methylsulfonic acid (MSA) (0.7 μM) dissolved in a 99.5:1 acetone: water mixture. MCC is hygroscopic and will absorb moisture allowing a localized aqueous-type environment to form. The eosin is also absorbed in this aqueous-type environment and can respond to changes in the pH surrounding it. MSA (pKa = –1.9) is used to ensure that the local pH is below the pKa (2.9) of eosin. This protonates eosin and keeps it in the low-fluorescent state. Upon spin-coating (600 rpm, 50 s), the solvent evaporates readily leaving a thin (70 micron) sensor film on the support. The sensor film is left to dry at room temperature for 24 hours and is then placed overnight in 0.7 μM MSA solution to ensure protonation of the eosin dye.

To prevent the sensor film from sensing the bulk solution pH, a gas permeable silicone layer (25% w/w in n-heptane) is spin coated (1200 rpm, 50 s) on top (100 micron nominal). Dissolved ammonia gas can diffuse through this layer, and ions and other liquid species cannot cross this gas-permeable membrane. In this manner, the sensor measures the local pH on the film side of the membrane and not the bulk solution pH outside the membrane.

One chemical pathway can be thought of as follows. This ammonia reacts with the water surrounding the eosin and MCC to form ammonium ions and hydroxide ions. This raises the local pH resulting in a change from the low-fluorescent to the high-fluorescent state.

$$\overset{\text{bulk}}{NH_{3(g)}}\ \text{diffuse}\ \overset{\text{sensor}}{NH_{3(g)} + H_2O_{(l)} \rightleftharpoons NH^+_{4(aq)} + OH^-_{(aq)}} \quad (4)$$

Equation (4) shows the reaction scheme showing the interaction of dissolved ammonia in the bulk phase with the local pH and is illustrated in Fig. 3.

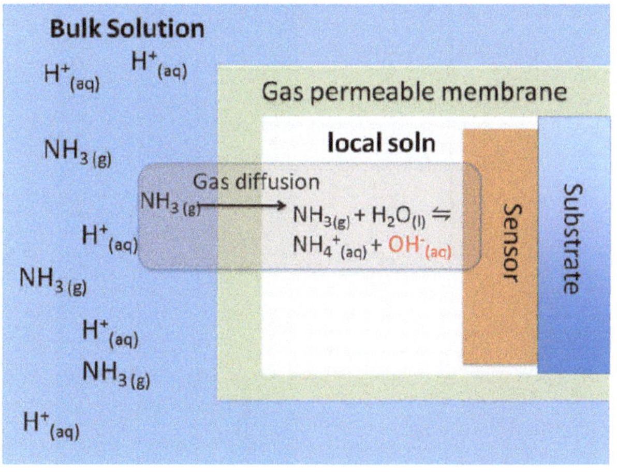

Fig. 3 Schematic of sensor mechanism illustrating (in boxed region) the diffusion of dissolved $NH_{3(g)}$ through the gas-permeable membrane and changing local pH (increase in OH-concentration).

Another chemical pathway could be the direct deprotonation of eosin by dissolved ammonia. In either way of both chemical pathways, the overall effect is the same, and eosin is deprotonated to form the highly-fluoresecent anion.

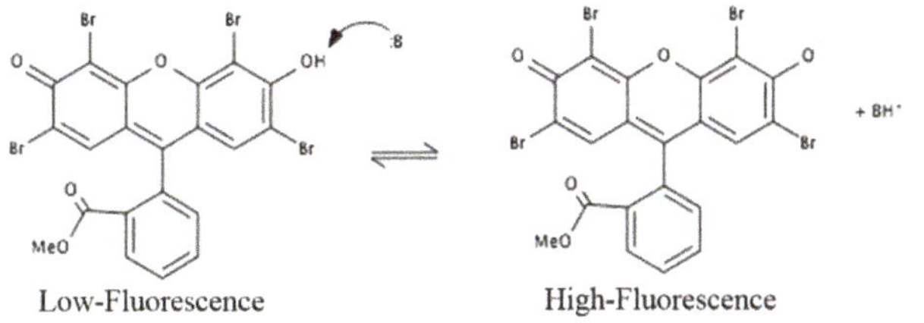

Low-Fluorescence High-Fluorescence

Fig. 4 Deprotonation of eosin. In the case of ammonia B=NH_3 and $BH^+ = NH_4^+$. In the case of hydroxide ion pathway, simply substitute B with OH^- and BH^+ with.

Standard solutions of ammonium ion to produce to required concentration of ammonia are prepared as follows.

Buffer solutions at pH7 are prepared using the 10:6 ratio of 0.01 M sodium dihydrogen phosphate and 0.01 M disodium hydrogen phosphate solutions.

Using the Hendersen-Hasselbalch equation the concentration of $[NH_3]$ can be calculated:

$$[NH_3] = [NH_4^+] \times 10^{(pH-pKa)} \quad (5)$$

or to obtain a given concentration of $[NH_3]$, the concentration of $[NH_4^+]$ can be determined:

$$[NH_4^+] = [NH_3] \times 10^{(pKa-pH)} \quad (6)$$

or given that the pKa is 9.24 and the pH is 7

$$[NH_4^+] = [NH_3] \times 10^{2.24}. \quad (7)$$

Table 1 outlines the concentrations of NH_4Cl made up in the (phosphate) pH7 buffer solutions.

Table 1 Concentrations and masses of NH_4Cl in 1 L of pH7 phosphate buffer solution to produce 1, 2, 3, 5 and 10 ppm NH_3 solutions, using (5).

$[NH_3]$ (ppm)	$[NH_3]$ (mM)	$[NH_4Cl]$ (mM)	$[NH_4Cl]$ (mgL^{-1})
1	0.059	10.20	545.8
2	0.117	20.41	1091.6
3	0.176	30.61	1637.4
5	0.294	51.02	2729.0
10	0.587	102.04	5458.1

2.2 Development of optics and electronics

The optical components and sensor module designed is shown in Fig. 5(a). The size of the module is approximately 38 mm×38 mm×30 mm.

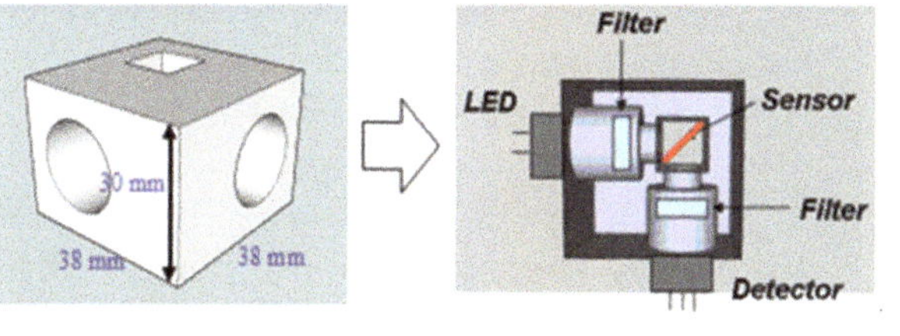

(a) (b)

Fig. 5 Optical components and sensor module designed: (a) optical components and sensor holder and (b) schematic diagram of the optical components within module.

The incident light and emitted fluorescence paths are divided into 2 channels. The LED has a peak emission wavelength of 490 nm, and the detector has a spectral range of 400 nm to 800 nm. An optical band-pass filter (475 ± 25 nm) is placed after the LED, and an optical long-pass filter (530 nm) is placed before the detector [Fig. 5(b)]. A hole (13 mm×13 mm×30 mm) in the module allows water to flow through to fill the region where the ammonia sensor is fixed across the diagonal. For this work, a cuvette is used with the sensor placed within a fixed volume of water and varying amounts of ammonia solution added to the water. The position of the ammonia sensor is fixed within the cuvette.

The electronics design is shown in Fig. 6. The main board is a micro-controller based PCB which consists of a micro-controller chip, power management circuit, A/D converter (ADC), Op-amp, and radio frequency (RF) transceiver. Another transceiver board is used to receive data sent from the main board and push the data to one of the serial COM ports on the PC through a USB cable.

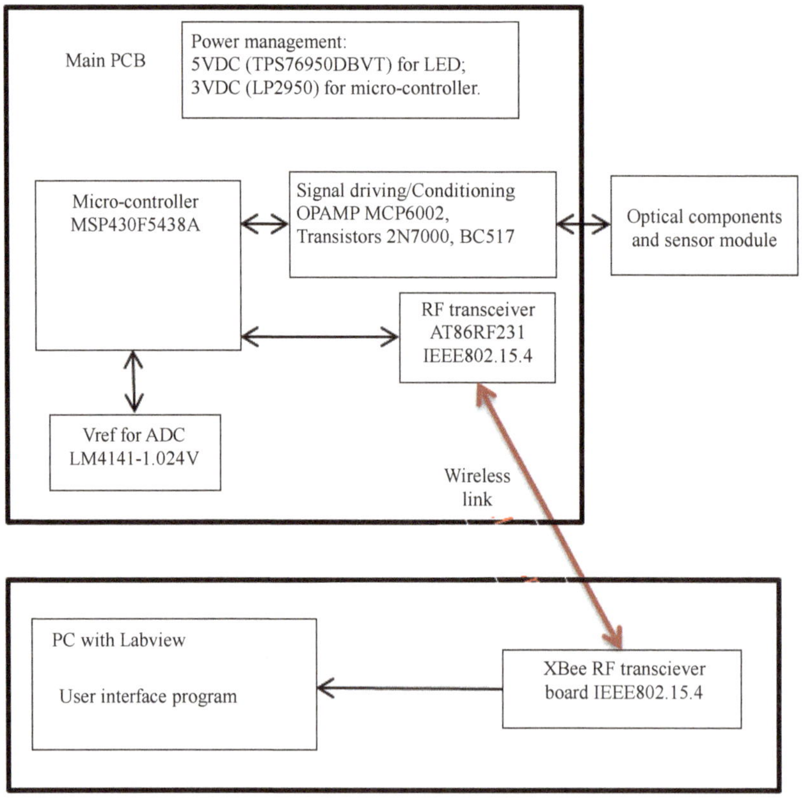

Fig. 6 Electronics design of the sensing system.

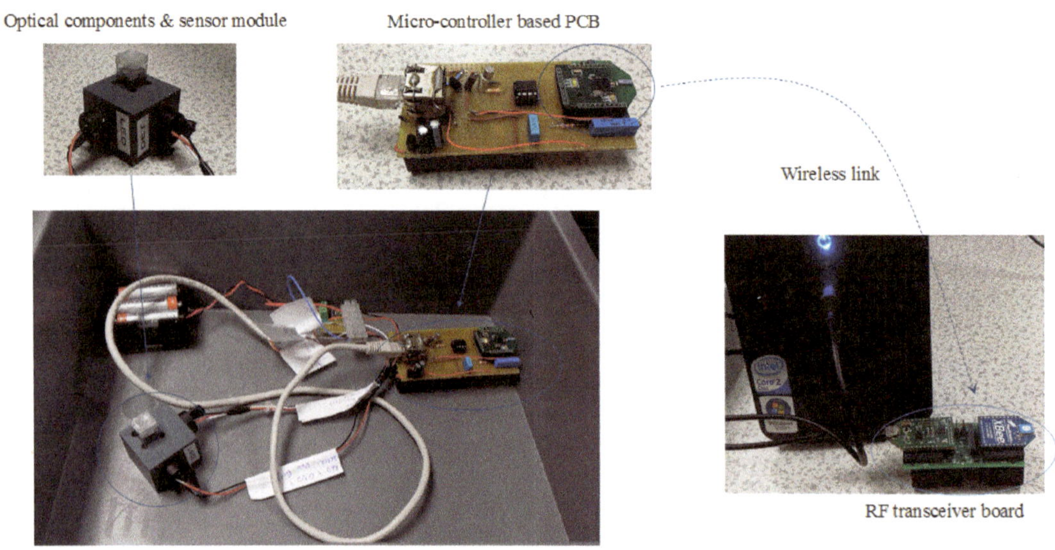

Fig. 7 Photograph of the ammonia sensing system developed.

To avoid photo bleaching, the micro-controller is programmed to modulate the excitation light from the LED. The LED is on for 400 micro-seconds in every 15 s. The photodiode detects the fluorescence signal emitted from the ammonia sensor, which is collected by the micro-controller through an A/D converter. The micro-controller then calculates the peak-to-peak (pk-pk) amplitude and sends the results through an RF transceiver on the board to a second RF transceiver that transmits the data to a PC through a USB cable. The results can be displayed using a custom-written Labview routine on the PC.

3. Experimental results

A photograph of the system developed is shown in Fig. 7. The system is powered by 3 AA batteries, and the measurement dataare transmitted wirelessly to a PC using RF transceivers and processed using a custom-written Labview routine.

Figure 8 shows the response of the system for varies ammonia concentrations. During the measurement, the sensor is rinsed in a zero buffer (0 ppm of ammonia, pH7) and placed inside the cuvette, calibrated volumes of dissolved ammonia are pipetted into the cuvette. In the system, as there is an overlap between the LED's spectral output (490 nm) and the photo-detector's detection range (400 nm ~ 800 nm), the optical filters used do not block all the LED's light which is scattered to the photo-detector. This "leaked" excitation light from the LED is then picked up by the photo-detector that forms a baseline which is constant. To clearly demonstrate the changes of the fluorescence intensity emitted from the sensor for varies ammonia concentrations, the baseline has been deducted in the plots.

For the calibration, an average of the last 100 data points for each dissolved ammonia concentration are used to generate the calibration curve, as shown in Fig. 9. A three-parameter exponential decay function ($y=A_1\exp(-x/t_1)+y_0$) is fitted to the sensor response which gives an accurate fit of r^2=0.999.

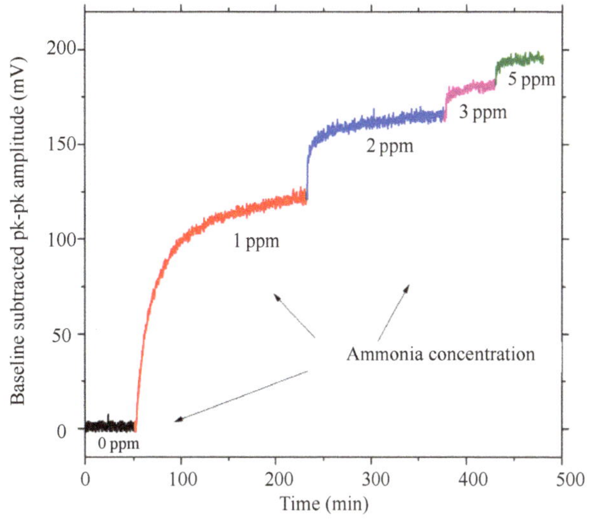

Fig. 8 Response curve of the system for various concentrations of dissolved ammonia in a buffer solution (pH7).

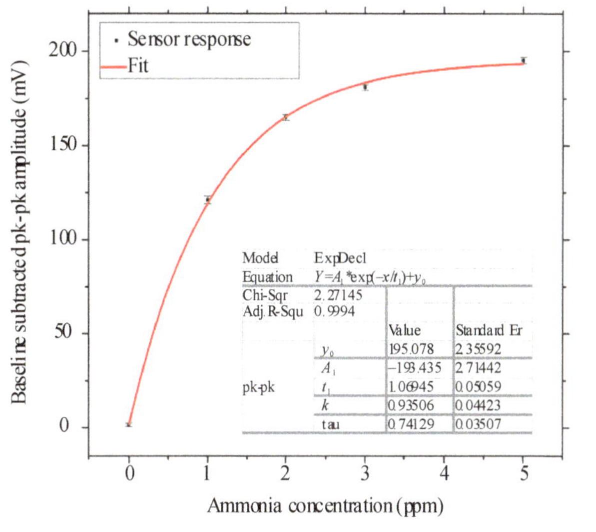

Fig. 9 Calibration curve, generated from the average of the last 100 data points of different dissolved ammonia concentrations. Error bars represent the standard deviation.

Figure 10 shows the response time of the sensor for a variety of dissolved ammonia concentrations. The typical response time from 0 ppm to 0.1 ppm, 1 ppm, and 10 ppm ammonia concentrations are 200 minutes, 80 minutes, and 20 minutes, respectively. This response is a convolution of diffusion process across the membrane and numbers of eosin molecules which have been deprotonated into the higher fluorescent state. As the numbers of fluorescent molecules are determined by the local pH of eosin, this exponential curve is expected. By varying the concentration of eosin and/or the amount of MSA, the response curve can be changed to give a stronger response or a more sensitive response. Larger concentrations of MSA would require larger concentrations of ammonia to affect the requisite change in local pH to enable a fluorescent state to form. Hence the applicable sensing range can be fine-tuned. Response time is mainly due to diffusion controlled process across the gas-permeable membrane. To achieve faster time, a thinner and structurally complete membrane is required.

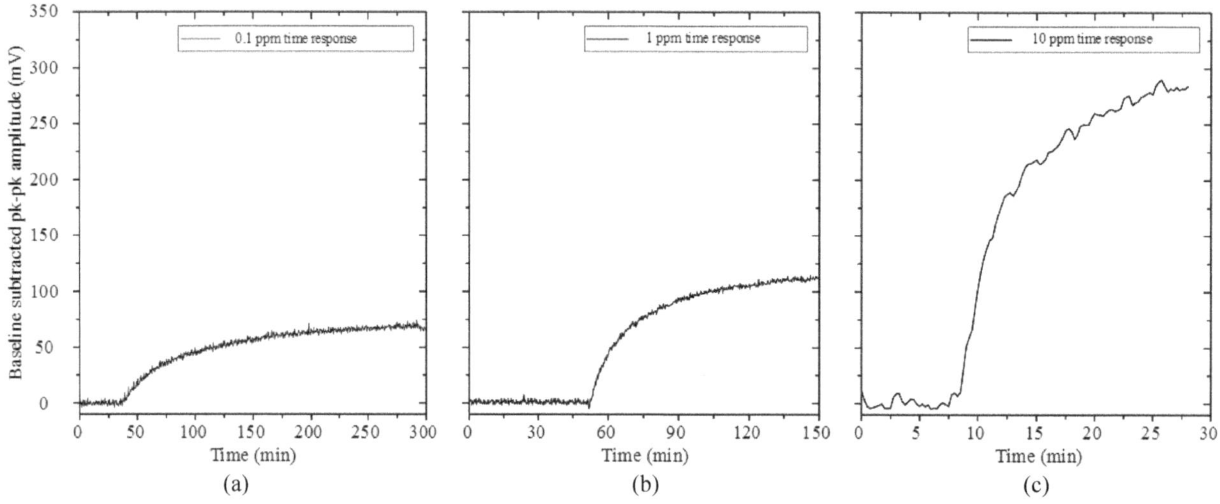

Fig. 10 Response time of the sensor for varies dissolved ammonia concentrations: (a) 0.1 ppm, (b) 1 ppm, and (c) 10 ppm.

4. Conclusions

In this work, an optical chemical sensor based ammonia detection system is developed, which is capable of detecting dissolved ammonia up to 5 ppm. To fabricate the optical chemical sensor in the system, eosin is used as the fluorescence dye which is immobilized on the glass substrate by a gas-permeable protection layer. The sensing system uses a compact module for housing the optical components and a battery powered micro-controller based circuit for the control of the incident light signal from the LED and readout of the fluorescence signal from the sensor. This allows the system to operate independently of the laboratory instruments that makes it low-cost and highly portable, and enables its usefulness for compact applications. Moreover, the system can communicate with a PC through wireless transceivers that makes it possible for real-time and remote monitoring of dissolved ammonia.

Acknowledgment

This work was supported by Enterprise Ireland (EI) and Science Foundation Ireland (SFI). The authors would like to thank Pat O'Leary from Faaltech Technologies Ltd and Jean-Michel Rubillon from Cork Institute of Technology for their support and help.

References

[1] L. M. Nollet, *Handbook of water analysis*. Japan: Food Science and Technology, 2000.

[2] B. Timmer, W. Olthuis, and A. V. D. Berg, "Ammonia sensors and their applications—a review," *Sensors and Actuators B: Chemical*, 2005, 107(2): 666–677.

[3] C. McDonagh, C. S. Burke, and B. D. MacCraith, "Optical chemical sensors," *Chemical Reviews*, 2008, 108(2): 400–422.

[4] O. S. Wolfbeis, "Fiber-optic chemical sensors and biosensors," *Analytical Chemistry*, 2008, 80(12): 4269–4283.

[5] A. R. Firooz, M. Movahedi, and A. A. Ensafi,

"Selective and sensitive optical chemical sensor for the determination of Hg (II) ions based on tetrathia-12-crown-4 and chromoionophore I," *Sensors and Actuators B: Chemical*, 2012, 171–172(8): 492–498.

[6] Y. Wang, B. Li, L. Zhang, L. Liu, O. Zuo, and P. Li, "A highly selective regenerable optical sensor for detection of mercury (II) ion in water using organic–inorganic hybrid nanomaterials containing pyrene," *New Journal of Chemistry*, 2010, 34(9): 1946–1953.

[7] K. T. Lau, S. Edwards, and D. Diamond, "Solid-state ammonia sensor based on Berthelot's reaction," *Sensors and Actuators B: Chemical*, 2004, 98(1): 12–17.

[8] S. Tao, L. Xu, and J. C. Fanguy, "Optical fiber ammonia sensing probes using reagent immobilized porous silica coating as transducers," *Sensors and Actuators B: Chemical*, 2006, 115(1): 158–163.

[9] T. Abel, B. Ungerböck, I. Klimant, and T. Mayr, "Fast responsive, optical trace level ammonia sensor for environmental monitoring," *Chemistry Central Journal*, 2012, 6(1): 21777–21778.

[10] K. Waich, T. Mayr, and I. Kilmant, "Fluorescence sensors for trace monitoring of dissolved ammonia," *Talanta*, 2008, 77(1): 66–72.

[11] R.W. Sabnis, *Handbook of biological dyes and stains*. New York: Wiley, 2010.

11

Polymer Integrated Waveguide Optical Biosensor by Using Spectral Splitting Effect

Xiaonan HAN[1], Xiuyou HAN[1*], Yuchen SHAO[1], Zhenlin WU[1], Yuxin LIANG[1,4], Jie TENG[2], Shuhui BO[3], Geert MORTHIER[4], and Mingshan ZHAO[1]

[1]*School of Physics and Optoelectronic Engineering, Dalian University of Technology, Dalian, 116024, China*

[2]*No. 38 Research Institute, China Electronics Technology Group Corporation, Hefei, 230088, China*

[3]*Technical Institute of Physics and Chemistry, Chinese Academy of Sciences, Beijing, 100190, China*

[4]*Photonics Research Group, Department of Information Technology, Ghent University-IMEC, 9000 Ghent, Belgium*

[*]Corresponding author: Xiuyou HAN E-mail: xyhan@dlut.edu.cn

Abstract: The polymer waveguide optical biosensor based on the Mach-Zehnder interferometer (MZI) by using spectral splitting effect is investigated. The MZI based biosensor has two unequal width sensing arms. With the different mode dispersion responses of the two-arm waveguides to the cladding refractive index change, the spectral splitting effect of the output interference spectrum is obtained, inducing a very high sensitivity. The influence of the different mode dispersions between the two-arm waveguides on the spectral splitting characteristic is analyzed. By choosing different lengths of the two unequal width sensing arms, the initial dip wavelength of the interference spectrum and the spectral splitting range can be controlled flexibly. The polymer waveguide optical biosensor is designed, and its sensing property is analyzed. The results show that the sensitivity of the polymer waveguide optical biosensor by using spectral splitting effect is as high as 10^4 nm/RIU, with an improvement of 2–3 orders of magnitude compared with the slot waveguide based microring biosensor.

Keywords: Optical biosensor; integrated waveguide; spectral splitting; sensitivity

1. Introduction

Integrated waveguide optical biosensors have broad applications in environmental monitoring, food safety, and healthcare with the advantages of the low cost, compact structure, real-time detection, and label-free sensing [1–4]. Among various structures for integrated waveguide optical biosensors [5–8], Mach-Zehnder interferometers (MZIs) have been attracting much attention due to their simple structure, easiness for fabrication. MZI based biosensors make use of the optical interference effect for sensing. When the phase difference between two arms is altered by the ambient analyte, the intensity or the resonant wavelength of the output interference signal changes, which reflects the information of the analyte, such as the kind or the concentration [9–11].

The sensitivity of the MZI based integrated waveguide optical biosensor can be improved by increasing the cumulative amount of phase change. For example, in [12], the two arms of the MZI based biosensor were set to be the spiral structure to increase the arm length. It increases the interaction distance between the analyte and lightwave, and improves the sensitivity. However, this structure increases the size of the biosensor. In [13], the sensing arm of the MZI based biosensor was set to be a slot waveguide structure to increase the interaction degree between the analyte and lightwave, aiming at obtaining large phase change. However, this approach increases the complexity of the waveguide design and the difficulty of the device fabrication.

In [14], the spectral splitting characteristic of the integrated waveguide interference structure was studied. If the phase difference between two optical paths satisfies a certain condition, there is a minimum power in the output spectrum at the dip wavelength, which is called as the critical wavelength. When the index of the analyte changes, the minimum power of the output spectrum is split into two dip wavelengths towards the opposite directions. The index change in the analyte can be determined by measuring the splitting wavelength, and then the information of the analyte is obtained. Because of the different mode dispersions of the two optical paths, a large splitting wavelength can be obtained with a small change of ambient refractive index, thus the very high sensitivity can be obtained. In [15], the silicon on the isolator (SOI) waveguide based MZI sensor with two unequal width arms was prepared. Laser irradiation was used to simulate the sensing process. The sensor could work at the critical wavelength by controlling the temperature, and the spectral splitting phenomenon was observed experimentally. In [16], the porous silicon based MZI sensor with two unequal width arms which worked at 1550 nm was simulated and analyzed. Different optical field distributions can be obtained since the two arms have different widths, thus the condition of spectral splitting can be satisfied. Because the position of the critical wavelength varies with the change in phase difference, the different phase periods with different phase differences can be distinguished. It solves the phase-ambiguity problem of the interference sensors and increases the dynamic range of sensing. The basic principle of the interference spectrum is analyzed in above-mentioned literatures, and the spectrum splitting phenomenon has been observed experimentally. However, as far as we know, the optimal design of the biosensor structure with a controllable initial critical wavelength and a spectral splitting range to meet the practical sensing and interrogation applications haven't been reported in detail.

In this paper, the polymer waveguide MZI based biosensor by using the spectral splitting effect is investigated. The relationships between the phase difference and the cladding refractive indices or the length of the sensing arms are analyzed. The widths of the sensing arms are chosen properly to obtain a large phase difference and to realize the spectral splitting. For the practical sensing and interrogation applications, the two unequal width arms of the MZI are chosen to be different lengths. By adjusting the length difference, the initial critical wavelength of the interference spectrum can be designed flexibly, and the spectral splitting range can be controlled easily. The polymer waveguide MZI based biosensor fabricated by the micro-nano imprinting technique is designed. The results show that the controllable initial critical wavelength and spectral splitting range can be realized with the proposed sensing arm structure. Compared with the slot waveguide microring biosensor, the sensitivity of the spectral splitting based MZI biosensor can be enhanced by 2 to 3 orders of magnitude.

2. Operation principle

Figure 1 illustrates the structure of the spectral splitting based waveguide biosensor. It is mainly composed of two multimode interference (MMI) couplers and two sensing arms. The widths of the two sensing arms are unequal in order to obtain the different mode dispersion responses to the refractive index change in the analyte. The MMI structure is adopted instead of the usual Y branch structure to split and combine the lightwave, which is helpful to reduce the transmission loss caused by the big Y branch angle and has a large tolerance to the imperfection of waveguide fabrication. Figure 2 shows the waveguides' cross section of the two sensing arms. Polymer Ormocore [17, 18] is used as the waveguide core, which is on the SiO_2 under the cladding. The biological sample under test, namely the analyte, covers both sensing arms, which avoids etching an extra sensing pool [19, 20]. Therefore, the complexity of the waveguide device fabrication is reduced, and the sensor array integration can be easily realized. The micro-nano imprinting technique is used to fabricate the polymer waveguide biosensor [21], thus the waveguide core may have the residual flat layer after demolding. The thickness of the residual flat layer is H_s. The structural parameters of the integrated waveguide biosensor are as follows: W_1 and W_2 are the waveguide widths of the sensing arms, H is the waveguide height, L_{arm1} and L_{arm2} are the longitudinal lengths of the two arms, G_1 and G_2 are the lateral spacings between one sensor arm and the straight waveguide of the MMI structure, l_1 and l_2 are the lengths of two bend waveguides, L_{bend} is the longitudinal length of the bend waveguides, L_{MMI} is the length of the MMI, W_{MMI} is the width of the MMI, D is the spacing between the two output straight waveguides of the MMI, and L_{taper} and W_{taper} are the length and width of the taper section, respectively.

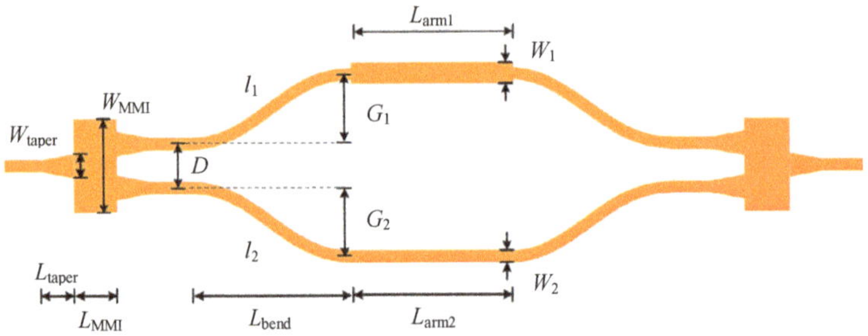

Fig. 1 Structure of the waveguide biosensor with two unequal width arms.

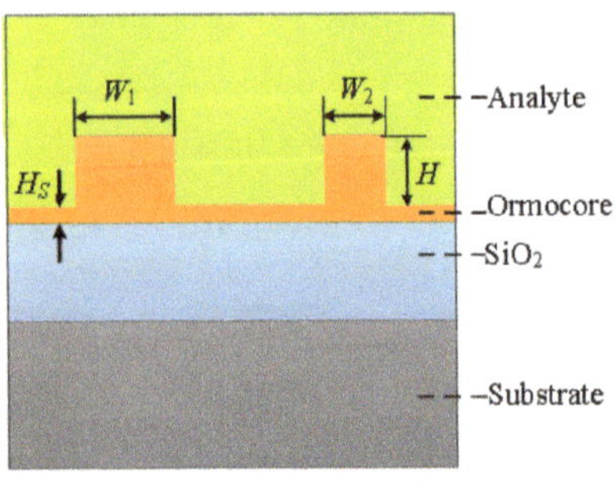

Fig. 2 Waveguides' cross section of two sensing arms.

According to the optical interference principle of the MZI, the output power of the biosensor can be expressed as

$$P_{out} = 0.5P_{in}\left(1+\cos\Delta\varphi\right) \quad (1)$$

where P_{in} and P_{out} are the input power and output power of the MZI based biosensor, respectively. The phase difference between the two sensing arms represented by $\Delta\varphi$ can be expressed as

$$\Delta\varphi = \frac{2\pi}{\lambda}L_{arm}\left(n_{eff1} - n_{eff2}\right) \quad (2)$$

where L_{arm} is the length of each arm, assuming L_{arm1}=L_{arm2}=L_{arm}, and n_{eff1} and n_{eff2} are the effective refractive indices of the guided modes in the two sensing arms, respectively.

Figure 3 illustrates the spectral splitting principle of the interference biosensor. When the total phase difference between the two arms of the MZI based biosensor is an odd times of π, the output power will reach the minimum, which is shown as the curves with dot mark in Fig. 3. The corresponding wavelength is called the critical wavelength. When the refractive index of the upper cladding changes, the phase difference will change accordingly, then the output power at the critical wavelength will not be the minimum. The unique minimum power is split into two dip wavelengths towards the opposite directions, which leads to the spectral splitting shown as the curves with triangle mark in Fig. 3. Due to the different mode dispersions of the guided modes in the two arms, the splitting wavelength can be very large with a small refractive index change in the upper cladding. As a result, the high sensitivity of the biosensor can be obtained.

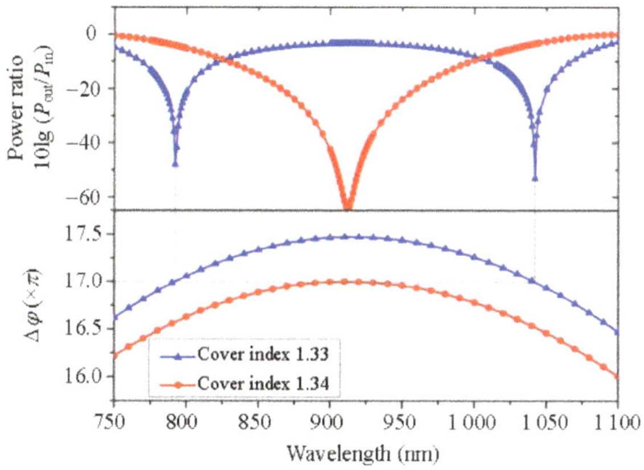

Fig. 3 Spectral splitting principle of the MZI based biosensor.

3. Biosensor structure design

According to the waveguides' cross section structure shown in Fig. 2, the finite difference beam propagation method (FD-BPM) is used to design the MZI waveguide biosensor structure. With the need to inject the phosphate buffer solution (PBS) in the microfluidic channel as the reference in the optical sensing applications [21], the PBS is chosen as the original material of the upper cladding in the analysis of waveguide. The transverse electric (TE) polarization state is chosen, and the refractive indices of each waveguide layer at 850 nm are as follows: upper cladding n_{PBS} =1.340, core n_{core} = 1.543, and lower cladding n_{cladding} =1.453. With the micro-nano imprinting technique, the waveguide height H =1 μm and flat layer thickness H_s=0.2 μm are chosen.

According to the optical interference principle and spectral splitting characteristic, the effective refractive indices of the guided modes in the two sensing arms affect the phase difference directly. The two sensing arms of the MZI based biosensor have different widths and only transmit the fundamental modes. The requirement of the single-mode condition for the waveguide width is analyzed. The effective refractive index with different waveguide widths is shown in Fig. 4, where Mode 0 means the fundamental mode and Mode 1 means the first-order guided mode. When the waveguide width is less than 1.5 μm, the single-mode condition is met. The narrow width will enhance the aspect ratio of waveguides' cross section, which increases the difficulty of the waveguide device preparation. So the widths of two arms are chosen as W_1=1.4 μm and W_2=1 μm, respectively.

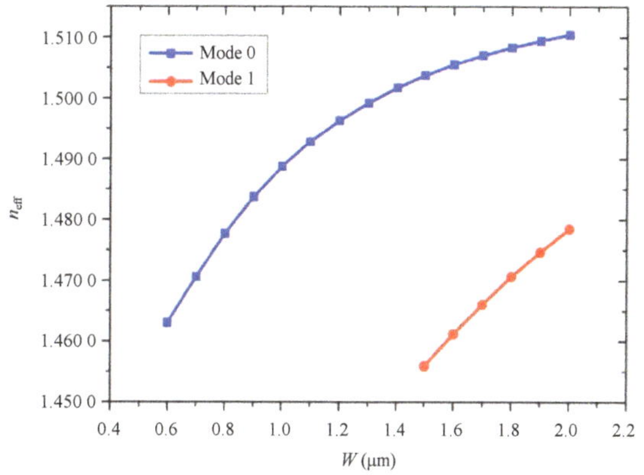

Fig. 4 Effective refractive index of the guided mode versus waveguide width.

Besides the effective refractive index of the arm waveguide, the lengths of sensing arms (L_{arm1} and L_{arm2} in Fig. 1) also affect the phase difference response greatly. When the lengths of two bend waveguides are equal (l_1 = l_2), the lengths of two sensing arms are also equal (L_{arm1} = L_{arm2}), and both are set to be 624 μm, then the total phase difference will be an odd times of π calculated with (2). The concentration change in the biological sample solution is simulated by changing the refractive index of upper cladding. Figure 5 shows the transmission spectrum of the biosensor. The dip shifts away from the critical wavelength in the opposite directions, thus the spectral splitting is obtained. We plan to use the super luminescent diode (SLED, Fiberer Corp.) with a wavelength range of 820 nm – 860 nm in the subsequent experiment, that is to say, the detectable wavelength range is 820 nm – 860 nm. So the spectral splitting range due to the concentration change in the biological sample solution should be between 820 nm and 860 nm. According to Fig. 5, if we choose the left side of the splitting spectrum for sensing, the initial critical wavelength should be adjusted. The initial critical wavelength can be changed by altering the widths of two sensing arms. Because the adjustable range of the waveguide width is very small, the adjustive accuracy should be very high, which increases the difficulty of waveguide device fabrication. Here, two unequal length sensing arms are adopted, and the initial critical wavelength can be flexibly adjusted by changing the length difference.

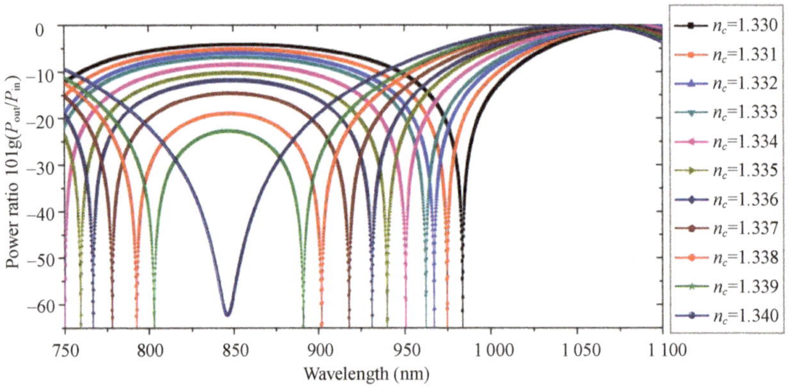

Fig. 5 Transmission spectrum under different indices of upper cladding ($L_{arm1} = L_{arm2}$ = 624 μm).

Figure 6 illustrates the modified biosensor structure with two unequal length sensing arms. The longitudinal lengths of both the two sensing arms are L_{arm}. The lateral spacing between the narrower arm and straight waveguide of the MMI structure is G_3. The narrow sensing arm consists of two identical bend waveguides, and the length of each section is $\sqrt{\left(G_3^2 + \left(L_{arm}/2\right)^2\right)}$. When the lengths of the two sensing arms are not equal, (1) can be expressed as

$$P_{out} = 0.5P_{in}\left(1 + \cos\left(\Delta\varphi_w - \Delta\varphi_l\right)\right) \tag{3}$$

where $\Delta\varphi_w$ and $\Delta\varphi_l$ are the phase difference induced by the equal length of two arms and phase difference induced by the length difference between two arms, respectively.

$$\Delta\varphi_w = \frac{2\pi}{\lambda} L_{arm}\left(n_{eff1} - n_{eff2}\right) \tag{4}$$

$$\Delta\varphi_l = \frac{2\pi}{\lambda} n_{eff2}\Delta L \tag{5}$$

where ΔL is the length difference between two arms of the MZI biosensor:

$$\Delta L = 2\sqrt{\left(G_3^2 + \left(L_{arm}/2\right)^2\right)} - L_{arm}. \tag{6}$$

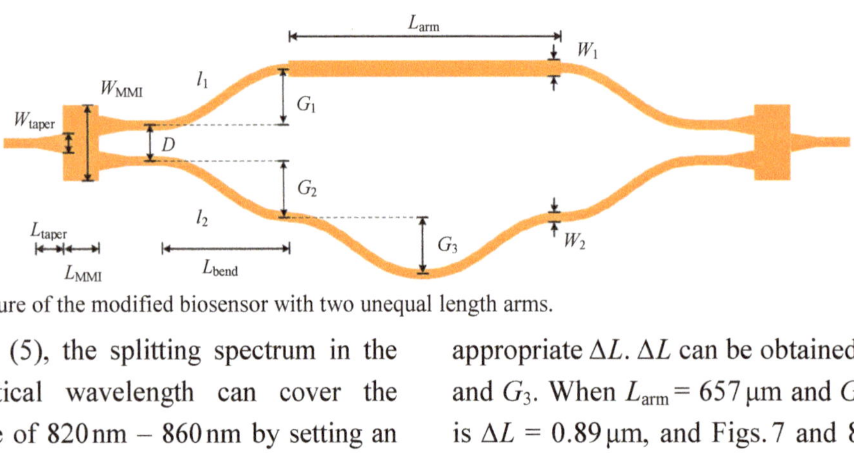

Fig. 6 Total structure of the modified biosensor with two unequal length arms.

According to (5), the splitting spectrum in the left of the critical wavelength can cover the wavelength range of 820 nm – 860 nm by setting an appropriate ΔL. ΔL can be obtained by adjusting L_{arm} and G_3. When L_{arm} = 657 μm and G_3 = 17.1 μm, there is ΔL = 0.89 μm, and Figs. 7 and 8 show the power transmission curves and phase difference curves, respectively. The spectrum corresponding to $n_{cladding}$= 1.340 will obtain the minimum power value at the critical wavelength. And the left side of the splitting spectrum will cover the wavelength range of 820 nm–860 nm, which meets the design requirements.

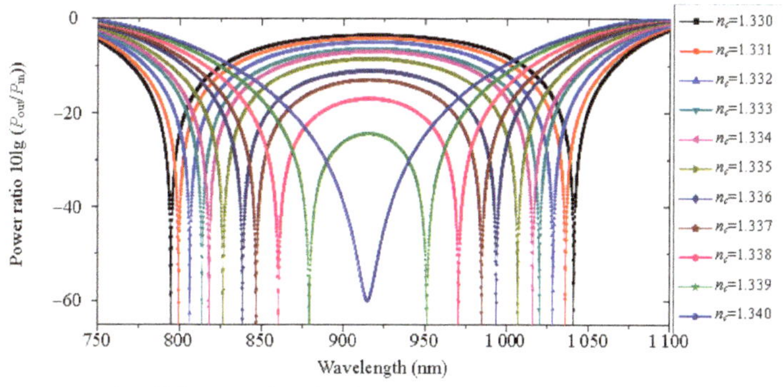

Fig. 7 Power transmission curves under different refractive indices of upper cladding (L_{arm}=657 μm and G_3=17.1 μm).

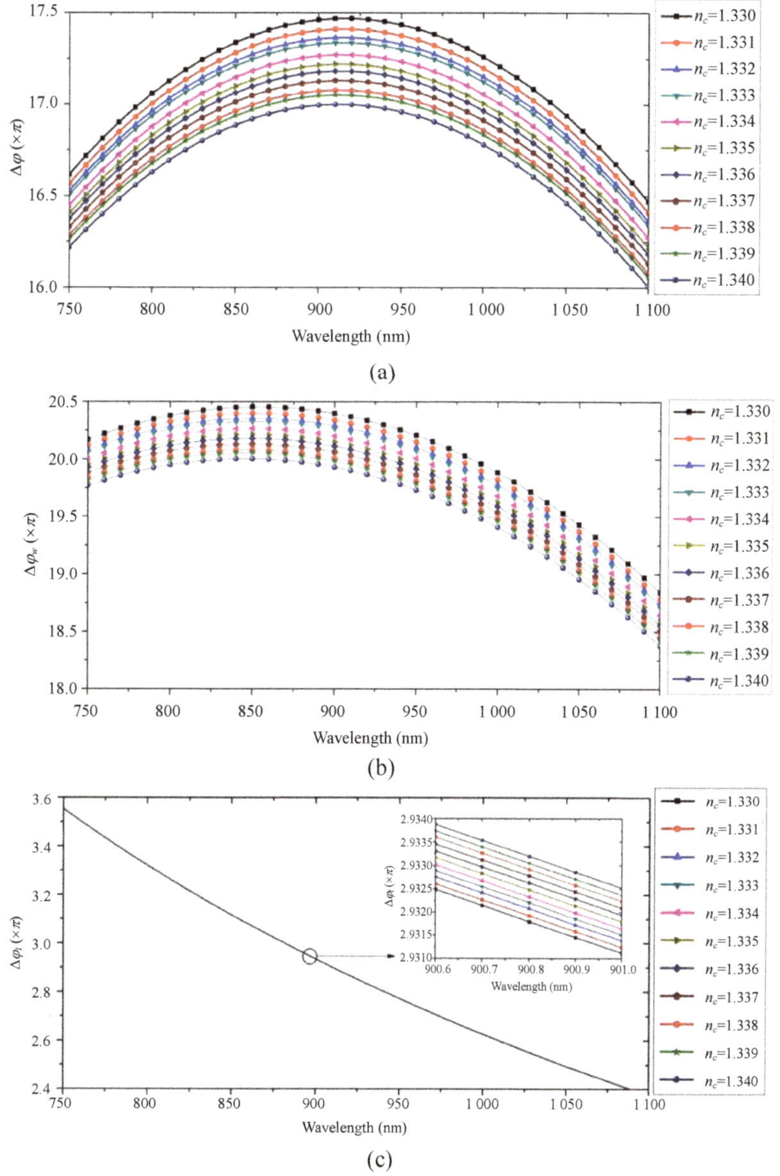

(c)

Fig. 8 Phase difference versus wavelength (L_{arm} = 657 μm and G_3 = 17.1 μm): (a) total phase difference $\Delta\varphi$, (b) phase difference $\Delta\varphi_w$ induced by the equal length of two arms, and (c) phase difference $\Delta\varphi_l$ induced by the length difference between two arms.

The next is to design the MMI based coupler. Figure 9 illustrates the MMI waveguide structure. The MMI structure is designed from the viewpoint of reducing the insertion loss. According to the principle of self-imaging of the MMI structure [22, 23], the optimized structural parameters of the MMI are as follows: the width of the MMI structure is $W_{MMI} = 9\ \mu m$, the length of the MMI structure is $L_{MMI} = 74\ \mu m$, the spacing between two output straight waveguides is $D = 4.6\ \mu m$, the length of the taper section is $L_{taper} = 62\ \mu m$, and the width of the taper section is $W_{taper} = 2.2\ \mu m$. Figure 10 shows the power transmission of the MMI based coupler. The output power of both arms is 49.3%, corresponding to the extra insertion loss of each arm of 0.07 dB. Finally, the optimized structural parameters of the MZI based biosensor are as follows: $H = 1\ \mu m$, $H_s = 0.2\ \mu m$, $W_1 = 1.4\ \mu m$, $W_2 = 1\ \mu m$, $G_1 = G_2 = 3\ \mu m$, $G_3 = 17.1\ \mu m$, $L_{arm} = 657\ \mu m$, $L_{bend} = 100\ \mu m$, $L_{taper} = 62\ \mu m$, $W_{taper} = 2.2\ \mu m$, $D = 4.6\ \mu m$, $W_{MMI} = 9\ \mu m$, $L_{MMI} = 74\ \mu m$.

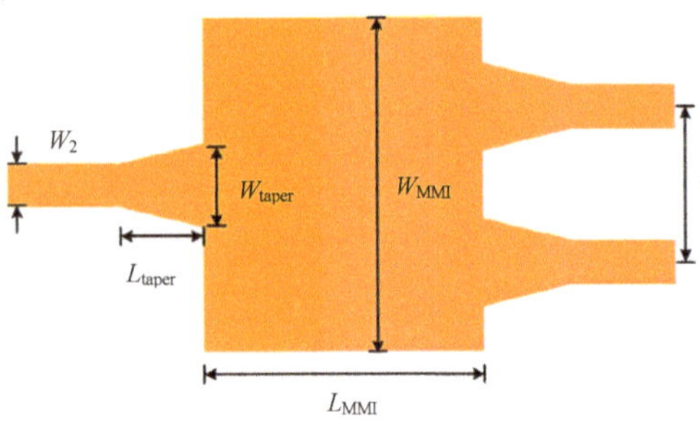

Fig. 9 Structure of the MMI waveguide.

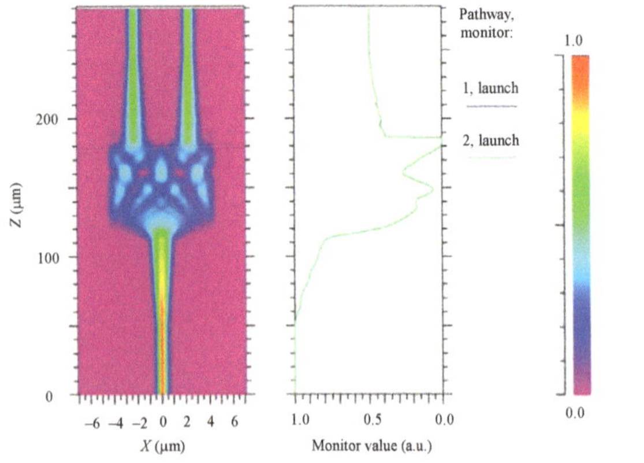

Fig. 10 Field distribution of the optical wave in the MMI waveguide structure.

4. Sensitivity analysis

From Fig. 7, it can be seen that the critical wavelength with the cladding refractive index of 1.34 is 912 nm, which is not within the range of the detectable wavelength 820 nm – 860 nm from the SLED. Because the wavelength difference between the shifted dip wavelength and critical wavelength is too large with a small upper cladding index change, the left dip wavelength of the splitting spectrum with the cladding index of 1.339 is set as the reference to determine the sensitivity of the MZI based biosensor for the practical interrogation. In Fig. 7, the splitting spectrum value $\Delta\lambda$, namely the wavelength difference between the shifted dip wavelength with different cladding refractive indices and reference dip wavelength with the cladding index of 1.339 can be obtained, and the data are listed in Table 1. According to Table 1, the sensitivity of the MZI biosensor with spectral splitting is about 10^4 nm/RIU, which is improved by 2 – 3 orders of magnitude compared with the slot waveguide microring sensor [24]. The relationship curve between the cladding index variation and spectral splitting value is plotted in Fig. 11. By fitting the curve of Fig. 11 with polynomial, the relationship equation can be obtained as (7). According to (7), the analyte refractive index variation Δn_c can be obtained by the measured splitting spectrum value $\Delta\lambda$.

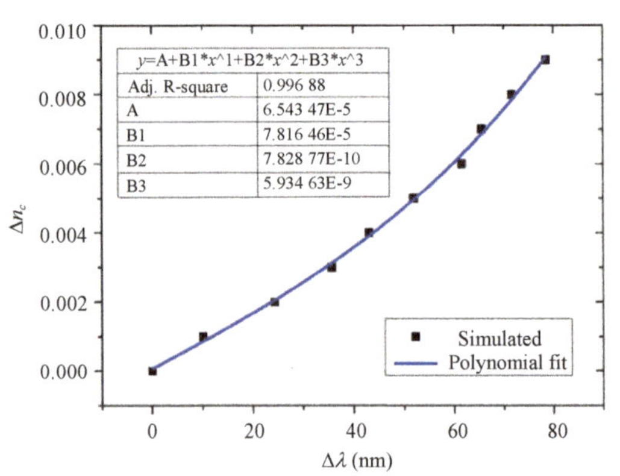

Fig. 11 Upper cladding refractive index variation versus splitting spectrum value.

$$\Delta n_c = 5.93463\times10^{-9}\Delta\lambda^3 + 7.82877\times10^{-1}\ \Delta\lambda^2 + 7.81646\times10^{-5}\Delta\lambda + 6.54347\times10^{-5} \quad (7)$$

Table 1 Spectral splitting value and ratio between the spectral splitting value and upper cladding index change.

n_c	1.33	1.331	1.332	1.333	1.334	1.335	1.336	1.337	1.338	1.339
$\Delta\lambda$ (nm)	78.6	71.6	65.5	61.5	51.9	43.0	35.6	24.3	10.1	0.0
$\Delta\lambda/\Delta n_c$ ($\times10^4$ nm/RIU)	0.87	0.90	0.94	1.03	1.04	1.08	1.19	1.22	1.01	

5. Conclusions

In this paper, the polymer waveguide optical biosensor based on the MZI structure by using spectral splitting effect has been investigated. With the different mode dispersion responses of the guided modes in the two unequal width waveguides, the high sensitivity can be achieved by properly choosing the waveguide widths of the two sensing arms. In order to meet the requirement of the practical sensing and interrogation application, the two sensing arms with different lengths are chosen. By adjusting the length difference between the two sensing arms, the initial critical wavelength and splitting spectrum range can be controlled easily. The polymer waveguide biosensor with the micro-nano imprinting technique is designed, and the sensing property is analyzed. The results show that the sensitivity of the polymer waveguide biosensor as high as 10^4 nm/RIU can be achieved.

Acknowledgment

This work was supported in part by the International Science & Technology Cooperation Program of China (No. 2014DFG32590), National Natural Science Foundation of China (No. 61307040), National R&D Program (No. 2012AA040406), National Research Foundation of China (No. 6140450010305), Natural Science Foundation of Liaoning Province (No. 2014020002), and Fundamental Research Funds for the Central Universities (DUT15ZD231 and DUT2015TD47).

References

[1] A. L. Washburn and R. C. Bailey, "Photonics-on-a-chip: recent advances in integrated waveguides as enabling detection elements for real-world, lab-on-a-chip biosensing applications," *Analyst*, 2011, 136(2): 227-236.

[2] V. J. Cadarso, A. Llobera, M. Puyol, and H. Schift, "Integrated photonic nanofences: combining subwavelength waveguides with an enhanced evanescent field for sensing applications," *ACS Nano*, 2016, 10(1): 778-785.

[3] L. G. Carrascosa, C. S. Huertas, and L. M. Lechuga, "Prospects of optical biosensors for emerging label-free RNA analysis," *Trac Trends in Analytical Chemistry*, 2016, 80: 177-189.

[4] E. Makarona, P. Petrou, S. Kakabakos, K. Misiakos, and I. Raptis, "Point-of-need bioanalytics based on planar optical interferometry," *Biotechnology Advances*, 2016, 34(3): 209-233.

[5] X. Fan, I. M. White, S. I. Shopova, H. Zhu, J. D. Suter, and Y. Sun, "Sensitive optical biosensors for unlabeled targets: a review," *Analytica Chimica Acta*, 2008, 620(1-2): 8-26.

[6] H. N. Daghestani and B. W. Day, "Theory and applications of surface plasmon resonance, resonant mirror, resonant waveguide grating, and dual polarization interferometry biosensors," *Sensors*, 2010, 10(11): 9630-9646.

[7] M. Sathish and S. Talabattula, "Polarization analysis of an asymmetrically etched rib waveguide coupler for sensing applications," *Photonic Sensors*, 2013, 3(2): 178-183.

[8] J. H. Zhang, F. S. Chen, B. Sun, and K. X. Chen, "Integrated optical waveguide sensor for lighting impulse electric field measurement," *Photonic Sensors*, 2014, 4(3): 215-219.

[9] Z. Qi, N. Matsuda, K. Itoh, M. Murabayashi, and C. R. Lavers, "A design for improving the sensitivity of a Mach-Zehnder interferometer to chemical and biological measurands," *Sensors and Actuators B: Chemical*, 2002, 81(2-3), 254-258.

[10] S. Dante, D. Duval, B. Sepúlveda, A. B. González-Guerrero, J. R. Sendra, and L. M. Lechuga, "All-optical phase modulation for integrated interferometric biosensors," *Optics Express*, 2012, 20(7): 7195-7205.

[11] T. F. Yang, "Optimal design and fabrication of polymeric planar optical waveguide bio-chemical sensor," Ph.D. dissertation, Jilin University, Jilin, 2014.

[12] A. Densmore, D. X. Xu, S. Janz, P. Waldron, T. Mischki, G. Lopinski, *et al.*, "Spiral-path high-sensitivity silicon photonic wire molecular sensor with temperature-independent response," *Optics Letters*, 2008, 33(6): 596-598.

[13] Q. Liu, X. Tu, K. W. Kim, J. S. Kee, Y. Shin, K. Han, *et al.*, "Highly sensitive Mach-Zehnder interferometer biosensor based on silicon nitride slot waveguide," *Sensors and Actuators B: Chemical*, 2013, 188(11): 681-688.

[14] R. Levy and S. Ruschin, "Critical sensitivity in hetero-modal interferometric sensor using spectral interrogation," *Optics Express*, 2008, 16(25): 20516-20521.

[15] R. Levy, S. Ruschin, and D. Goldring, "Critical sensitivity effect in an interferometer sensor," *Optics Letters*, 2009, 34(19): 3023-3025.

[16] T. Hutter, S. R. Elliott, and S. Ruschin, "Dynamic range enhancement and phase-ambiguity elimination in wavelength-interrogated interferometric sensor," *Sensors and Actuators B: Chemical*, 2013, 178(3): 593-597.

[17] Ormocore Products, Micro Resist Technol. GmbH, Berlin, Germany, 2014.

[18] R. Morarescu, P. K. Pal, X. Han, M. Zhao, P. Bienstman, and G. Morthier, "Polymer microring resonators for biosensing applications by nanoimprint lithography," in *The 17th International Conference on Transparent Optical Networks*, Hungary, Sept. 5-9, pp. 1-4, 2015.

[19] R. Bruck, E. Melnik, P. Muellner, R. Hainberger, and M. Lämmerhofer, "Integrated polymer-based Mach-Zehnder interferometer label-free streptavidin biosensor compatible with injection molding," *Biosensors and Bioelectronics*, 2011, 26(9): 3832-3837.

[20] B. Sepúlveda, J. S. D. Río, M. Moreno, F. J. Blanco, K. Mayora, C. Domínguez, *et al.*, "Optical biosensor microsystems based on the integration of highly sensitive Mach-Zehnder interferometer devices," *Journal of Optics A: Pure and Applied Optics*, 2006, 8(7): S561-S566.

[21] L. H. Wang, J. Ren, X. Y. Han, T. Claes, X. G. Jian, P. Bienstman, *et al.*, "A label-free optical biosensor built on a low-cost polymer platform," *IEEE Photonics Journal*, 2012, 4(3): 920-930.

[22] L. B. Soldano and E. C. M. Pennings, "Optical multi-mode interference devices based on self-imaging: principles and applications," *Journal of Lightwave Technology*, 1995, 13(4): 615-627.

[23] Y. C. Shao, X. Y. Han, X. N. Han, Z. L. Lu, Z. L. Wu, J. Teng, *et al.*, "Optimal design of 850 nm 2×2 multimode interference polymer waveguide coupler by imprint technique," *Photonic Sensors*, 2016, 6(3): 234-242.

[24] X. N. Han, X. Y. Han, Y. C. Shao, Z. L. Lu, J. Teng, Z. L. Wu, *et al.*, "Study on polymer microring optical biosensor based on slot waveguide," *Acta Optical Sinica*, 2016, 38(4): 04130011-04130018.

12

Improving Smoothing Efficiency of Rigid Conformal Polishing Tool Using Time-Dependent Smoothing Evaluation Model

Chi SONG[1,3*], Xuejun ZHANG[1,2], Xin ZHANG[1,2], Haifei HU[1,2], and Xuefeng ZENG[1,2]

[1]*Changchun Institute of Optics, Fine Mechanics and Physics, Chinese Academy of Sciences, Changchun, 130033, China*

[2]*Key Laboratory of Optical System Advanced Manufacturing Technology, Chinese Academy of Sciences, Changchun, 130033, China*

[3]*University of Chinese Academy of Sciences, Beijing, 100049, China*

[*]Corresponding author: Chi SONG E-mail: songchi7787@126.com

Abstract: A rigid conformal (RC) lap can smooth mid-spatial-frequency (MSF) errors, which are naturally smaller than the tool size, while still removing large-scale errors in a short time. However, the RC-lap smoothing efficiency performance is poorer than expected, and existing smoothing models cannot explicitly specify the methods to improve this efficiency. We presented an explicit time-dependent smoothing evaluation model that contained specific smoothing parameters directly derived from the parametric smoothing model and the Preston equation. Based on the time-dependent model, we proposed a strategy to improve the RC-lap smoothing efficiency, which incorporated the theoretical model, tool optimization, and efficiency limit determination. Two sets of smoothing experiments were performed to demonstrate the smoothing efficiency achieved using the time-dependent smoothing model. A high, theory-like tool influence function and a limiting tool speed of 300 RPM were obtained.

Keywords: Optics design and fabrication; optics fabrication; polishing

1. Introduction

Large aspheric optical surfaces can be precisely manufactured using computer-controlled optical surfacing (CCOS). For next-generation large-aperture and high-resolution imaging optical systems such as the thirty meter telescope (TMT) and giant magellan telescope (GMT) [1, 2], correcting mid-spatial-frequency (MSF) errors on the optical surfaces is very important. Failure to control the MSF characteristics yields reduced optical performance due to the resultant MSF errors, which are directly related to the point-spread-function sharpness [3].

Two different approaches to controlling MSF errors exist: directed figuring and natural smoothing. The directed figuring of small-scale errors requires small tools, which in turn requires a long polishing run time, high-accuracy optical metrology, and high-accuracy tool positioning. A large polishing tool can naturally correct MSF errors smaller than the tool size, while also removing large-scale errors within a short time period. However, smoothing MSF errors from aspheric mirrors using large tools is challenging, because the curvature changes on the surface require tools with sufficient compliance to fit the surface, but with sufficient rigidity to realize natural smoothing.

In 2010, Kim and Burge developed a rigid conformal (RC) lap using a visco-elastic non-Newtonian fluid to overcome the conflict between the desired rigidity and flexibility for large polishing laps [4]. Compared with active stress laps [5] and semi-flexible tools [6], an RC lap provides numerous advantages, such as ease of large-scale manufacture, superior surface roughness, and a highly stable theory-like tool influence function (TIF). However, the RC-lap smoothing efficiency performance is poorer than expected [7].

To quantify the smoothing effect, various mathematical models have been developed. For example, Brown *et al.* proposed a smoothing model for an elastic-backed flexible lapping belt in 1981 [8]. Subsequently, Jones simulated MSF error evolution based on a simple linear parametric model [9], and Mehta and Reid later proposed the classic Bridging model to study the smoothing effect, which is based on the theory of elasticity [10]. Tuell *et al.* then improved the Bridging model using the spatial Fourier decomposition method [11]. Later, Kim *et al.* introduced a parametric smoothing model based on the simplified Bridging model to describe smoothing effects [12]; this parametric smoothing model was further verified by Shu *et al.* using a correlation-based model [13]. Shu *et al.* also noted that the model presented by Kim *et al.* neglected the instantaneous property and therefore constructed a new model based on the Bridging model [14]. This new model discloses the exponential decay of the MSF errors with time during smoothing. Finally, Nie *et al.* constructed a generalized numerical pressure distribution model to solve the superposition of innumerable sinusoidal errors with different frequencies and amplitudes through finite element analysis (FEA) [15].

As indicated above, the majority of the past works have focused on the construction of models to quantitatively describe the smoothing effect. However, no explicit guidance has been developed to specify methods through which the smoothing efficiency can be improved. For instance, the parametric smoothing model reveals that the smoothing efficiency is related to the tool stiffness. However, if the smoothing time is considered, a greater number of factors can influence the smoothing efficiency.

In this paper, we present a time-dependent smoothing model containing specific factors directly related to the smoothing efficiency, which is derived from the parametric smoothing model and Preston equation. Based on this model, we can maximize the RC-lap smoothing performance. The remainder of the paper is organized as follows. In Section 2, we briefly introduce the general information on the RC lap and the parametric smoothing model. In Section 3, we propose a strategy to improve the smoothing efficiency, which incorporates the theoretical model, tool optimization, and efficiency limit determination. In Section 4, we apply the above principles to actual smoothing experiments, and present the results. The conclusions are given in Section 5.

2. RC lap and parametric smoothing model

(1) RC lap and smoothing effect

The RC lap uses a non-Newtonian fluid (Silly-Putty™) as a medium, which conforms to an aspheric surface while generating a naturally smoothing effect. Having rigidity and viscosity intermediate between a solid and a liquid, the RC lap has the advantages of both rigid and compliant tools. When the tool is slowly travelling along the tool path, it acts like a compliant tool over a large time scale. However, in the short term, the RC lap behaves like a rigid tool, as a result of the tool motion and bumpy surface.

The basic principles of the RC-lap smoothing effect are straightforward. As the RC lap has a flow characteristic, it can fully deform in response to MSF errors. The MSF-error peaks have higher pressure than the valleys, because of the elastic property of the non-Newtonian fluid. The MSF errors then converge, because of the nonuniform material removal. The polishing pressure distribution between the RC lap and workpiece induced by a particular spatial frequency feature for the one-dimensional (1D) case is shown in Fig. 1. Note that the specific tool structure is presented in Section 3(3).

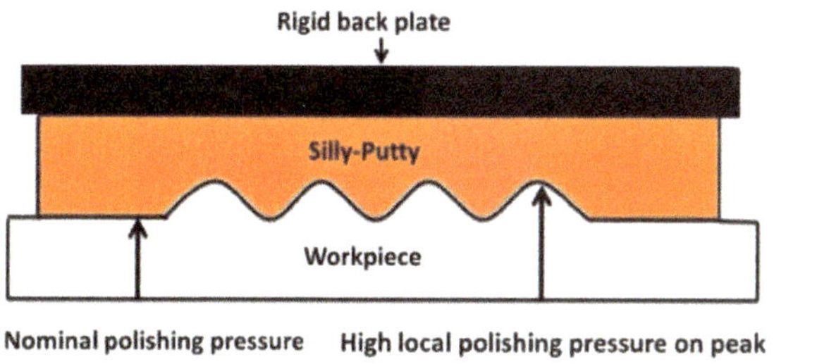

Fig. 1 RC-lap polishing pressure distribution induced by sinusoidal error.

(2) Parametric smoothing model

The parametric smoothing model can be used to describe the smoothing effect for visco-elastic polishing tools [12]. The dimensionless smoothing factor *SF* is defined as the ratio of $\Delta\varepsilon$ to ΔZ, where $\Delta\varepsilon$ is the difference between the peak-to-valley (PV) magnitude of a particular spatial frequency error before and after smoothing, and ΔZ is the corresponding change in the nominal removal depth. The linear relationship between *SF* and the initial surface error ε_{ini} can be expressed as

$$SF = \frac{\Delta\varepsilon}{\Delta Z} = k \cdot (\varepsilon_{ini} - \varepsilon_0) \tag{1}$$

where ε_0 is a fixed value indicating the final smoothing limit, and k is the sensitivity to ε_{ini}.

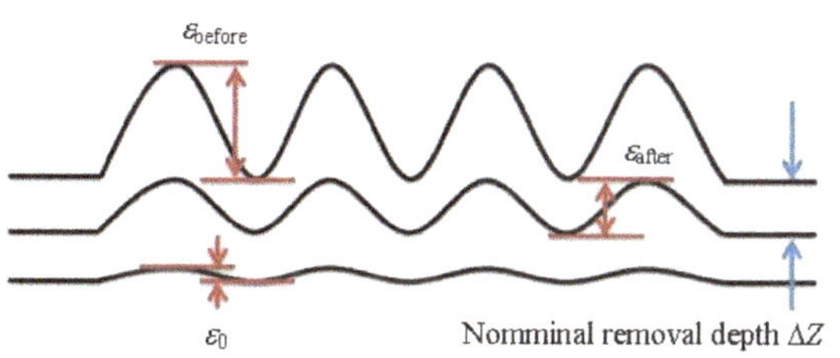

Fig. 2 Before ε_{before}, after ε_{after}, and final ε_0 smoothing profiles for sinusoidal error.

As shown in Fig. 2, ε_0, $\Delta\varepsilon$, and ΔZ can be measured experimentally during smoothing, and the smoothing efficiency k can be obtained by fitting the measurement results. The *SF* function slope is proven to be related to the tool stiffness κ_{total}. However, the k defined in this model neglects the instantaneous property. It has been shown that the MSF errors decay exponentially with time during smoothing [14]. Unfortunately, the smoothing rate is not yet well defined.

3. Smoothing-efficiency *k* improvement strategy

(1) Theoretical model

In order to derive an explicit equation for k improvement, we now present a time-dependent smoothing model, which contains specific factors related to the smoothing rate.

For the orbital motion, the instantaneous lap velocity V (mm/s) can be obtained from the stroke speed Ω (RPM) and the orbital radius A (mm), such that

$$V = \frac{2\pi \cdot \Omega \cdot A}{60}. \tag{2}$$

Based on the dynamic mechanical properties of Silly-Putty, the storage modulus E' is a function of the applied stress frequency f (Hz), as shown in Fig. 3 [16]. Because f is determined by the local features under the tool motion, it can be expressed as

$$f = V \cdot \xi = 2\pi \cdot \Omega \cdot A \cdot \xi / 60 \tag{3}$$

where ξ (mm^{-1}) is the spatial frequency of a sinusoidal error.

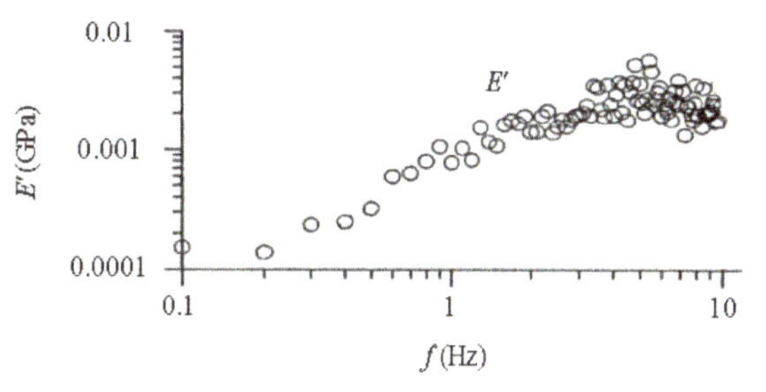

Fig. 3 Silly-putty storage modulus E' as a function of applied stress frequency f.

According to the parametric smoothing model derivation process, the compressive stiffness of the entire tool κ_{total} and the slope of the *SF* function, which corresponds to *k*, can be expressed as

$$\frac{1}{\kappa_{total}} = \frac{1}{\kappa_{elastic}(f)} + \frac{1}{\kappa_{others}} \quad (4)$$

$$k = \frac{1}{P_{nominal} \cdot [\frac{1}{\kappa_{elastic}(f)} + \frac{1}{\kappa_{others}}]} \quad (5)$$

where $\kappa_{elastic}$ is the elastic stiffness of the Silly-Putty, κ_{others} is the combined stiffness of all other structures, e.g., the polishing pad and polishing compound fluid, and $P_{nominal}$ is the nominal polishing pressure under the tool.

Here, we assume that E' is equivalent to the Young's modulus, because E' represents the elastic property of the Silly-Putty. The elastic stiffness $\kappa_{elastic}$ is proportional to E'. The entire structure of the RC lap remains unchanged, except for variation of the elastic stiffness; therefore, the combined stiffness of all other components is a constant. Further, κ_{total} can be expressed as a function of *V* with

$$k = \frac{\kappa_{total}(V)}{P_{nominal}}. \quad (6)$$

Inspired by Burge, Kim, and Martin [17], the following differential equation can be derived directly from (1):

$$\frac{d\varepsilon(Z)}{dZ} = \frac{d[\varepsilon(Z) - \varepsilon_0]}{dZ} = -k \cdot [\varepsilon(Z) - \varepsilon_0] \quad (7)$$

having the solution

$$\varepsilon(Z) = (\varepsilon_{ini} - \varepsilon_0) \cdot e^{-k \cdot Z} + \varepsilon_0 \quad (8)$$

where ε is the PV magnitude of a particular spatial frequency error as a continuous function of the nominal removal depth *Z*. In order to transform $\varepsilon(Z)$ to a function of the polishing time *t*, we combine (8) and the well-known Preston equation

$$Z = K_p \cdot P \cdot V \cdot t. \quad (9)$$

Hence, we obtain

$$\varepsilon(t) = (\varepsilon_{ini} - \varepsilon_0) \cdot e^{-k \cdot K_p \cdot P_{nominal} \cdot V \cdot t} + \varepsilon_0 \quad (10)$$

where K_p is the coefficient of the Preston equation, *P* is the polishing pressure, and *t* is the polishing time. Using (6) and (10), the final mathematical model is expressed as

$$\varepsilon(t) = (\varepsilon_{ini} - \varepsilon_0) \cdot e^{-\kappa_{total}(V) \cdot K_p \cdot V \cdot t} + \varepsilon_0. \quad (11)$$

Equation (11) clearly indicates that the smoothing rate is related to κ_{total}, K_p, and *V*. We can divide the contribution of *V* to the smoothing rate into two terms: $\kappa_{total}(V)$ and the removal term $K_p \cdot V$. The former is a non-analytic function of *V*, and the contribution of *V* decreases if *V* increases, based on the E' trend shown in Fig. 3. For the latter, it is apparent that $K_p \cdot V$ is related to *V* directly. In other words, we can assume that an infinite smoothing rate will be obtained using an infinite *V*. Thus, $K_p \cdot V$ dominates *k*, and determining the limitation *V* is the first task of our strategy. Our second task is to apply this limitation *V* to actual smoothing experiments, so as to calculate κ_{total} by fitting the data to the *SF* function. Finally, we evaluate the combined influence on *k* of κ_{total}, K_p, and *V*, based on our smoothing model. In addition, (11) is also very valuable for quantitative prediction of the smoothing effect for a real fabrication process. For a CCOS process, the dwell map is based on the de-convolution of the target removal map using a TIF. If we obtain this dwell map, we can estimate the MSF errors before polishing and the resultant change.

(2) Tool optimization

The smoothing is not an independent process, as it must be accompanied by the figuring. In other words, no technique to correct MSF errors without removing any large-scale errors has been developed in the history of CCOS. The use of a low-quality TIF to figure the surface is one of the MSF-error sources. Thus, the TIF is one of the key factors influencing the smoothing process, which further explains why we employ the RC lap as the polishing tool in this study.

The use of a lowered drive pin hole allows the original RC lap to overcome the gradient pressure effect, which is induced by the moment from the shear force on the workpiece surface [4]. When polishing hard ceramic materials such as RB-SiC, the polishing pressure is higher (>1 psi) than the conventional range (0.2 psi–0.6 psi). An unstable TIF occurs when higher polishing pressure is applied to the RC lap, which is caused by the local high pressure at the tool edge. A new RC-lap structure is designed herein, in order to mitigate the high nominal polishing pressure at the edge using a lowered back plate. Schematic 3-dimentional (3D) models of the old and new RC-lap designs are shown in Fig. 4.

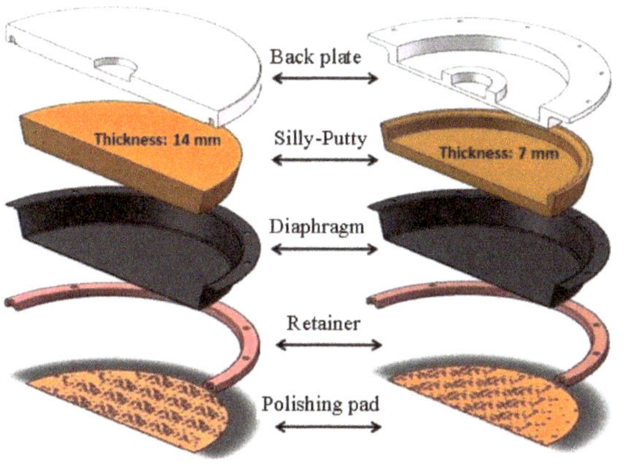

Fig. 4 3D schematic structures of old (left) and new (right) RC-lap structures (exploded and halved).

In order to understand the source of the high edge pressure, we consider an example using the static FEA to analyze the nominal contact pressure distribution between the tool and workpiece. The static elastic modulus of Silly-Putty is assumed to be 2 MPa at 7 Hz, based on the dynamic mechanical properties of this material [15]. For a 68-N vertical force applied to the drive pin hole, symmetrical FEA models are established for the old and new RC laps (100 mm), using the ABAQUS software. The old (new) assembly model consists of 41592 (57797) nodes and 36292 (51874) elements, 35472 (50874) and 820 (1000) of which are linear hexahedral elements of type C3D8I and linear wedge elements of type C3D6, respectively. The FEA models and boundary conditions are shown in detail in Fig. 5. The RC-lap assembly-component material properties are listed in Table 1.

(a) (b)

Fig. 5 Symmetrical FEA models for: (a) old and (b) new RC laps.

Table 1 Assembly-component material properties.

Component	Material name	Elasticmodulus (MPa)	Poisson's ratio
Back plate	Aluminum	7.0e4	0.38
Non-Newton fluid	Silly-Putty	2 at 7 Hz	0.45
Diaphragm	Woven fabric reinforced elastomer	10	0.46
Pad	Universal LP-66	48	0.40
Mirror	RB-SiC	3.4e5	0.26

The nominal contact pressure distribution between the RC lap and the workpiece is clearly shown in Fig. 6. An abrupt pressure change occurs at the edge in both cases. Because the RC lap consists of flexible materials, such as Silly-Putty, elastomer, and polyurethane, the high pressure at the edge is caused by the deformation of these materials when the vertical force to the drive pin hole becomes excessively high. Our new model minimizes this effect using a lowered back plate. The high edge pressure generates an undesirable TIF shape. Figure 7 depicts the orbital motion; the lap orbits the TIF center with a certain orbit radius. Ideally, the TIF peak region lies in the center, because the lap constantly covers the peak region in the figure. However, the line 1 in Fig. 7 indicates high pressure at the tool edge, which never affects the TIF peak region. Thus, the TIF peak region no longer corresponds to the TIF peak, and the line 2 indicates the new peak with a sharp cliff. Hence, a volcano-like TIF is obtained. Again, this effect is reduced by our proposed model, as shown below. Note that the real measured TIF may differ from the static FEA result, because the real motion is a dynamic problem. In addition, we have ignored the polishing compound fluid.

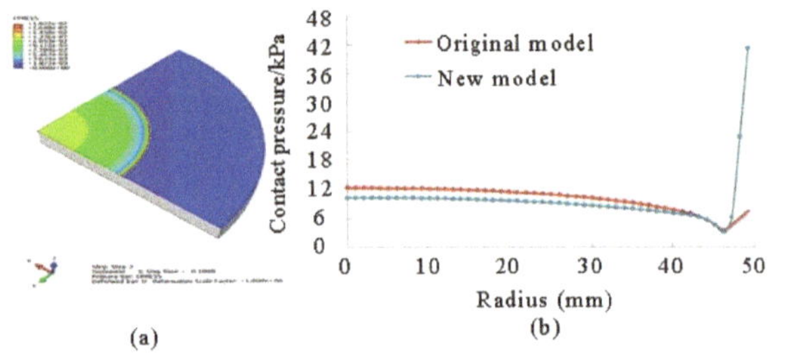

(a) (b)

Fig. 6 Nominal contact pressure P distribution: (a) 3D P contour for new model and (b) P distribution along workpieceradius for both old and new models.

In this study, we conducted two experiments using a 100-mm RC lap at 300-RPM orbital speed, with a 15 mm-orbital radius and under 1.47-psi polishing pressure; the results are shown in Fig. 8. The shape of the real measured TIF of our new structure was beyond our expectation, and a high, theory-like TIF was obtained.

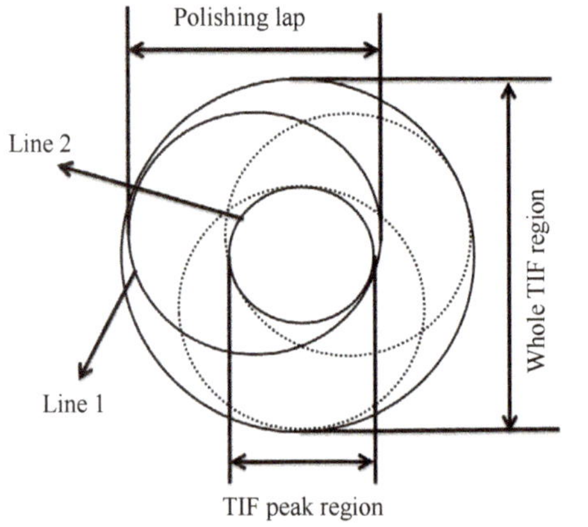

Fig. 7 Schematic picture of orbital motion (note: the line 1 and line 2 represent the high pressure at the lap edge and the edge of the TIF peak region, respectively; the dashed line indicates the lap position distribution region).

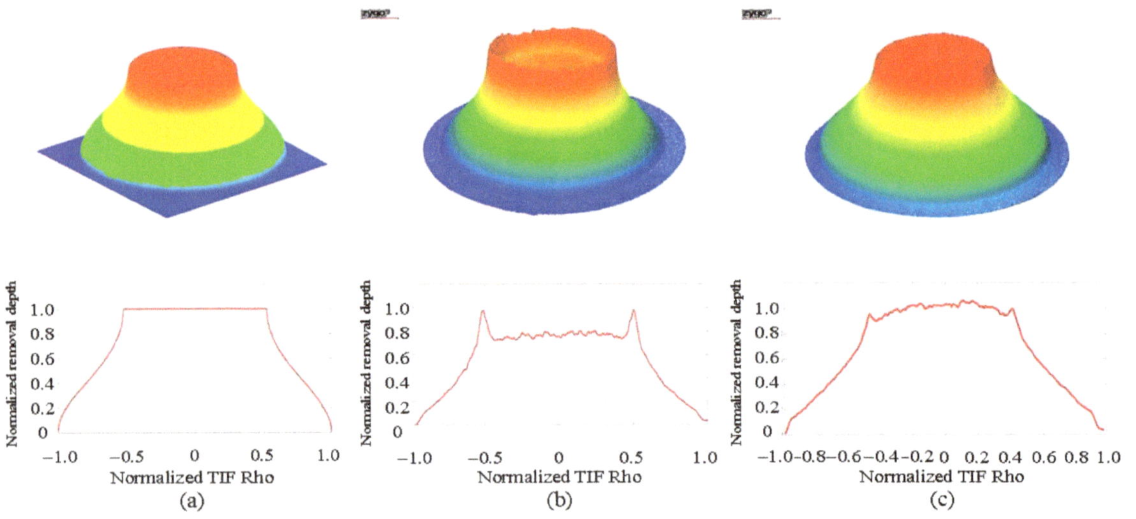

(a) (b) (c)

Fig. 8 3D measured TIFs (top) and their normalized radial profiles (bottom) for (a) theoretical, (b) old, and (c) new models.

(3) Preston coefficient vs. tool speed

As discussed in Section 3(1), the influence of $K_p{\cdot}V$ on the smoothing rate is greater than that of $\kappa_{\text{total}}(V)$. In this section, we discuss the $K_p{\cdot}V$ limitation using actual removal experiments.

For many CCOS processes, the material removal amount is calculated based on the well-known Preston equation given in (9). In order to determine the $K_p{\cdot}V$ limitation, the K_p values for a wide range of V (50 RPM–400 RPM) were measured based on the Preston equation. Forty experiments were performed to measure the TIFs, with the K_p then being calculated from the TIFs. The detailed experimental conditions are listed in Table 2. The actual 100-mm RC lap and a computer numerically controlled (CNC) polishing machine are shown in Figs. 9 and 10, respectively.

Table 2 Overall experimental conditions.

Polishing tool	100-mm RC lap
Workpiece	150-mm RB-SiC
Polishing compound	Diamond slurry poly (2-μm particle size)
Tool motion	Orbital tool motion (15-mm orbital radius)
Tool motion speed range	50 RPM–400 RPM (50-RPM interval)
Polishing pressure	1.47 psi
Repetitions	5

(a) (b)

Fig. 9 Workpiece and polishing tool for experiments: (a) 150-mm RB-SiC and (b) 100-mm RC lap.

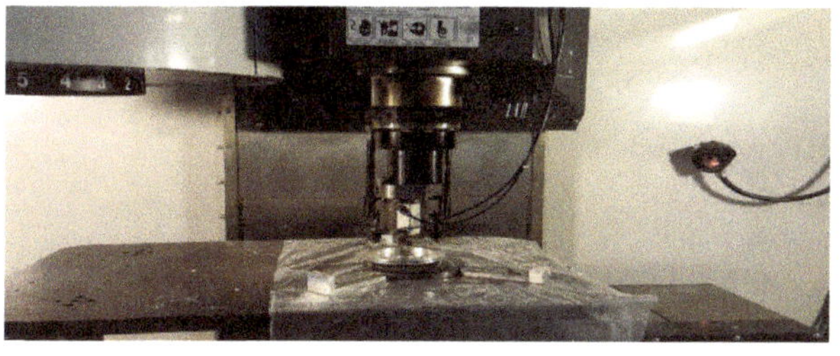

Fig. 10 CNC polishing machine used for experiments.

To ensure that there is no adverse impact on the surface roughness *Ra* from the high-speed orbital motion, it is necessary to measure the *Ra* values for all experiments. The average *Ra* values are plotted in Fig. 11, and the typical macro roughness of RB-SiC is shown in Fig. 12. The results indicate that *Ra* is insensitive to the tool speeds for all cases.

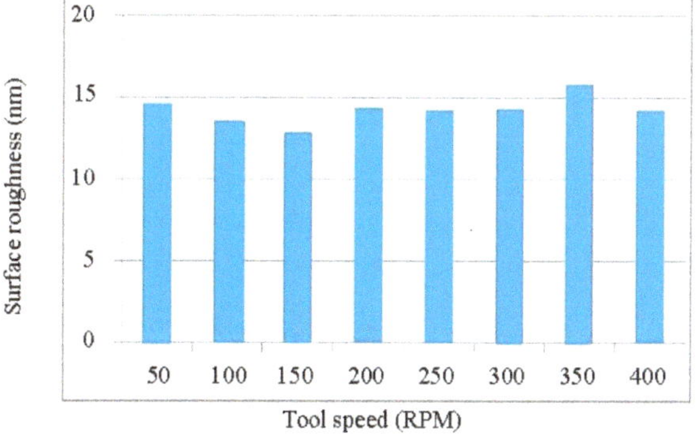

Fig. 11 Average surface roughness *Ra* values for all experiments.

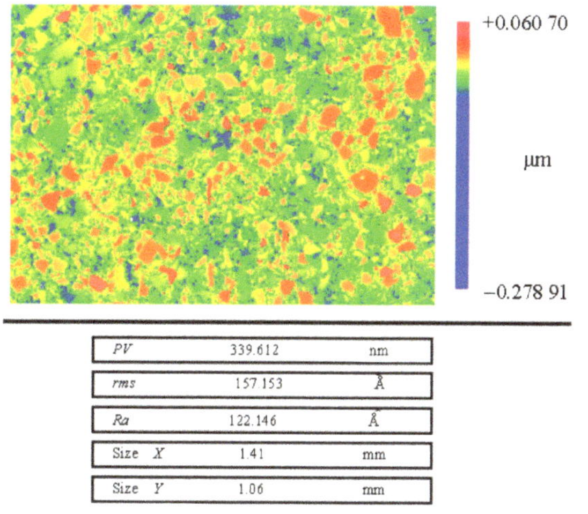

Fig. 12 RB-SiC roughness for 150-RPM orbital motion under 1.47-psi polishing pressure *P*.

The measured K_p is plotted in Fig. 13. Each marker is the average value, and the standard deviation is indicated by the error bar. For a V of 50-RPM, the K_p is comparatively low. As V increases in the 100-RPM – 300-RPM range, K_p increases exponentially. However, for the V range of 300 RPM – 400 RPM, K_p decreases linearly. In order to clearly illustrate the relationship between the material removal and V, the ΔZ per hour is calculated (solid line, Fig. 13) using the averaged K_p values. Hence, it is apparent that the material Z increases only slightly as V increases from 300 RPM to 400 RPM, indicating that the smoothing rate $K_p \cdot V$ has reached a limitation. As stated above, our second task is to apply the limiting V to actual smoothing experiments, so as to calculate κ_{total} by fitting the data to the *SF* function. As mentioned in Section 3(1), $\kappa_{\text{total}}(V)$ is a non-analytic function.

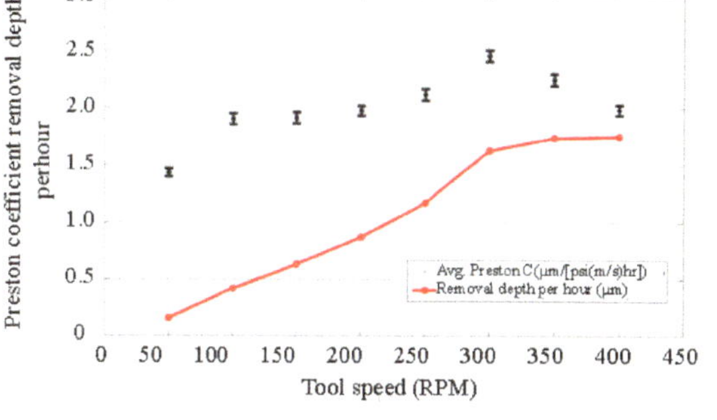

Fig. 13 Non-linearity of Preston coefficient K_p over tool-speed V range.

Therefore, its contribution must be verified via smoothing experiments; these experiments are discussed in Section 4.

4. Smoothing experiments and results

(1) Experimental setup

In order to quantitatively calculate the compressive $\kappa_{total}(V)$, the parametric smoothing model in (1) was used. The linear trend was fitted with the experimental data (ε_0, $\Delta\varepsilon$, and ΔZ) to obtain the slope of the *SF* function, i.e., k. Hence, κ_{total} could be calculated from (6).

A sinusoidal ripple with ξ = 0.2 mm^{-1} and PV=~0.1 µm was generated on the central areas of 150-mm diameter RB-SiC workpieces using magneto rheological finishing (MRF), as shown in Fig. 14. The reference area at the edge was intentionally left for the ΔZ measurement. Smoothing experiments were then conducted on these ripples.

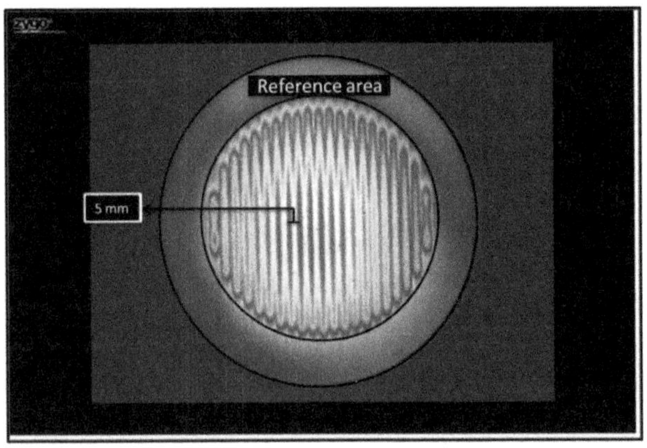

Fig. 14 Intensity map of MRF-generated sinusoidal ripples and reference area to measure nominal removal depth ΔZ.

To facilitate a more informative comparison, two sets of experiments comparing the κ_{total} obtained using 300-RPM and 400-RPM RC-lap orbital tool motions were performed. A Zygo Verifire™ QPZ interferometer was used to monitor $\Delta\varepsilon$ and ΔZ. The smoothing process was repeated until there was no obvious surface error reduction (ε_0). Details of the experimental setup are provided in Table 3.

Table 3 Experimental conditions for smoothing.

	300-RPM tool speed	400-RPM tool speed
Polishing tool	100-mm RC lap	100-mm RC lap
Workpiece	150-mm RB-SiC	150-mm RB-SiC
Tool motion	Orbital tool motion (15-mm orbital radius)	Orbital tool motion (15-mm orbital radius)
Polishing pressure	1.47 psi	1.47 psi
Polishing compound	Diamond slurry poly (2-µm particle size)	Diamond slurry poly (2-µm particle size)

(2) Experimental results

Because some errors were induced during the smoothing processes, a band-pass (wavelength: 2 mm – 6.5 mm) fast Fourier transform (FFT) filter was applied to separate the MSF error information from the measured map. Some measured surface errors and the filtered data are presented as examples in Fig. 15. The averaged *PV* value was determined from the filtered data to obtain the ε_0, $\Delta\varepsilon$, and ΔZ values, and then to fit the *SF* function.

The *SF* function was successfully fit using the experimental data, as shown in Fig. 16. From k, i.e., the slope of the *SF* function, and (6), the RC-lap compressive κ_{total} was calculated as listed in Table 4.

It is clearly apparent that the 400-RPM V yields higher compressive κ_{total}, which is caused by the non-linear visco-elastic behavior of the Silly-Putty. In Section 3(1), it was explained that the final k is related to κ_{total}, V, and K_p. Applying (11) to the K_p measured in Section 3(3) and the κ_{total} calculated from the *SF* function, we can express the surface error as a function of the smoothing time t directly. An example having a 0.3-µm ε_{ini} is shown in Fig. 17 to facilitate a comparison of the k obtained for 300-RPM and 400-RPM V values.

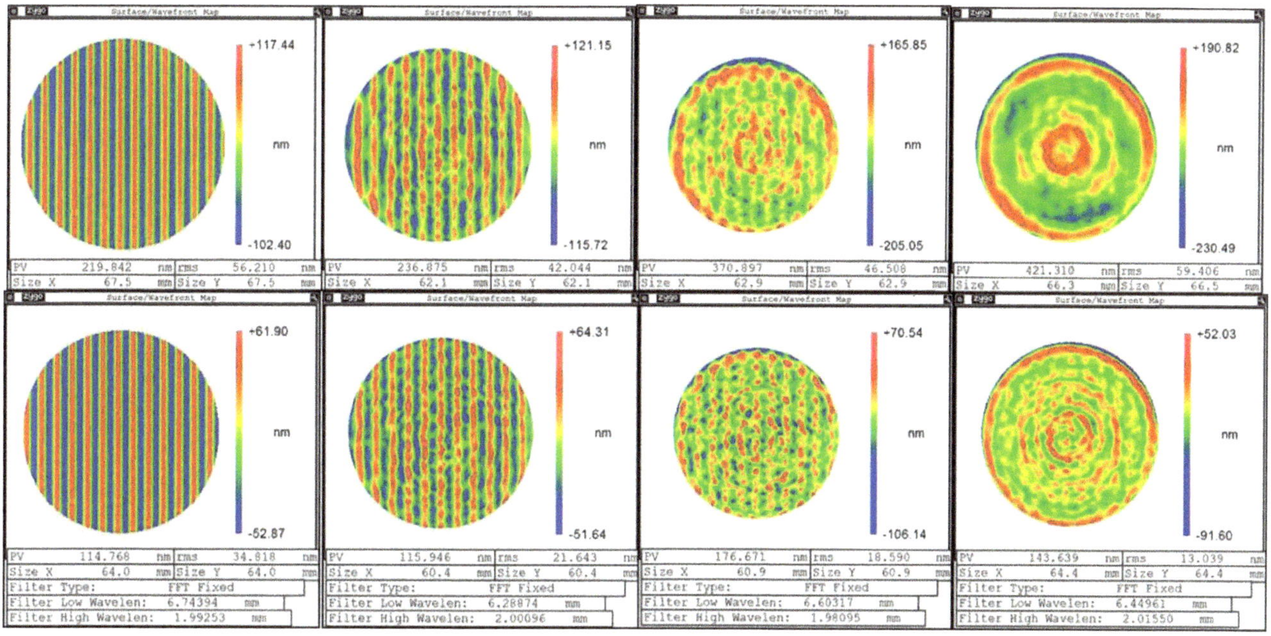

(a)

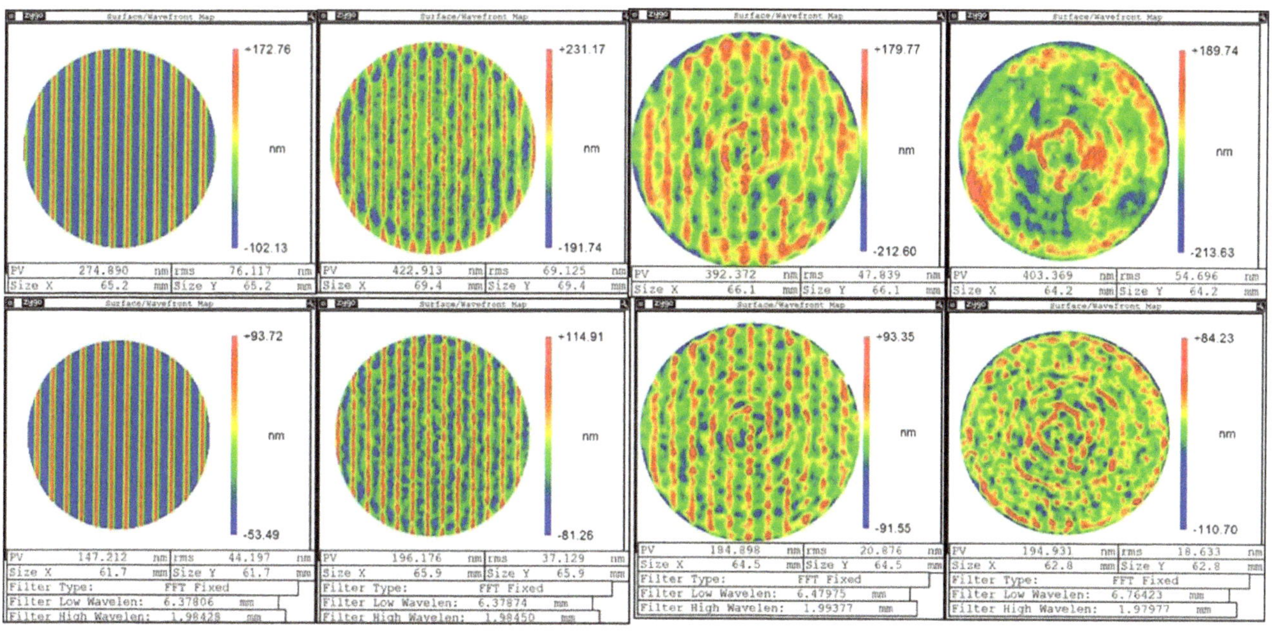

(b)

Fig. 15 Selected measured surface errors (top) and filtered data (bottom) in timed sequence: (a) 300-RPM and (b) 400-RPM tool-speed V results.

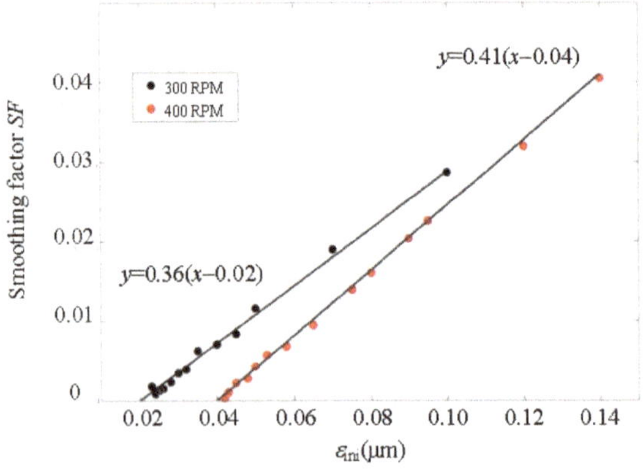

Fig. 16 Measured smoothing factor *SF vs.* initial magnitude ε_{ini} and linear fitting results for 300-RPM and 400-RPM tool speed *V*.

Table 4 Slope of *SF* function *k* and compressive tool stiffness κ_{total}.

RPM	300	400
k	0.36	0.41
κ_{total}(Pa/μm)	3649	4054

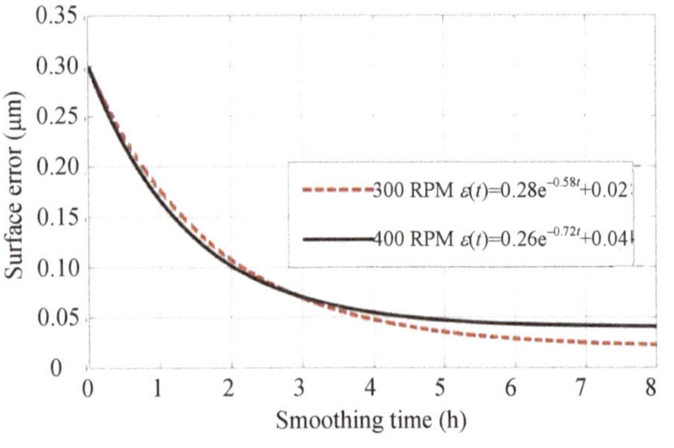

Fig. 17 Surface error ε_{ini} vs. smoothing time t for 300-RPM and 400-RPM tool speed *V*.

The result is straightforward. The ripples are almost fully smoothed out for both cases. The convergence rate is higher for the 400-RPM tool speed, but ε_0 for the 300-RPM case is lower. Our actual data in Fig. 15 also explain this result. The error map for the 300-RPM case is smoother, having a 13.039-nm root-mean-squared (RMS) value in the final filtered data. In contrast, the RMS value of the 400-RPM case is 18.633 nm. The smoothing rates for the 300-RPM and 400-RPM cases are almost identical, which verifies our assumption: the $K_p \cdot V$ term is more important than the κ_{total} term for the RC laps. In addition, the induced errors of the 400-RPM case, which may be caused by the higher κ_{total} or the higher V, yield a higher (i.e., less favorable) ε_0. Thus, the optimal smoothing speed V considering t is 300 RPM.

Note that the errors induced by the polishing parameters will be studied in a separate investigation in the future. Further, as shown in Fig. 18, the 300-mm RC lap was successfully employed in the Key Laboratory of Optical System Advanced Manufacturing Technology at the Changchun Institute of Optics, Fine Mechanics, and Physics (CIOMP).

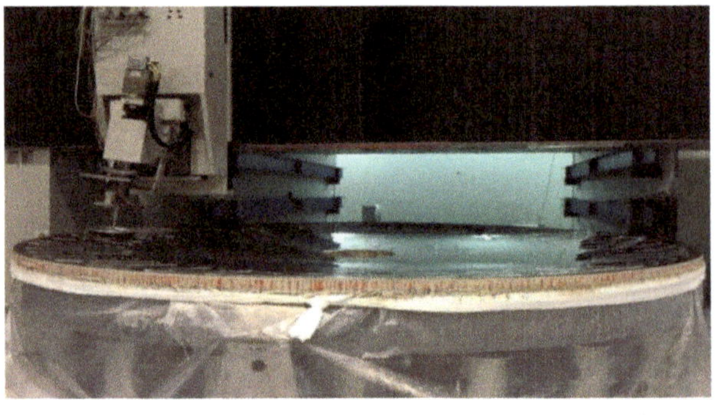

Fig. 18 300-mm diameter RC lap on 4-m-diameter RB-SiC mirror at CIOMP.

5. Conclusions

A time-dependent smoothing evacuation model that contains specific smoothing parameters was successfully derived from the parametric smoothing model and the Preston equation. Based on the time-dependent model, we proposed a strategy to improve the smoothing rate. A new RC-lap structure was designed to overcome the extreme polishing pressure, while providing a highly stable, theory-like TIF. Using our new structure, the limiting speed of 300-RPM was determined via a series of experiments. The surface roughness values were also measured for all experiments, being stable at 12 nm −15 nm for all cases. Two sets of smoothing experiments were performed using 300-RPM and 400-RPM tool speeds. The *SF* function was successfully fit against the experimental data and the tool stiffness was calculated. Based on the time-dependent model, the smoothing rates for various tool speeds were compared. Hence, the optimal smoothing tool speed was defined as being 300-RPM.

Acknowledgment

This research is financially supported by the National Natural Science of China (NSFC) (61210015) and Youth Foundation of National Natural Science Foundation (61605202).

References

[1] J. Nelson and G. H. Sanders, "The status of the thirty meter telescope project," *SPIE*, 2008, 7012: 70121A-1–70121A-18.

[2] M. Johns, P. Mccarthy, K. Raybould, A. Bouchez, A. Farahani, J. Filgueira, *et al*., "Giant magellan telescope: overview," *SPIE*, 2012, 8444: 84441H-1–84441H-16.

[3] T. Hull, M. J. Riso, J. M. Barentine, and A. Magruder, "Mid-spatial frequency matters: examples of the control of the power spectral density and what that means to the performance of imaging systems," *SPIE*, 2012, 8353: 835329-1–835329-17.

[4] D. W. Kim and J. H. Burge, "Rigid conformal polishing tool using non-linear visco-elastic effect," *Optics Express*, 2010, 18(3): 2242–2257.

[5] H. M. Martin, D. S. Anderson, J. R. P. Angel, R. H. Nagel, S. C. West, and R. S. Young, "Progress in the stressed-lap polishing of a 1.8-mf/1 mirror," *SPIE*, 1990, 1236: 682–690.

[6] H. Burge, B. Anderson, S. Benjamin, M. K. Cho, K. Z. Smith, and M. J. Valente, "Development of optimal grinding and polishing tools for aspheric surfaces," *SPIE*, 1990, 4451: 153–164.

[7] Y. Shu, X. Nie, F. Shi, and S. Li, "Compare study between smoothing efficiencies of epicyclic motion and orbital motion," *Optik – International Journal for Light and Electron Optics*, 2014, 125(16): 4441–4445.

[8] N. J. Brown, P. C. Baker, and R. E. Parks, "Polishing-to-figuring transition in turned optics," *Contemporary Methods of Optical Fabrication*, 1981, 306(6): 58–65.

[9] R. A. Jones, "Computer simulation of smoothing during computer-controlled optical polishing," *Applied Optics*, 1995, 34(7): 1162–1169.

[10] P. K. Mehta and P. B. Reid, "Mathematical model for optical smoothing prediction of high-spatial-frequency surface errors," *SPIE*, 1999, 3786: 447–459.

[11] M. T. Tuell, J. H. Burge, and B. Anderson. "Aspheric optics: Smoothing the ripples with semi-flexible tools," *Optical Engineering*, 2002, 15(2): 1473–1474.

[12] D. W. Kim, W. H. Park, H. K. An, and J. H. Burge, "Parametric smoothing model for visco-elastic polishing tools," *Optics Express*, 2010, 18(21): 22515–22526.

[13] Y. Shu, D. W. Kim, H. M. Martin, and J. H. Burge, "Correlation-based smoothing model for optical polishing," *Optics Express*, 2013, 21(23): 28771–28782.

[14] Y. Shu, X. Nie, F. Shi, and S. Li, "Smoothing evolution model for computer controlled optical surfacing," *Journal of Optical Technology*, 2014, 81(3): 164–167.

[15] X. Nie, S. Li, F. Shi, and H. Hu, "Generalized numerical pressure distribution model for smoothing polishing of irregular midspatial frequency errors," *Applied Optics*, 2014, 53(6): 1020–1027.

[16] A. C. Fischer-Cripps, "Multiple-frequency dynamic nanoindentation testing," *Journal of Materials Research*, 2004, 19(19): 2981–2988.

[17] J. H. Burge, D. W. Kim, and H. M. Martin, "Process optimization for polishing large aspheric mirrors," *SPIE*, 2014, 9151: 91512R-1–91512R-13.

A Study on Refractive Index Sensors Based on Optical Micro-Ring Resonators

Georgios N. TSIGARIDAS*

Department of Physics, School of Applied Mathematical and Physical Sciences, National Technical University of Athens, Zografou Campus, GR-15780 Zografou, Athens, Greece

*Corresponding author: Georgios N. TSIGARIDAS E-mail: gtsig@mail.ntua.gr

Abstract: In this work, the behavior of refractive index sensors based on optical micro-ring resonators is studied in detail. Using a result of waveguide perturbation theory in combination with numerical simulations, the optimum design parameters of the system, maximizing the sensitivity of the sensor, are determined. It is found that, when optimally designed, the sensor can detect relative refractive index changes of the order of $\Delta n/n \approx 3\times10^{-4}$, assuming that the experimental setup can detect relative wavelength shifts of the order of $\Delta\lambda/\lambda \approx 3\times10^{-5}$. The behavior of the system as bio-sensor has also been examined. It is found that, when optimally designed, the system can detect refractive index changes of the order of $\Delta n \approx 10^{-3}$ for a layer thickness of t=10 nm, and changes in the layer thickness of the order of $\Delta t \approx 0.24$ nm, for a refractive index change of Δn=0.05.

Keywords: Optical micro-ring resonators; refractive index sensors; bio-sensors; nano-photonic sensors

1. Introduction

Optical ring resonators are interesting optical devices with a plethora of applications especially in optical switching [1–4], routing [4–7], and sensing [8–10]. An optical ring resonator usually consists of a straight waveguide coupled to a circular one, as shown in Fig. 1(a), or two straight waveguides coupled through a circular one, as shown in Fig. 1(b).

In the first case the transmittance is given by the formula [11]

$$T_n = \frac{a^2 + r^2 - 2ar\cos\varphi}{1 + r^2 a^2 - 2ar\cos\varphi} \quad (1)$$

where r is the self-coupling coefficient of light between the straight and circular waveguides, and a is a loss parameter related to the power attenuation coefficient α through the equation $a^2=\exp(-\alpha L)$, with L being the length of the circular waveguide. Here, φ is the single-pass phase shift, defined as

$$\varphi = \beta L = \frac{2\pi n_{\mathrm{eff}}}{\lambda_0} L \quad (2)$$

where β is the propagation constan, n_{eff} the effective refractive index of the propagating mode in the circular waveguide, and λ_0 the free space wavelength. Setting $\cos\varphi \to y$, (1) takes the form:

$$T_n = \frac{a^2 + r^2 - 2ary}{1 + r^2 a^2 - 2ary}. \quad (3)$$

The derivative of T_n with respect to y is as follows:

$$\frac{dT_n}{dy} = -\frac{2a\left(-1+a^2\right)r\left(-1+r^2\right)}{\left(1+a^2r^2-2ary\right)^2}. \quad (4)$$

It is obvious that $dT_n/dy \le 0$ since both a and r

are less than or equal to unity. Thus, the transmittance is minimized for y=1, which implies that the condition for minimum transmittance, as far as the phase shift is concerned, is

$$\cos\varphi = 1 \quad \text{or} \quad \varphi = 2m\pi \tag{5}$$

where m is an integer. From (2) and (5), it follows that the values of the wavelength minimizing the transmittance (resonant wavelengths) are given by the formula

$$\lambda_{0,\mathrm{res}} = \frac{L}{m} n_{\mathrm{eff}} = \frac{2\pi R}{m} n_{\mathrm{eff}} \tag{6}$$

where R is the radius of the circular waveguide. In this case the transmittance takes the form

$$T_{n,\mathrm{res}} = \left(\frac{a-r}{1-ar}\right)^2 \tag{7}$$

which implies that for $a = r$, critical coupling, the transmittance is zero.

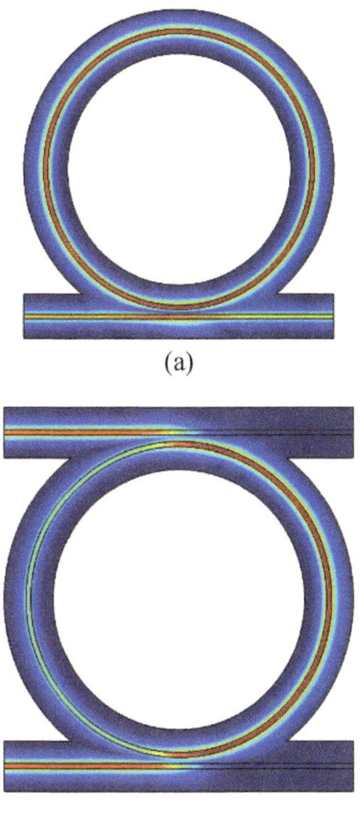

(a)

(b)

Fig. 1 Two basic configurations of an optical micro-ring resonator: (a) a linear waveguide is coupled to a circular one, and (b) two linear waveguides are coupled to each other through a circular one.

As far as the arrangement shown in Fig. 1(b) is concerned, the transmittance through the lower and upper waveguides are given by the formulae

$$T_p = \frac{r_1^2 + a^2 r_2^2 - 2ar_1r_2y}{1 + a^2r_1^2r_2^2 - 2ar_1r_2y} \tag{8a}$$

and

$$T_d = \frac{a\left(1-r_1^2\right)\left(1-r_2^2\right)}{1 + a^2r_1^2r_2^2 - 2ar_1r_2y} \tag{8b}$$

respectively. Here r_1 and r_2 are the coupling coefficients between the lower-circular and upper-circular waveguides respectively and $y = \cos\varphi$. Following a similar analysis it can be shown that the condition for the minimum transmittance through the lower waveguide, and simultaneously the maximum transmittance through the upper waveguide, is described again by (6). In this case (8a) and (8b) take the form

$$T_{p,\mathrm{res}} = \left(\frac{r_1 - ar_2}{1 - ar_1r_2}\right)^2 \tag{9a}$$

and

$$T_{d,\mathrm{res}} = \frac{a\left(1-r_1^2\right)\left(1-r_2^2\right)}{\left(1-ar_1r_2\right)^2}. \tag{9b}$$

Equation (9a) implies that for $r_1 = ar_2$, T_p=0. In this case the transmittance through the upper waveguide takes the form

$$T_{d,\mathrm{res}}\Big|_{r_1=ar_2} = \frac{a\left(r_2^2-1\right)}{a^2r_2^2-1}. \tag{10}$$

It is obvious that for $a=1$, $T_d=1$. Thus, in the case of a symmetrical optical ring resonator, shown in Fig. 1(b), the light is perfectly coupled to the upper waveguide, when the wavelength is given by (6), the loss parameter is equal to unity $(a=1)$, and the coupling coefficients are equal to each other $(r_1 = r_2)$.

As far as the use of micro-ring resonators as refractive index sensors is concerned, the most important result of the above analysis is that, in both configurations, the resonant wavelength is given by (6), which practically implies that it is proportional to the effective refractive index of the propagating mode in the circular waveguide. However, the value of n_{eff} depends on the refractive index of the environment. Therefore, the system can function as a refractive index sensor. In more detail, supposing that the effective refractive index changes by Δn_{eff} due to a change in the refractive index of the environment by Δn, (6) takes the form

$$\lambda_{0,\text{res}} + \Delta\lambda_{0,\text{res}} = \frac{2\pi R}{m}\left(n_{\text{eff}} + \Delta n_{\text{eff}}\right) \tag{11}$$

where $\Delta\lambda_{0,\text{res}}$ is the corresponding change in the resonant wavelength. Dividing (11) and (6) by parts follows that:

$$\frac{n_{\text{eff}} + \Delta n_{\text{eff}}}{n_{\text{eff}}} = \frac{\lambda_{0,\text{res}} + \Delta\lambda_{0,\text{res}}}{\lambda_{0,\text{res}}} \quad \text{or} \quad \frac{\Delta n_{\text{eff}}}{n_{\text{eff}}} = \frac{\Delta\lambda_{0,\text{res}}}{\lambda_{0,\text{res}}}. \tag{12}$$

Consequently, the relative change in the resonant wavelength is equal to the relative change in the effective refractive index.

According to the above analysis, the most important parameter regarding the performance of an optical ring resonator when used as a refractive index sensor is the dependence of the effective refractive index of the propagating mode on the refractive index of the environment. Mathematically, this is described by the quantity as

$$S = \frac{1}{n_{\text{eff}}}\frac{dn_{\text{eff}}}{dn} \quad \text{or} \quad S = \frac{1}{n_{\text{eff}}}\frac{\Delta n_{\text{eff}}}{\Delta n} \tag{13}$$

where in the second form of (13), it is supposed that the refractive index change of the environment is sufficiently small, so that the dependence of Δn_{eff} on Δn is linear. Using the parameter S, (12) takes the form:

$$\frac{\Delta\lambda_{0,\text{res}}}{\lambda_{0,\text{res}}} = S\Delta n. \tag{14}$$

From (14), it is obvious that S can be regarded as a measure of the sensitivity of the system when used as a refractive index sensor. In practice, the parameter S can be numerically calculated quite easily as described in the following section.

2. Numerical analysis

The profile and effective refractive indexes of the propagating modes in the circular waveguide are calculated using the finite element analysis [12]. The simulations were assisted by the Comsol Multiphysics software package. The analysis is made on a two dimensional cross-section of the waveguide, assuming that its profile and properties remain constant throughout its length. Firstly, the electromagnetic field is written in the form

$$\begin{aligned} E(x,y,z) &= A(x,y)\exp(-\alpha_0 z)\exp(\mathrm{i}\beta z) \\ &= A(x,y)\exp(-\gamma z) \end{aligned} \tag{15a}$$

$$\begin{aligned} H(x,y,z) &= B(x,y)\exp(-\alpha_0 z)\exp(\mathrm{i}\beta z) \\ &= B(x,y)\exp(-\gamma z) \end{aligned} \tag{15b}$$

where $E(x,y,z)$ and $H(x,y,z)$ are the electric and magnetic field components of the propagating wave, respectively. The two-dimensional functions $A(x,y)$ and $B(x,y)$ describe the mode profile, assuming that the electromagnetic wave is propagating in the $+z$ direction. Here α_0 is the linear loss coefficient and β is the propagation constant, related to the effective refractive index through the formula as

$$\beta = k_0 n_{\text{eff}} \tag{16}$$

where $k_0 = 2\pi/\lambda_0$ is the free space wavenumber. The parameter γ corresponds to the generalized propagation constant, defined as $\gamma = \alpha_0 - \mathrm{i}\beta$.

Writing the Maxwell equations in matrix form and eliminating the longitudinal field components, an eigenvalue problem of the form

$$\mathbf{\Omega}^2 \begin{bmatrix} \mathbf{A}_x \\ \mathbf{A}_y \end{bmatrix} = \left(\frac{\gamma}{k_0}\right)^2 \begin{bmatrix} \mathbf{A}_x \\ \mathbf{A}_y \end{bmatrix} \tag{17}$$

is obtained, where

$$\mathbf{\Omega}^2 = \begin{bmatrix} \mathbf{D}^e_{x'}\varepsilon^{-1}_{zz}\mathbf{D}^h_{y'} & -\left(\mathbf{D}^e_{x'}\varepsilon^{-1}_{zz}\mathbf{D}^h_{x'} + \mu_{yy}\right) \\ \mathbf{D}^e_{y'}\varepsilon^{-1}_{zz}\mathbf{D}^h_{y'} + \mu_{xx} & \mathbf{D}^e_{y'}\varepsilon^{-1}_{zz}\mathbf{D}^h_{x'} \end{bmatrix} \times \begin{bmatrix} \mathbf{D}^h_{x'}\mu^{-1}_{zz}\mathbf{D}^e_{y'} & -\left(\mathbf{D}^h_{x'}\mu^{-1}_{zz}\mathbf{D}^e_{x'} + \varepsilon_{yy}\right) \\ \mathbf{D}^h_{y'}\mu^{-1}_{zz}\mathbf{D}^e_{y'} + \varepsilon_{xx} & \mathbf{D}^h_{y'}\mu^{-1}_{zz}\mathbf{D}^e_{x'} \end{bmatrix}. \tag{18}$$

where $\mathbf{D}^e_{x'(y')}$ and $\mathbf{D}^h_{x'(y')}$ are matrix differential operators with respect to $x'(y')$ for the electric and magnetic field respectively, and ε_{xx}, ε_{yy}, ε_{zz}, μ_{xx}, μ_{yy}, and μ_{zz} are matrices containing the diagonal elements of the relative permittivity and permeability tensors at each position. The vector matrix $\begin{bmatrix} \mathbf{A}_x \\ \mathbf{A}_y \end{bmatrix}$ corresponds to the transverse distribution of the electric field, $A(x, y)$, with $\mathbf{A}_x$ and $\mathbf{A}_y$ containing in single column format the x and y components of the electric field at each position. The coordinates (x', y', z') are related to (x, y, z) through the transformation $(x' \to k_0 x, y' \to k_0 y, z' \to k_0 z)$.

Thus, solving (17) for the eigenvalues $\lambda = (\gamma/k_0)^2$ and the eigenvectors $\begin{bmatrix} \mathbf{A}_x \\ \mathbf{A}_y \end{bmatrix}$, the generalized propagation constant and consequently the effective refractive index as well as the transverse electric field distribution (profile) for each mode are obtained.

It should be noted that in most applications of practical interest, the single mode operation is preferable, because the existence of more than one propagating modes would complicate the behavior of the system. For example, in the case of multi-mode operation, (6) implies that more than one resonant wavelengths exist for each configuration, which would result in multiple dips (or peaks, if measured through the upper waveguide) in the transmittance spectrum, complicating both the theoretical and experimental analysis of the system. Therefore, in the simulations, special care is taken in order to ensure single mode operation.

3. Simulations, results, and discussion regarding design optimization

According to a result of waveguide perturbation theory [13], the change in the effective refractive index of the propagating mode due to a refractive index change in some part of the waveguide is proportional to the fraction of the mode power in this specific part of the waveguide, defined as

$$\eta_p = \frac{\iint_{S_p} |A(x,y)|^2 \, dxdy}{\iint_{S_{\text{tot}}} |A(x,y)|^2 \, dxdy} \tag{18}$$

where S_p is the area where the refractive index has changed and S_{tot} is the total area of the waveguide. Further, according to (13), the sensitivity S is proportional to Δn_{eff} and consequently it will also be proportional to η_p. This practically means that the sensitivity increases as the overlapping of the propagating mode with its environment becomes more extensive.

Therefore, in the simulations the core of the waveguide is considered to be in direct contact with the environment. Further, the core dimensions are chosen to be small, close to the limit of waveguiding, in order to increase the mode spreading and maximize the parameter η_p. The cross section of the core of the waveguide is considered to be rectangular, as shown in Fig. 2, where the mode profile is also depicted, for two different heights of the waveguide core. The material of the core is chosen to be silicon nitride (SiN), which has the advantage that its refractive index is relatively insensitive to temperature fluctuations [14, 15]. It is supposed that the core is developed on a layer of SiO_2, as usually happens in practice. The environment is chosen to be water, which is the most common solvent, and it is also suitable for biological applications. Thus, the simulation model consists of a vertical column of SiN with small dimensions (the core of the waveguide) developed on a horizontal layer of SiO_2, while all the system is submersed in water. The free space wavelength is supposed to be λ_0=1550 nm, which is a common wavelength emitted by fiber lasers, and further corresponds to the minimum of the fiber losses [16].

The refractive indices of the materials used in the simulations at the aforementioned wavelength are [17] $n_{\text{SiN}} = 2.463$, $n_{\text{SiO}_2} = 1.443$, and $n_{\text{water}} = 1.318$ for the silicon nitride, silicon dioxide, and water, respectively. The dimensions of the waveguide core are chosen to be $h = \lambda_{\text{env}}$ and $w = \lambda_{\text{env}}/4$, where h is the height and w the width of the waveguide, while $\lambda_{\text{env}} = \lambda_0/n_{\text{env}}$ is the wavelength in the environment of the sensor, water in our case. This choice for the dimensions of the waveguide core ensures single mode operation and sufficient mode spreading. In order to optimize further the design of the refractive index sensor the parameter η_p is calculated as a function of the core width w with the height fixed to $h = \lambda_{env}$. The results are shown in Fig.3, where it is clear that the maximum value of η_p is obtained for $w = 0.175\ \lambda_{env}$.

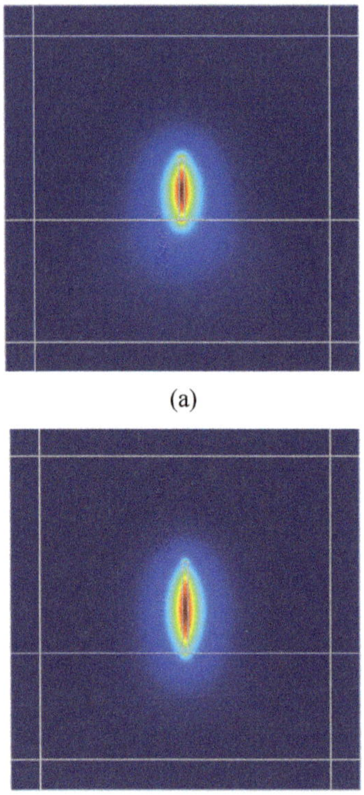

(a)

(b)

Fig. 2 Profile of the fundamental mode in a single core waveguide in the case of (a) $h = \lambda_{env}$ and (b) $h = 3\lambda_{env}/2$. In both cases the waveguide width is $w = 0.175\lambda_{env}$.

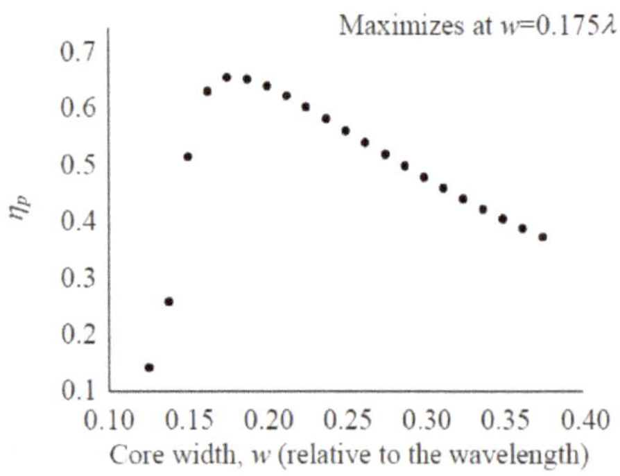

Fig. 3 Values of the parameter η_p as function of the core width w in the case of $h = \lambda_{env}$.

In order to optimize the height of the waveguide, the parameter η_p is calculated as the function of the waveguide height, with the width fixed to $w = 0.175\lambda_{env}$. The results are shown in Fig.4. It is clear that the value of η_p does not change much for $h \geq \lambda_{env}$. Further, for $h > 3\lambda_{env}/2$, the multimode behavior is exhibited, which is not desirable for practical applications, as explained in the previous section. Thus, the optimum width of the waveguide core is $w_{opt} = 0.175\ \lambda_{env}$ while the optimum height lies in the region $\lambda_{env} \leq h_{opt} \leq 3\lambda_{env}/2$.

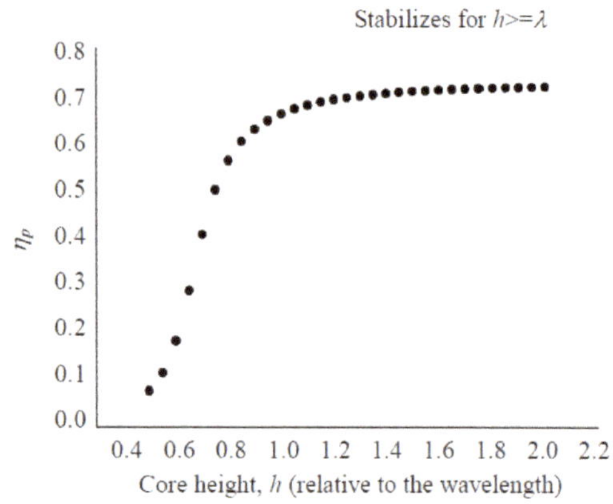

Fig. 4 Values of the parameter η_p as function of the core height h in the case of $w = 0.175\lambda_{env}$.

A double core configuration has also been studied, as shown in Fig.5, where the mode profile is depicted. In this case, the dimensions of each core are chosen to be $w = 0.175\ \lambda_{env}$ and $h = \lambda_{env}$. Higher values of h have not been considered because the system exhibits multi-mode operation. In order to find the optimum distance between the fiber cores, the parameter η_p has been calculated as function of the core separation d. The results are shown in Fig.6.

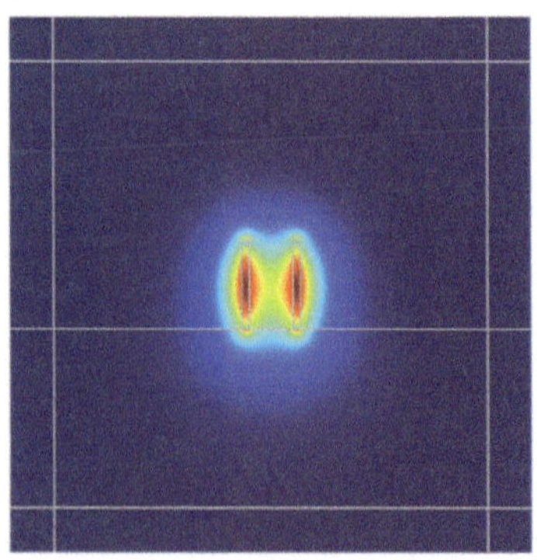

Fig. 5 Profile of the fundamental mode in a dual core waveguide in the case of $h = \lambda_{env}$, $w = 0.175\lambda_{env}$, and $d = \lambda_{env}/2$.

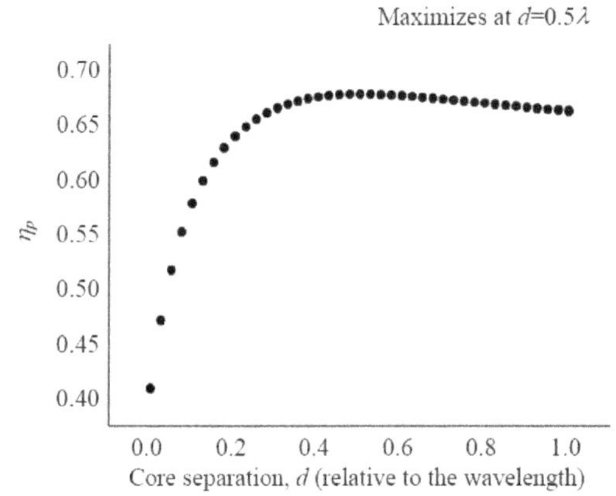

Fig. 6 Values of the parameter η_p as function of the core separation d in the case of $w = 0.175\lambda_{env}$ and $h = \lambda_{env}$.

It is clear that η_p is maximized for $d = \lambda_{env}/2$. Thus, the optimum conditions for a dual core design are $w_{opt} = 0.175\ \lambda_{env}$, $h_{opt} = \lambda_{env}$, and $d_{opt} = \lambda_{env}/2$. It should be noted that although these optimal values refer to an incident wavelength of 1 550 nm, they can be used as guidelines for other wavelengths. This is due to the fact that the Maxwell equations are invariant under scale transformations and consequently the behavior of the system does not change if its dimensions are scaled according to the wavelength ratio. For example, if the wavelength is multiplied by a factor x and all the dimensions of the waveguide are multiplied by the same factor, then its behavior will remain unchanged. Of course, in practice this cannot be achieved because the optical properties of the materials change due to dispersion. However, in any case, these values can be used as good starting points for the design optimization. Next, the sensitivity is calculated for the three optimal configurations, specifically $w = 0.175\ \lambda_{env}$, $h = \lambda_{env}$, single core, $w = 0.175\ \lambda_{env}$, $h = 3\lambda_{env}/2$, single core, and $w = 0.175\ \lambda_{env}$, $h = \lambda_{env}$, $d = \lambda_{env}/2$, dual core. The results are shown in Table 1.

Table 1: Basic characteristics of the three optimal designs described above.

No. of cores	w/λ_{env}	h/λ_{env}	d/λ_{env}	n_{eff}	dn_{eff}/dn	S
1	0.175	1	-	1.46951	0.317	0.216
1	0.175	1.5	-	1.50435	0.330	0.219
2	0.175	1	0.5	1.50773	0.365	0.242

For these values of the sensitivity, (14) implies that the relative change in the resonant wavelength is $\Delta\lambda_{0,res}/\lambda_{0,res} = 0.216\Delta n$ for a single core design $(h = \lambda_{env})$ and further increases by a factor of 1.014 in the case of $h = 3\lambda_{env}/2$ and 1.120 for a dual core configuration. This practically means that if the experimental setup is able to resolve a wavelength shift of the order of $\Delta\lambda/\lambda \approx 3\times10^{-5}$, which can be easily achieved by modern high-end optical spectrum analyzers, then the system can detect a refractive index change of $\Delta n_{min} \approx 1.4\times10^{-4}$ for a single core configuration $(h = \lambda_{env})$, which is further reduced by a factor of 1.014 in the case of $h = 3\lambda_{env}/2$ and 1.120 in the case of a dual core design.

It should be noted that the gain in the sensitivity for a dual core design is not that significant in order to justify the complexity of the design and the manufacturing challenges that entails. Therefore, it is suggested that in most practical situations, the design of choice would be a single core waveguide with dimensions $w = 0.175\ \lambda_{env}$ and $\lambda_{env} \le h \le 3\lambda_{env}/2$. Furthermore, as it will be shown in the following section, the choice of $w = 0.175\ \lambda_{env}$ and $h = 3\lambda_{env}/2$ for the core dimensions provides better sensitivity than the dual core design as far as the behavior of the system as biosensor is considered.

4. Study of the bio-sensing properties of the system

In recent years, the use of photonic nano-sensors as bio-sensors is a hot topic of research, with a plethora of designs and applications [18–23]. Therefore, in the following, the behavior of the system described in the previous section as biosensor is considered. For this purpose, a thin layer of biomaterial is supposed to be attached on the surface of the waveguide core. The dependence of the parameter η_p, namely the fraction of the mode power propagating within the layer, as function of the layer thickness, is shown in Fig. 7. It is assumed that the refractive index within the bio-layer increases by Δn=0.01 compared to the environment, although the parameter η_p is not very sensitive to the refractrive index of the biolayer.

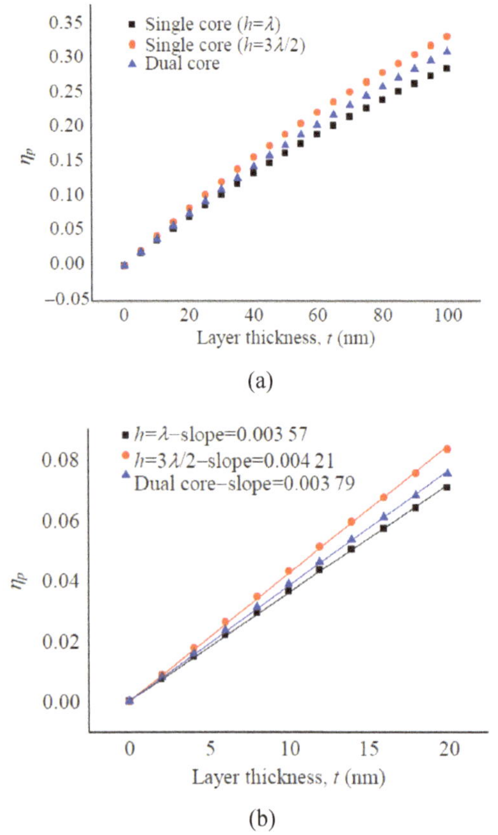

Fig. 7 Values of the parameter η_p as function of the layer thickness t for a range of (a) 100 nm and (b) 20 nm (zooming).

It is clear that η_p and consequently the sensitivity increase linearly for small values of the layer thickness but exhibit a trend for saturation as t increases beyond ~40 nm. It should also be noted that in this case the sensitivity of the single core waveguide with height $h=3\lambda_{env}/2$ is greater than that of the dual core one. In order to study the sensing ability of the system as the refractive index of the bio-layer varies, the relative change in the effective refractive index has been calculated as function of the refractive index difference of the layer material. Two values of the layer thickness are considered, namely t=10 nm and t=20 nm. The results are shown in Fig. 8.

It is clear that the dependence is linear, and further the single core waveguide with height $h=3\lambda_{env}/2$ exhibits the highest sensitivity, with a slope of 0.026 in the case of t=10 nm and 0.050 when the layer thickness doubles. This practically means that if the experimental setup can resolve a relative change in the resonant wavelength of the order of 3×10^{-5}, then the system can sense refractive index changes within the bio-layer of the order of $\Delta n_{min}\approx1.15\times10^{-3}$ for a layer thickness of 10 nm, which is reduced by a factor of two when the layer thickness doubles. Consequently, the system is able to detect very subtle changes in the composition or in the environment of the bio-layer. Finally, the sensing ability of the system regarding the thickness of the bio-layer has been examined. For this purpose, the values of the relative change in the effective refractive index have been calculated as function of the layer thickness, supposing that the refractive index change within the layer is $\Delta n=0.01$ and $\Delta n=0.05$. The results are shown in Fig. 9.

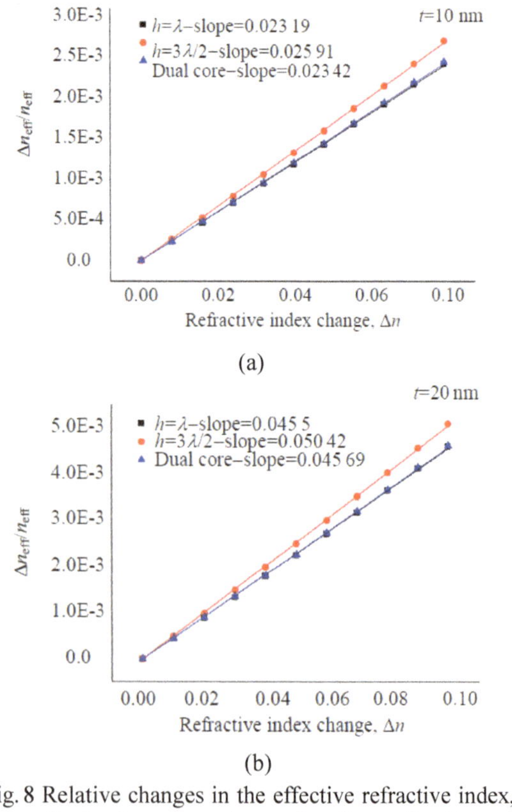

Fig. 8 Relative changes in the effective refractive index, and consequently in the resonant wavelength, as function of the refractive index change Δn in the bio-layer, when its thickness is (a) t=10 nm and (b) t=20 nm.

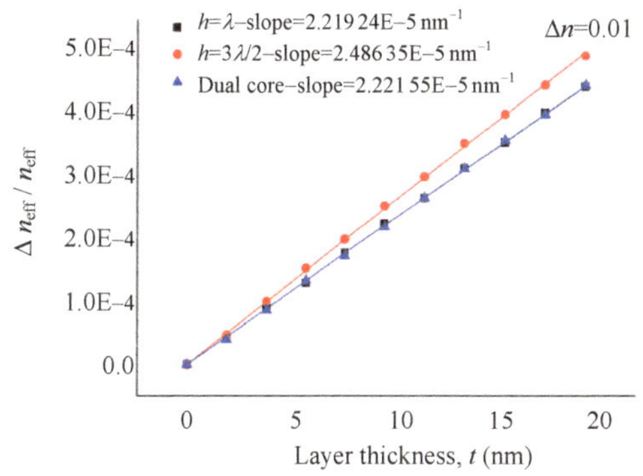

(a)

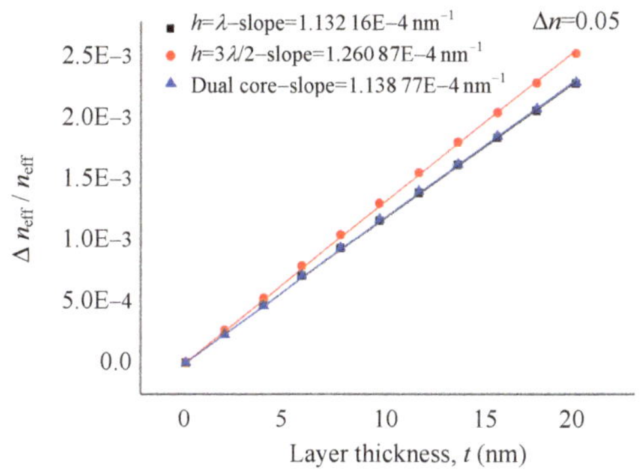

(b)

Fig. 9 Relative change in the effective refractive index, and consequently in the resonant wavelength, as function of the thickness t of the bio-layer, when the change in its refractive index is (a) $\Delta n = 0.01$ and (b) $\Delta n = 0.05$.

It is clear that the dependence is again linear, and the highest sensitivity is exhibited by the $h = 3\lambda_{\mathrm{env}}/2$ single core waveguide. Indeed, the slope in this case is $2.49\times10^{-5}\,\mathrm{nm}^{-1}$ for $\Delta n = 0.01$ and increases to $1.26\times10^{-4}\,\mathrm{nm}^{-1}$ in the case of $\Delta n = 0.05$. Consequently, supposing that the minimum relative wavelength shift that can be measured is 3×10^{-5}, the minimum change in the layer thickness that can be detected is $\Delta t_{\min} \approx 1.2$ nm in the case of $\Delta n = 0.01$, which is further reduced to $\Delta t_{\min} \approx 0.24$ nm for $\Delta n = 0.05$. This impressive result leads to the conclusion that the system can detect even the slightest changes in the configuration of the biomolecules.

5. Conclusions

In this work the behavior of optical micro-ring resonators when functioning as refractive index sensors has been studied in detail. Detailed guidelines for maximizing the sensitivity of the sensors are provided. It is found that, when optimally designed, the system can detect relative refractive index changes of the order of $\Delta n/n \approx 10^{-4}$, assuming that the experimental setup is able to resolve relative wavelength shifts of the order of $\Delta\lambda/\lambda \approx 3\times10^{-5}$. The performance of the systems as bio-sensor has also been examined. It is found that, when optimally designed, the system can detect refractive index changes of the order of $\Delta n \approx 10^{-3}$ for a layer thickness of t=10 nm, and changes in the layer thickness of the order of $\Delta t \approx 0.24$ nm, for a refractive index change of $\Delta n \approx 0.05$.

References

[1] R. Q. Ji, L. Yang, L. Zhang, Y. H. Tian, J. F. Ding, H. T. Chen, *et al.*, "Microring-resonator-based four-port optical router for photonic networks-on-chip," *Optics Express*, 2011, 19(20): 18945–18955.

[2] R. Q. Ji, J. Xu, and L. Yang, "Five-port optical router based on microring switches for photonic networks-on-chip," *IEEE Photonics Technology Letters*, 2013, 25(5): 492–495.

[3] T. Hu, H. Shao, L. Z. Yang, C. Xu, M. Yang, H. Yu, *et al.*, "Four-port silicon multi-wavelength optical router for photonic networks-on-chip," *IEEE Photonics Technology Letters*, 2013, 25(23): 2281–2284.

[4] S. J. Emelett and R. Soref, "Design and simulation of silicon microring optical routing switches," *Journal of Lightwave Technology*, 2005, 23(4): 1800–1807.

[5] Q. Xu and M. Lipson, "All-optical logic based on silicon micro-ring resonators," *Optics Express*, 2007, 15(3): 924–929.

[6] V. Van, T. A. Ibrahim, K. Ritter, P. P. Absil, F. G. Johnson, R. Grover, *et al.*, "All-optical nonlinear switching in GaAs-AlGaAs microring resonators," *IEEE Photonics Technology Letters*, 2002, 14(1): 74–76.

[7] J. N. Xia, "Microring-resonator-based switch architectures for optical networks," Ph.D. dissertation, Politecnico di Torino, Turin MA, 2014.

[8] I. M. White and X. Fan, "On the performance quantification of resonant refractive index sensors," *Optics Express*, 2008, 16(2): 1020–1028.

[9] M. Gabalis, D. Urbonas, and R. Petruskevicius, "A perforated microring resonator for optical sensing applications," *Journal of Optics*, 2014, 16(10): 105003.

[10] A. Yalcin, K. C. Popat, J. C. Aldridge, T. A. Desai, J. Hryniewicz, N. Chbouki, *et al.*, "Optical sensing of biomolecules using microring resonators," *IEEE Journal of Selected Topics in Quantum Electronics*, 2006, 12(1): 148–155.

[11] W. Bogaerts, P. D. Heyn, T. V. Vaerenbergh, K. D. Vos, S. K. Selvaraja, T. Claes, *et al.*, "Silicon microring resonators," *Laser & Photonics Reviews*, 2015, 6(1): 47–73.

[12] J. M. Jin, *The finite element method in electromagnetics*. Piscataway, U.S.: Wiley, 2014.

[13] A. W. Snyder and J. D. Love, *Optical waveguide theory*. New York, U.S.: Springer US, 1983, pp. 1–37.

[14] K. B. Gylfason, C. F. Carlborg, A. Kazmierczak, F. Dortu, H. Sohlstom, L. Vivien, *et al.*, "On-chip temperature compensation in an integrated slot-waveguide ring resonator refractive index sensor array," *Optics Express*, 2010, 18(4): 3226–3237.

[15] V. Raghunathan, W. N. Ye, J. J. Hu, T. Izuhara, J. Michel, and L. Kimerling, "Athermal operation of silicon waveguides: spectral, second order and footprint dependencies," *Optics Express*, 2010, 18(17): 17631–17639.

[16] G. P. Agrawal, *Nonlinear Fiber Optics*, Pittsburgh: Academic Press, 2001, pp: 1–61.

[17] Optical constants for a variety of materials. Available online: http://refractiveindex.info/.

[18] M. Consales, M. Pisco, and A. Cusano, "Lab-on-fiber technology: a new avenue for optical nanosensors," *Photonic Sensors*, 2012, 2(4): 289–315.

[19] S. Olyaee, S. Najafgholinezhad, and H. Alipour-Banaei, "Four-channel label-free photonic crystal biosensor using nanocavity resonators," *Photonic Sensors*, 2013, 3(3): 231–236.

[20] M. Nejadebrahimy, L. Halimi, and H. Alipour-Banaei, "Design and simulation of ultrasensitive nano-biosensor based on OFPC," *Photonic Sensors*, 2015, 5(1): 43–19.

[21] S. Robinson and N. Dhanlaksmi, "Photonic crystal based biosensor for the detection of glucose concentration in urine," *Photonic Sensors*, 2017, 7(1): 11–19.

[22] X. N. Han, X. Y. Han, Y. H. Shao, Z. L. Wu, Y. X. Liang, J. Teng, *et al.*, "Polymer integrated waveguide optical biosensor by using spectral splitting effect," *Photonic Sensors*, 2017, 7(2): 131–139.

[23] H. Heng and R. Wang, "Electromagnetic resonant properties of metal-dielectric-metal (MDM) cylindrical microcavities," *Photonic Sensors*, 2017, 7(2): 148–156.

Potential of SPR Sensors Based on Multilayer Interfaces With Gold and LHM for Biosensing Applications

Abdellatif CHERIFI* and Benamar BOUHAFS

University of Tlemcen, Theoretical Physics Laboratory, Physics of Department, Tlemcen, 13000, Algeria

*Corresponding author: Abdellatif CHERIFI E-mail: Cherifi_physics@yahoo.com

Abstract: Recently, the subject on "plasmonics" has received significant attention in designing surface plasmon resonance (SPR) sensors. In order to achieve extremely high-sensitivity sensing, multilayered configurations based on a variety of active materials and dielectrics have been exploited. In this work, a novel SPR sensor is proposed and investigated theoretically. The structure, analyzed in attenuated total reflection (ATR), consists of multilayer interfaces between gold and a metamaterial (LHM) separated by an analyte layer as a sensing medium. By interchanging between gold and LHM, under the effect of the refractive index (RI) of analyte set to be in the range of 1.00 to 1.99, the sharp peak reflectivity at the SPR angle takes two opposite behaviors predicted from the transfer matrix method. At the threshold value of 1.568 of the refractive index of analyte and when the LHM is the outer medium, the layered structure exhibits a giant sharp peak located at 43° of intensity up to 10^5 due to the Goos-Hànchen effect. With respect to the refractive index (RI) change and thickness of analyte, the characteristics (intensity, resonance condition, and quality factor) of the SPR mode, which make the proposed device have the potential for biosensing applications, have been analytically modelized.

Keywords: Plasmonic; goos-Hànchen (GH) effect; SPR sensor systems; sub-wavelength scale

1. Introduction

The surface plasmon resonance (SPR) phenomenon known for a long time has been used in large practical applications such as terahertz (THz) filter [1], plasmon enhanced solar cell [2, 3], plasmonic detectors [4], optical antenna [5], biological analysis [6], and chemical sensors [7]. This resonant phenomenon corresponds to a coupling process between a monochromatic light and conduction electrons on the boundary of a metal surface and a dielectric. With an attenuated total reflection method (ATR), the SPR mode is strongly sensitive to both thickness and refractive index of the dielectric. Due to this characteristic, SPR becomes a highly efficient mechanism in designing plasmonic sensor systems. In 1982, the first application of SPR was done by Liedberg *et al.* [8, 9]. However, the efficient way of manipulating the characteristics of the electromagnetic field correlated to SP in a nanoscale, remains related particularly to the involved interface properties. Therefore, a multilayer SPR sensor leads to achieving other limits that can be preferable to those obtained according to the conventional Kretschmann configuration [10]. In the above configuration, when a p-polarized light is incident on the prism-metal interface and the reflectance is measured versus the incident angle, a sharp dip occurs at the resonance angle beyond the critical angle. Within the medium to be sensed in contact with the metal layer [11], the SP field evolutes in an evanescent character into both mediums of the interface. The resolution of an SPR-sensor in terms of line width and dip position of SPR-reflectivity depends on the environmental index change [12, 13]. As an example of SPR-sensor, we cite the optical sensing based on chalcogenide prism and graphene layers in addition to the conventionally used Au-layer for the detection of bimolecular analyte [14], periodic nanostructures Au nano-gratings [15], etc.

In recent years, plasmonic metamaterials hold a new promise for biosensing with unprecedented sensitivity and specificity. Metamaterials are composite materials whose permittivity ε and permeability μ are both negative. Thus, metamaterials are described by the negative refraction index $n = -(\mu\varepsilon)^{1/2}$. In this topic, the first discussion on metamaterials type was advanced in 1968 by Veselago [16]. In 1999, Pendry *et al.* identified a practical way to construct left-handed metamaterials (LHMs) of a different behavior which did not follow the conventional right hand rule [17].

Recently, researchers have successfully put forward many different biosensor configurations with greatly enhanced sensitivity to the LHM variation. For example, Schueler *et al.* [18] reviewed the LHM inspired composite transmission line microwave sensors. Chen *et al.* [19] reviewed the LHM application in sensing with an emphasis on split ring resonator-based sensors. Yang *et al.* studied LHM sensors based on a single metamaterials particle with the advantages of an easy fabrication and experimental robustness [20]. Upadhyar *et al.* [21] used LHM instead of metals to investigate SPs in low-wavelength sensors [22], and Zheludev [23] analyzed the future development of LHMs and pointed out that the sensing application represents a growing area.

Very recently, Pal *et al.* [24] proposed SPR sensors based on bi-LHMs with gold as an inner core and compared the above structure to another with two dielectric media instead of LHM. On the analysis of the reflectivity profile, the authors [24] pointed out the advantage of using the metamaterials to further enhance SPR sensor sensitivity. Some researchers have demonstrated typical multilayer devices as biosensors that provide sensitivity almost 6-fold higher than that of the conventional SPR sensor [25]. Sensitivity enhancement of the SPR-sensorwas the discussion subject in [26], in which a bi-metallic structure was established and a comparative analysis with the conventional SPR configuration to that of an SPR sensor of four and five layered LHM media was made.

The main objective of this work focuses on exploring SPR on a tunable nanocavity where an analyte (sensing medium) is embedded between symmetric interfaces gold/LHM. To interpret the SPR response, the transfer matrix method, applied as a repeated Fresnel's equations, conducts to express the reflection coefficient within the designed structure. The above response is calculated in the angular mode at 738-nm wavelength. By optimizing the thicknesses of the media to get a minimum dip, we investigate particularly the effect of the refractive index n_s on the SPR profile. The later spectrum leads to evaluating angular-sensitivity defined as the ratio between the SPR angle shift and refractive index, n_s and Q-factor. Finally, the best parameter conditions can be concluded to identify the best Q-factor and sensitivity.

2. Theory and characteristics of the considered sensor

As shown in Fig. 1, we consider a multilayer wave guide geometry with N layers. Each medium is subscribed by of a thickness d_k, permittivity ε_k, and permeability μ_k. This multilayer is disposed onto the base of a prism coupler. In the whole work, two arrangements have been analyzed, one is

Au-analyte- LHM, and the other is LHM-analyte-Au for which the respective reflectivity is deduced by means of the transfer matrix method.

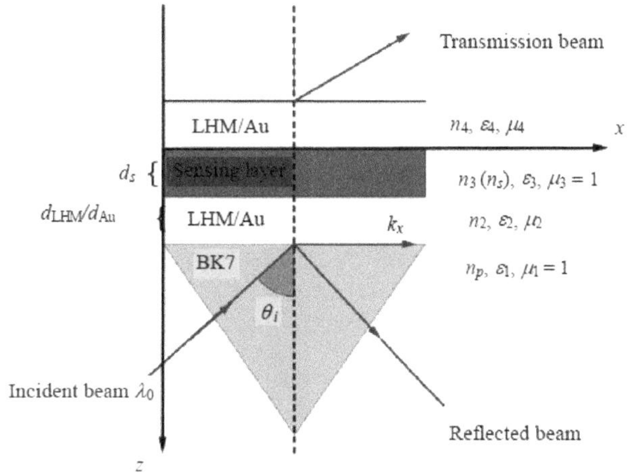

Fig. 1 Schematic illustration of the composite waveguide sensor under study.

In this method, the tangential fields at the first boundary z_1 are related to those at the last boundary z_n as it will be explained below. So, when the structure is irradiated with a light beam of transverse magnetic (TM or p) polarization under the incidence angle θ_i through a BK7-glass prism, consequently a field distribution occurs in the four media. The relationships between electromagnetic fields ($\mathbf{E}_1$, $\mathbf{B}_1$), ($\mathbf{E}_2$, $\mathbf{B}_2$), ($\mathbf{E}_3$, $\mathbf{B}_3$), and ($\mathbf{E}_4$, $\mathbf{B}_4$) in the four media obtained by using phase change due to light passing through the different layers denoted as k take the forms as

$$\begin{cases} E_k = (\mathrm{e}^{-\mathrm{i}\beta_k} \cos\beta_k) E_{k+1} - \left(\mathrm{i} \dfrac{\mathrm{e}^{-\mathrm{i}\beta_k}}{q_k} \sin\beta_k \right) B_{k+1} \\ B_k = (\mathrm{i} q_k \mathrm{e}^{-\mathrm{i}\beta_k} \sin\beta_k) E_{k+1} - \left(\mathrm{e}^{-\mathrm{i}\beta_k} \cos\beta_k \right) B_{k+1} \end{cases} \quad (1)$$

where β_k is the optically admittance, and q_k is the phase factor. For each medium subscribed by k, these quantities are given as follows:

$$q_k = (\mu_k / \varepsilon_k)^{1/2} \cos\theta_k = \frac{(\varepsilon_k \mu_k - n_1^2 \sin^2\theta_1)^{1/2}}{s_k} \quad (2)$$

and

$$\beta_k = \frac{2\pi d_k}{\lambda} (\varepsilon_k \mu_k - n_1^2 \sin^2\theta_1)^{1/2}. \quad (3)$$

For the proposed SPR sensor of four layers, the characteristics matrix is given as

$$\begin{bmatrix} \mathbf{E}_1 \\ \mathbf{B}_1 \end{bmatrix} = \mathbf{M} \begin{bmatrix} \mathbf{E}_4 \\ \mathbf{B}_4 \end{bmatrix} \quad (4)$$

where the matrix $\mathbf{M}$ takes the form as follows:

$$\mathbf{M} = \prod_{k=2}^{4} \begin{bmatrix} \cos\beta_k & -\mathrm{i}\sin\beta_k / q_k \\ -\mathrm{i} q_k \sin\beta_k & \cos\beta_k \end{bmatrix} = \begin{bmatrix} M_{11} & M_{12} \\ M_{21} & M_{22} \end{bmatrix}. \quad (5)$$

Finally, the amplitude of the reflection coefficient for p-polarized incident wave is given as [27]

$$r_p = \left| \frac{(M_{11} + M_{12} q_s) q_1 - (M_{21} + M_{22} q_s)}{(M_{11} + M_{12} q_s) q_1 + (M_{21} + M_{22} q_s)} \right|^2 \quad (6)$$

and the reflectivity for the p-polarized light can be analyzed with respect to the angle of incidence θ_i as follows:

$$R_p = | r_p |^2. \quad (7)$$

Using (7), the resolution of the SPR peak that exhibits the structure can be determined by adjusting thicknesses, d_k and refractive index, ε_k of the mediums (Cf. Fig. 1 for notations). From this SPR response, the determination of the resonance condition where the reflectivity, $R_p(\theta_i = \theta_{SPR})$ drops to a minimum value, provides the propagation wavelength, λ_{SPR} of the surface plasmon as:

$$\lambda_{SPR} = \frac{\lambda_0 / n_p}{\sin\theta_{SPR}} \quad (8)$$

where λ_0 is the free space wavelength in vacuum.

Generally, the performance of plasmonic sensors is evaluated in terms of angular sensitivity, resolution or detection limit (DL), and quality factor Q. Based on the reflectivity's profile, with a fixed wavelength λ_0, the angular sensitivity is the ratio between the changes in SPR angle $\partial\theta_{SPR}$ and the change ∂n_s [12] expressed as follows:

$$S = \frac{\partial\theta_{SPR}}{\partial n_s} \ or \ \frac{\partial\lambda_{SPR}}{\lambda n_s}. \quad (9)$$

For the quality factor, Q that can be evaluated from the linewidth at 50% of the SPR peak [28] noted as $(\Delta\theta_{0.5})^{-1}$, which is given as follows:

$$Q = \left(\frac{\Delta\theta_{0.5}}{\tan\theta_{SPR}} \right)^{-1} \text{ or } \left(\frac{\Delta\lambda_{0.5}}{\tan\lambda_{SPR}} \right)^{-1}. \quad (10)$$

3. Results and discussions

In order to outline the resolution of the proposed

SPR sensor as illustrated in Fig. 1, the reflectivity is calculated and plotted numerically for two arrangements between Au and LHM. Thus, three main parameters, namely intensity, detection accuracy, and quality factor, are evaluated particularly under the influence of the refractive index n_s of analyte. The refractive index of BK7-prism is set to be 1.52 [29].

Using (7), the angular profiles of SPR response are displayed in Figs. 2, 3, 4, and 5. It is noted that the specific arrangements that correspond to these data are reported in the insert of the figures. In these plots, we have taken for gold the complex permittivity expressed from Drude-Lorentz model, $\varepsilon_{Au} = -21.3+1.34i$ as tabulated in [30]. The electric permittivity and magnetic permeability for the LHM, are −33.5 and −11, respectively which are listed in [31]. The sensing medium termed analyte of thickness, d_s=385 nm, has the refractive index (RI) n_s, which varies from 1.00 to 1.66.

In the above conditions, under the effect of n_s of the sensing medium, the SPR curves are simulated in Fig. 2. These data indicate that an increase in RI contributes firstly to the trap single SPR mode or double SPR mode and secondly to reduce the linewidth, i.e. high quality factor.

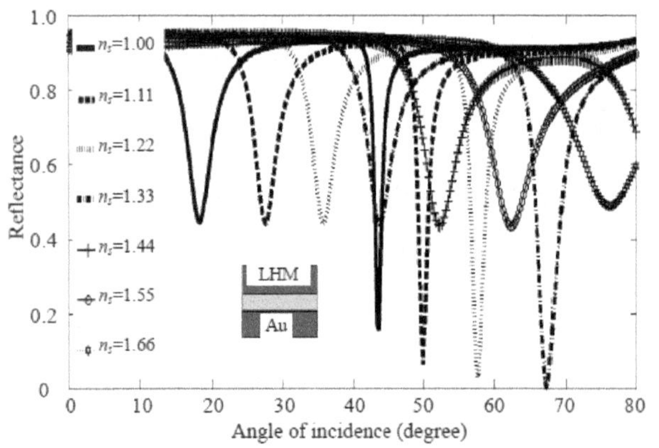

Fig. 2 Simulated SPR curves with different refractive indexes of sensing layer (thickness 385 nm) stacked between Au the layer (refractive index $\varepsilon_{Au} = -21.3 + 1.34i$ thickness 40 nm) and left-handed materials (LHM) of $\{\varepsilon_{LHM}, \mu_{LHM}\} = \{-33.5, -11\}$.

According to Fig. 3, it is seen that an increase in n_s from 1.33 to 1.35 leads to the best enhancement effect determined on the SPR-mode termed SP1 since we have already observed the zero-reflectance in a narrower angular range of 68°–70°.

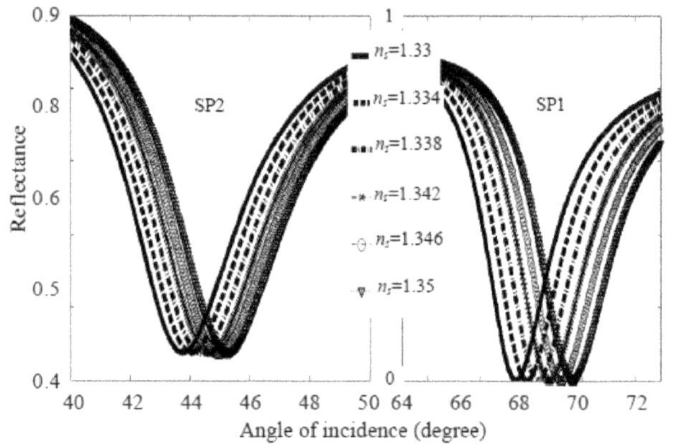

Fig. 3 Calculated p-reflectance with six different refractive indexes, n_s of the sensing layer. The other parameters are the same as in Fig. 2.

It is not the case of the SPR-mode termed SP2 which is excited at the lower angle than SP1. In other hand, when tuning the n_s, an apparent discrepancy in term of the resolution is seen on the minimum dips of SP1 and SP2.

The minimum of the intensity remains constant, i.e. $R_{SPR} = 0.44$ of the SP2 mode and where the characteristics of SP1 are $\theta_{SPR} = 67.42°$, $R_{SPR} = 3 \times 10^{-5}$, D.A.=0.313/°, FWHM=3.19°, and Q=43.23 at n_s=1.33. In addition, the best parameters of SPR1 are θ_{SPR}=27°, D.A.=11.11/°, and Q=325 which in turn measure the resonance angle with optimizing both Au and sensing layer thicknesses 70 nm and 240 nm, respectively. Guo *et al.* [28] obtained the angular sensitivity increasing as 103 °/RIU in a short range mode method, 97 °/RIU in a typical single mode method, and 70 °/RIU in a metallic method [32]. Additionally, the angular sensitivity of the SPR sensor based on prism was reported increasing from 94.66 °/RIU to 204.41 °/RIU, but it went with an increase in FWHM from 2.24° to 4.36° [33].

A similar study of the reflectance is simulated for another arrangement between gold and LHM, which is depicted in Fig. 4. The data show that with an active LHM of 720-nm thickness, the reflectance takes an opposite phase compared with the first arrangement Au/analyte/LHM structure. An increase in RI, n_s in the range of 1.00 to 1.99 causes an optical amplification of the SPR peak which is launched in a large angular range from 20° to 80°. Here we specify that an increase inthe RI, n_s in the restricted range of 1.44 to 1.66 gives rise to the Goos-Hànchen effect observed on the single SPR mode whose full width at half maximum is the narrowest. This structure exhibits one single SPR mode whose maximal intensity can be manipulated on the change in RI of the sensing from 1.00 to 1.55. The Goos-Hànchen peak shift has the best parameters as θ_{SPR} = 41.85 °, R_{SPR} = 3939, D.A. = 100 /° (FWHM = 0.01 °), and Q = 5134 at n_s = 1.55. For the particular RI, n_s=1.22, on the p-reflectance profile the structure exhibits two SPR modes with a large discrepancy between their respective intensities. The first dips denote SP1 (see left) with the parameters θ_{SPR} = 51.16 °, R_{SPR} = 0.0745, D.A. = 8.33 /° (FWHM = 0.12°), and Q = 593.1, and the second denote SP2 (see right) with θ_{SPR} = 18.55 °, R_{SPR} = 0.34, D.A. = 7.14 /° (FWHM = 0.14 °), and Q = 137.4. The inset shows the maximum resolution (detection limit) is D.A.$_{max}$ = 125 /° (FWHM = 0.008 °).

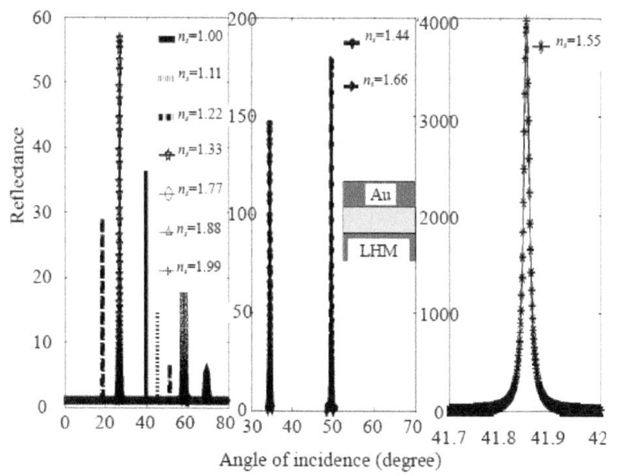

Fig. 4 Simulated SPR curves with different refractive indexes of the sensing layer (thickness 360 nm) stacked between left-handed materials (LHM), (refractive index{ε_{LHM}, μ_{LHM}} = {−33.5, −11}, thickness 720 nm) and the Au layer (refractive index ε_{Au} = −21.3+1.34i).

From Fig. 5, the reflection efficiency of the peak increases slightly with an increase in the refractive index of the sensing layer from 1.33 to 1.55. Additionally, we can see that n_s induces the shift of SPR from 27.04° towards a higher angle of the incident angle. One can also observe a particular influence on the reflectance intensity, as R (θ_{SPR} = 43.07 °) = 1.18×10^5 for n_s = 1.568, and a high quality factor is estimated to be 6000. The last issue that must be addressed is the question of the drop in the reflectance peaks for $\theta_{SPR} \geq 43.04$ ° for $n_s > 1.552$ of the sensing layer. This is simply due to the cut-off function of the RI associated to a nonabsorbent sensing layer. Finally, in the conditions of the parameters specified in left Fig. 5, a linear fit is predicted for SPR in RI change 1.33 to 1.42, i.e. the parameters linearity with equation as $\theta_{SPR} = 70.136 \times n_s - 66.244$ (°). And $Q = 1000 \times n_s - 1005$.

With error ± 0.009, ± 7.54 × 10^{-13} respectively and quadratic equation as $R_{SPR} = 3330.4 \times n_s^2 - 8548 \times n_s + 5535.4$ at $n_s > 1.43$ (right Fig. 5), the linearity equation of the resonance angle is $\theta_{SPR} = 67.72 \times n_s - 63.13$ with error ±0.0095, and other parameters of SPR are nonlinearity.

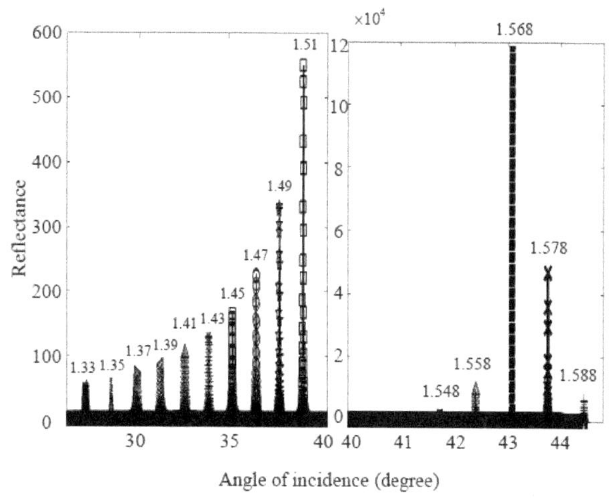

Fig. 5 Calculated reflectance for different refractive indexes of the sensing layer with different refractive indexes of the sensing layer, n_s=1.33, 1.35, 1.37, 1.39, 1.41, 1.43, 1.45, 1.47, 1.49, 1.51, 1.548, 1.558, 1.568, 1.578, and 1.588. The other parameters are the same with those in Fig. 4.

In the ATR technique with a BK7-prism, the response of the multilayer configuration based on Au as an active metal and LHM-bulk and mediated with GaAs predicts an optical amplification that can be turned on the variation of RI. This amplification measured versus the incident angle is correlated to the SPR mode excited around the incident angle of 41.830°. The above effect was observed by the authors of [34] where the obtained optical amplification did not exceed 10^3. The experimental use helps realize applications in precise processing and sensors. It was reported in [35] that the Goos-Hànchen shift was also sensitive to the change in refractive index and could be used as a different mechanism for sensing. In Figs.4 and 5, the shift in the resonance angle for a variation $\Delta n_s = 0.004$ remains constant in angular sensitivity that is in the order of 66 °/RIU.

4. Conclusions

The sensing properties evaluated from the reflectance curves on two arrangements between gold and LHM are investigated theoretically. Based on the simulated results, it has been shown that under the effect of RI of the sensing medium, gold as an active medium favors the presence of a high resolution required in biosensing. In appropriate conditions of the multilayer thicknesses, a single SPR of a giant intensity up to 10^4 is sustained by the structure through the sensing medium of RI as 1.558. The later expected due to the Goos-Hànchen effect confirms the functionality of the structure to operate as an optical amplifier and an active tunable filter.

References

[1] X. Y. Dai, L. Y. Jiang, and Y. J. Xiang, "Tunable THz angular/frequency filters in the modified Kretschmann-Raether configuration with the insertion of single layer graphene," *IEEE Photonics Journal*, 2015, 7(2): 1–8.

[2] Y. A. Akimov and W. S. Koh, "Resonant and nonresonant plasmonic nanoparticle enhancement for thin-film silicon solar cells," *Nanotechnology*, 2010, 21(23): 235201.

[3] Y. A. Akimov, K. Ostrikov, and E. P. Li, "Surface plasmon enhancement of optical absorption in thin-film silicon solar cells," *Plasmonics*, 2009, 4(2): 107–113.

[4] F. F. Ren, K. W. Ang, J. F. Song, Q. Fang, M. B. Yu, G. Q. Lo, *et al.*, "Surface plasmon enhanced responsivity in a waveguided germanium metal-semiconductor-metal photodetector," *Applied Physics Letters*, 2010, 97(9): 091102-1–091102-3.

[5] L. Tang, S. E. Kocabas, S. Latif, A. K. Okyay, D. S. Ly-Gagnon, K. C. Saraswat, *et al.*, "Nanometre-scale germanium photodetector enhanced by a near-infrared dipole antenna," *Nature Photonics*, 2008, 2(4): 226–229.

[6] B. Rothenhausler and W. Knoll, "Surface plasmon microscopy," *Nature*, 1988, 332(6165): 615–617.

[7] R. C. Jorgenson and S. S. Yee, "A fiber-optic chemical sensor based on surface plasmon resonance," *Sensors & Actuators B–Chemical*, 1993, 12(3): 213–220.

[8] C. Nylander, L. Bo, and T. Lind, "Gas detection by means of surface plasmons resonance," *Sensors & Actuators*, 1982, 3(82): 79–88.

[9] L. Bo, C. Nylander, and I. Lunstrom, "Surface plasmon resonance for gas detection and biosensing," *Sensors & Actuators*, 1983, 4(83): 299–304.

[10] E. Kretschmann and H. Raether, "Radiative decay of non-radiative surface plasmons excited by light," *Zeitschrift Für Naturforschung A*, 1968, 23(12): 2135–2136.

[11] J. Homola, S. S. Yee, and G. Gauglitz, "Surface plasmon resonance sensors: review," *Sensors & Actuators B–Chemical*, 1999, 54(1–2): 3–15.

[12] J. Homolar, *Surface plasmon resonance based sensors series on chemical sensors and biosensors*. Berlin, Germany: Springer-Verlag, 2006: 1–251.

[13] X. D. Hoa, A. G. Kirk, and M. Tabrizian, "Towards integrated and sensitive surface plasmon resonance biosensors: a review of recent progress," *Biosensors & Bioelectronics*, 2007, 23(2): 151–160.

[14] P. K. Maharana and R. Jha, "Chalcogenide prism and graphene multilayer based surface plasmon resonance affinity biosensor for high performance," *Sensors & Actuators B–Chemical*, 2012, 169(13): 161–166.

[15] H. S. Leong, J. Guo, R. G. Lindquisu, and Q. H. Liu, "Surface plasmon resonance in nanostructured metal films under the Kretschmann configuration," *Journal of Applied Physics*, 2009, 106(12): 124314-1–124314-5.

[16] V. G. Veselago, "The electrodynamics of substances with simultaneously negative values of ε and μ," *Soviet Physics Uspekhi*, 1968, 10(4): 509–514.

[17] J. B. Pendry, A. J. Holden, D. J. Robbins, and W. J. Stewart, "Magnetic from conductors and enhanced nonlinear phenomena," *IEEE Transactions on Microwave Theory Techniques*, 1999, 47(11): 2075–2084.

[18] M. Schueler, C. Mandel, M. Puentes, and R. Jakoby, "Metamaterial inspired microwave sensors," *IEEE Microwave Magazine*, 2012, 13(2): 57–68.

[19] T. Chen, S. Y. Li, and H. Sun, "Metamaterials application in sensing," *Sensors*, 2012, 12(3): 2742–2765.

[20] J. J. Yang, M. Huang, H. Tang, J. Zeng, and L. Dong, "Metamaterial Sensors," *International Journal of Antennas and Propagation*, 2013, 2013(4): 1–16.

[21] A. Upadhyay, Y. K. Prajapati, V. Singh, and J. P. Saini, "Sensitivity estimation of metamaterial loaded planar waveguide sensor," *Optical and Quantum Electronics*, 2015, 47(7): 2277–2287.

[22] A. Upadhyay, Y. K. Prajapati, V. Singh, and J. P. Saini, "Comprehensive study of reverse index waveguide based sensor with metamaterial as a guiding layer," *Optics Communications*, 2015, 348: 71–76.

[23] N. I. Zheludev, "The road ahead for metamaterials," *Science*, 2010, 328(5978): 582–583.

[24] S. Pal, Y. K. Prajapati, J. P. Saini, and V. Singh, "Resolution enhancement of optical surface plasmon resonance sensor using metamaterial," *Photonic Sensors*, 2015, 5(4): 330–338.

[25] S. Pal, Y. K. Prajapati, J. P. Saini, and V. Singh, "Sensitivity enhancement of metamaterial-based SPR biosensor for NIR," *Optica Applicata*, 2016, 46(1): 131–143.

[26] Y. K. Prajapati, A. Yadav, A. Verma, V. Singh, and J. P. Saini, "Effect of metamaterial layer on optical surface plasmon resonance sensor," *Optik–International Journal for Light and Electron Optics*, 2013, 124(18): 3607–3610.

[27] M. Yamamoto, "Surface plasmon resonance (SPR) theory: tutorial," *Encyclopedic Reference of Immunotoxicology*, 2010, 14(4): 388–398.

[28] J. Guo, P. D. Keathley, and J. T. Hastings, "Dual-mode surface-plasmon-resonance sensors using angular interrogation," *Optics Letters*, 2008, 33(5): 512–514.

[29] S. Szunerist, X. Castel, and R. Boukherroub, "Surface plasmon resonance investigation of silver and gold films coated with thin indium tin oxide layers: Influence on stability and sensitivity," *Journal of Physical Chemistry C*, 2008, 112(40): 15813–15817.

[30] A. D. Rakic, A. B. Djurisic, J. M. Elazar, and M. L. Majewski, "Optical properties of metallic films for vertical-cavity optoelectronic devices," *Applied Optics*, 1998, 37(22): 5271–5283.

[31] S. M. Xiao, V. P. Drachev, A. V. Kildishev, X. J. Ni, U. K. Chettiar, H. K. Yuan, *et al.*, "Loss-free and active optical negative-index metamaterials," *Nature*, 2010, 466(7307): 735–738.

[32] K. M. Byun, S. J. Kim, and D. Kim, "Grating-coupled transmission-type surface plasmon resonance sensors based on dielectric and metallic gratings," *Applied Optics*, 2007, 46(23): 5703–5708.

[33] G. Gupta and J. Kondon, "Tuning and sensitivity enhancement of surface plasmon resonance sensor," *Sensors & Actuators B–Chemical*, 2007, 122(2): 381–388.

[34] N. Goswami, A. Saha, and A. Ghosh, "Optical amplification with surface plasmon resonance and total internal reflection in gold nanostructure with BK7 parallel slab," *International Journal of Chemtech Research*, 2014, 7(3): 1148–1153.

[35] X. Yin, L. Hesselink, H. Chin, and D. A. B. Miller "Temporal and spectral nonspecularities in reflection at surface plasmon resonance," *Applied Physics Letters*, 2006, 89(4): 041102-1–041102-3.

15

Development of High Sensitivity Eight-Element Multiplexed Fiber Laser Acoustic Pressure Hydrophone Array and Interrogation System

Ming LI[1,2], Zhihui SUN[2], Xiaolei ZHANG[2], Shujuan LI[2], Zhiqiang SONG[2], Meng WANG[2], Jian GUO[2], Jiasheng NI[2], Chang WANG[2*], Gangding PENG[3], and Xiangang XU[1]

[1]*State Key Laboratory of Crystal Materials, Shandong University, Jinan, 250100, China*

[2]*Jinan Research Center, Laser Institute of Shandong Academy of Sciences, Jinan, 250014, China*

[3]*School of Electronic Engineering and Telecommunication, University of New South Wales, Sydney, 2052, Austrilia*

[*]Corresponding author: Chang WANG E-mail: ch_wangs@163.com

Abstract: Fiber laser hydrophones have got widespread concerns due to the unique advantages and broad application prospects. In this paper, the research results of the eight-element multiplexed fiber laser acoustic pressure array and the interrogation system are introduced, containing low-noise distributed feedback fiber laser (DFB-FL) fabrication, sensitivity enhancement packaging, and interferometric signal demodulation. The frequency response range of the system is 10 Hz–10 kHz, the laser frequency acoustic pressure sensitivity reaches 115 dB re Hz/Pa, and the equivalent noise acoustic pressure is less than $60\mu Pa/Hz^{1/2}$. The dynamic range of the system is greater than 120 dB.

Keywords: Fiber laser hydrophone; DFB-FL; packaging sensitization; eight-element multiplex; interferometric demodulation

1. Introduction

Optical fiber hydrophones have been used in the military, marine monitoring, resource exploration, and other aspects of the application because of strong advantages after several decades of continuous research and development. The United States of America and other developed countries have begun to equip and deploy fiber hydrophones. In China, the research of fiber hydrophones is limited by the optical fiber production technology and the development of the grating writing technology, which started late but goes fast. National University of Defense Technology, Institute of Acoustics, Chinese Academy of Sciences, and other units have done a lot of research on the optical fiber hydrophone and achieved fruitful results. Recently, the research on the fiber laser hydrophones gets a wider range of concerns, dues to unique advantages of small size, high sensitivity, dynamic range, and so on.

In 2012, Y. Leguillon discussed the factors that affect the accuracy of the sensor and the factors that affect the number of sensor arrays. The sensor array, consisting of 12 distributed feedback (DFB) fiber lasers with a pumping source of 200 mW of 1 480 nm, had very low intensity noise and frequency noise [1]. In 2013, Scott Foster developed a fiber laser hydrophone device, which was deployed to the south coast of Australia 33 m deep seabed for a field test. It was the first time that a fiber laser hydrophone tested the sea state zero (SS0) [2]. In 2014, F. X. Launay introduced a two-dimensional array of 12 fiber laser hydrophones with a pressure sensitivity of 103.5 dB (re Hz/Pa) at 1 kHz [3]. In China, the 23rd Institute of China Electronics Technology Group, Navy Engineering University, and National University of Defense Technology have done a lot of research on the systematic fiber laser hydrophone technology [4–6], and this work is mainly concentrated on the hydrophone basic technology and reuse tests. These results have laid an important foundation for further research on fiber laser hydrophone systems, but there is still a lot of work to do to achieve a comprehensive breakthrough in application technology. In 2008, W. T. Zhang reported an experiment of distributed feedback fiber laser (DFB-FL) hydrophone packaging [7].This package structure was relatively flat in frequency response under 1 kHz frequency range, which showed the fiber laser hydrophone application has taken a key first step. In 2012, W. T. Zhang *et al.* used hydrophone array of eight fiber lasers, to place an underwater sounder, which was 8 m underwater, and 5 km away for the field test in Qiandao Lake of Zhejiang Province. From the test, the authors founded that the response consistency of the hydrophone unit was good, and the test value conformed to the theoretical value [8]. In 2014, H. C. Gu who is from Institute of Acoustics, Chinese Academy of Sciences, introduced a linear dragging array containing four fiber laser hydrophones. The sensitivity of a single hydrophone was about –135 dB. In the range of 20 Hz to 1 kHz, the amplitude fluctuation was less than ±1.5 dB [9].

This paper systematically introduces the research results of the Laser Research Institute of Shandong Institute of Science and Technology in the high sensitivity of the eight-element multiplexed array fiber laser acoustic pressure hydrophone system. The laser frequency acoustic pressure sensitivity of the fiber laser hydrophone system is 115 dB re Hz/Pa in the frequency range of 10 Hz – 10 kHz, the equivalent noise acoustic pressure is less than 60 μPa/Hz$^{1/2}$, and the dynamic range of the system is greater than 120 dB.

2. Fabrication of distributed feedback fiber laser

Fiber laser is the key sensor of the fiber laser hydrophone. Firstly, the laser output power determines the signal to noise ratio of the sensor signal. Secondly, the laser mode competition, including vertical mode competition and polarization mode competition, will affect the demodulation stability of the fiber laser hydrophone system. Thirdly, the intensity of the laser noise and frequency noise will eventually be converted into hydrophone system's phase noise, which affects the system's minimum detectable signal and dynamic range. We have a grating fabrication and testing system based on argon ion laser. As shown in Fig.1, a phase-shifted fiber grating is fabricated by the dynamic phase mask method and scanning exposure method. The phase mask's position is controlled by the nano-precision piezoelectric actuator, which can be put in phase shift accurately in the grating at any position and produce a reliable asymmetric distributed feedback fiber laser (DFB-FL).

The DFB-FL uses a single grating written on the active fiber as the gain medium and the resonant cavity. The laser's structure is simple so that it can avoid laser mode hopping. The early research shows that the $\lambda/4$ phase shift gratings have a minimum pumping threshold and a stable single longitudinal mode operating performance when the grating factor (kl) is at 10 or so [10, 11]. When the erbium-doped phase-shifted fiber Bragg grating is fabricated, the length of erbium-doped fiber is 45 mm, on which the phase-shifted grating with the length of 42 mm is written, the grating factor is between 10 and 12, and the π phase shift point is in the scale position of 4:6 [12]. The output wavelengths of the asymmetric DFB-FL are 1528 nm – 1552 nm, the output optical power ratio is more than 100:1, the line width is less than 15 kHz, and the relative intensity noise is less than –100 dB/Hz (at 0 – 50 kHz). The DFB-FL can meet the requirements of the laser system for laser power, narrow line-width, and low noise. The power balance of eight asymmetric DFB-FLs multiplexing is less than 4 dBm.

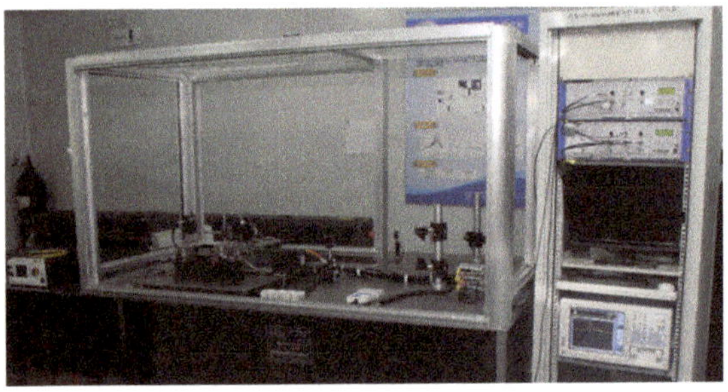

Fig. 1 Grating fabrication system of argon ion laser.

3. Fiber laser hydrophone sensitization package

For bare DFB-FL that has not been encapsulated, the acoustic pressure causes a small laser wavelength change (approximately 3.6 pm/MPa). To achieve the required acoustic pressure sensitivity requirements (>1 nm/MPa) for practical applications, DFB-FL needs sensitization package.

DFB-FL is a laser that has been inserted a π-phase shift to form a distributed feedback structure when fabricating a grating in an erbium-doped fiber. The central wavelength λ_B of the laser is determined by the Bragg equation of the fiber grating as follows:

$$\lambda_B = 2n_{\text{eff}}\Lambda \tag{1}$$

where n_{eff} is the effective refractive index of the fiber core, and Λ is the grating period.

The influence of the ambient environment (temperature and pressure) can cause the grating period Λ and the effective refractive index n_{eff} of the DFB-FL to change, resulting in a change in the laser central wavelength λ_B as follows:

$$\Delta\lambda_B = 2n_{\text{eff}}\Delta\Lambda + 2\Delta n_{\text{eff}}\Lambda \,. \tag{2}$$

The fiber laser hydrophone detects the acoustic wave signal emitted or reflected by the underwater target. The change in the sound wave in the underwater sound field leads to the change in the water pressure P. The sensitivity of the fiber laser hydrophone can be defined as follows:

$$K_p = \frac{\Delta\lambda_B}{\lambda_B P} = \left[\frac{n_{\text{eff}}^2}{2}(P_{11}+P_{12})-1\right]\frac{1-2\nu}{E} \tag{3}$$

where E is the Young's modulus and ν is the Poisson's ratio of the hydrophone.

It can be seen that if the Young's modulus E and Poisson's ratio of the hydrophone ν can be reduced by effective encapsulation, the acoustic pressure sensitivity of the hydrophone can be effectively improved [13]. By analyzing different materials, we used Divinycell H high performance polyurethane as the sandwich material for hydrophone packaging. The Young's modulus is about 320 MPa. It has excellent adhesion strength, superior chemical resistance, low water absorption, and excellent thermal insulation performance.

We made different structures of fiber laser hydrophone using the polyurethane material for encapsulation contrast. A water acoustic frequency response test system was set up as shown in Fig. 2 to carry out the frequency response sensitivity test of different package fiber laser hydrophones. A standard piezoelectric ceramic hydrophone RHSA-20 provided by 715th Research Institute of China Shipbuilding Industry Corporation was used to calibrate the measured sound field. Place the standard piezoelectric hydrophone RHSA-20 in the calibration field of the standing wave tube in parallel. The fiber laser hydrophone was pumped by 1480-nm semiconductor laser. Use the Michelson interferometer of 2.4 m arm length difference provided by the Optiphase Company for converting laser wavelength signal to phase signal. Then enter the phase signal into the OPD4000 signal demodulator for the signal demodulation provided by the Optiphase Company. The laser phase acoustic pressure sensitivity of the fiber laser hydrophone is defined as follows:

$$M_p = 20\lg(M_\phi / M_r) \quad (4)$$

$$M_\phi = \Delta\phi / P \quad (5)$$

where $M_r = 1\ \text{rad}/\mu\text{Pa}$.

The laser frequency acoustic pressure sensitivity of the fiber laser hydrophone is defined as follows:

$$M_F = 20\lg(M_f / M_r) \quad (6)$$

$$M_f = \Delta\nu / P \quad (7)$$

where $M_r = 1\ \text{Hz}/\text{Pa}$.

Through the test, we found that the cylindrical package structure's sensitivity frequency response curve was the most flat, shown in Fig. 3, whose outer diameter was 8 mm and wall thickness was 2 mm.

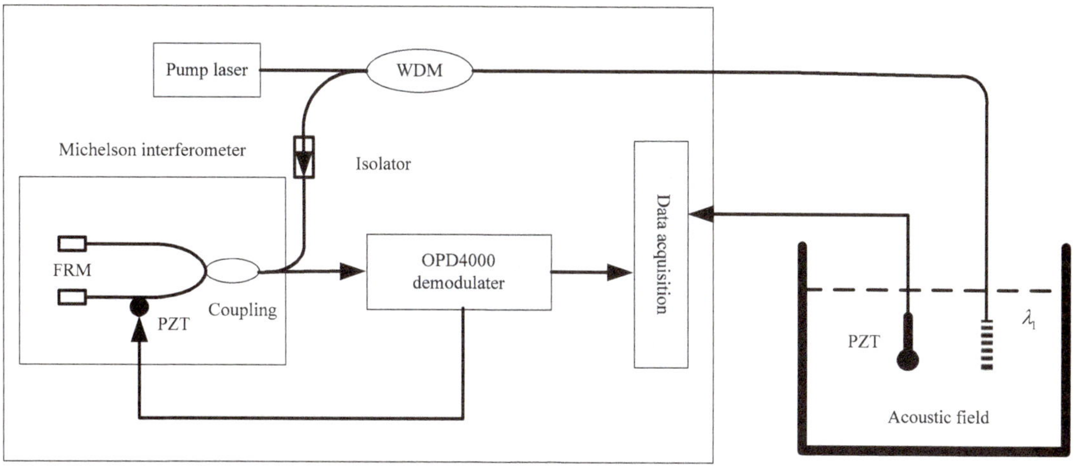

Fig. 2 Fiber laser hydrophone's sensitivity test system.

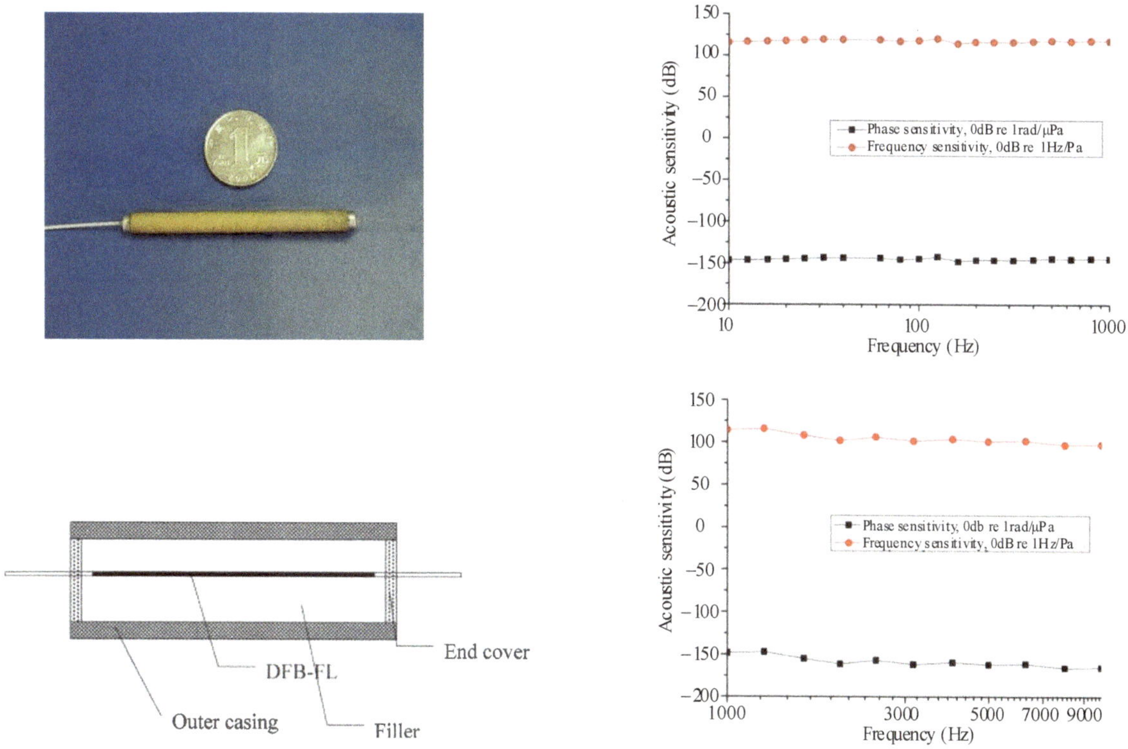

Fig. 3 Fiber laser hydrophone of cylindrical capsulation structure and its sensitivity frequency response curve.

4. Design of the signal interference demodulation system

The laser generated by the DFB-FL produces the interference fringe after the fiber interferometer, and the intensity of the interference fringe is defined as follows:

$$I = I_1 + I_2 + 2\sqrt{I_1 I_2}\cos(\Delta\phi) \qquad (8)$$

where I_1 and I_2 are the intensities of two interference light beams, and $\Delta\phi$ is the phase difference of two interference light beams.

The wavelength variation $\Delta\lambda_B$ of the DFB-FL is transformed into phase change $\Delta\phi$ by the fiber interferometer with the arm length difference nL :

$$\Delta\phi = \frac{2\pi nL}{\lambda_B^2}\Delta\lambda_B . \qquad (9)$$

Through the photoelectric detector for photoelectric conversion and signal demodulation, the phase change value can be got. The accurate acquisition of the hydrophone's optical phase shift is the key to the success of phase shift sensitivity detection. In this paper, the phase generated carrier (PGC) demodulation technique is used to measure the change in optical phase shift after the fiber laser hydrophone is modulated by the acoustic pressure signal. The PGC technology has the characteristics of low noise, ease to implement multiplexing, and large dynamic demodulation range [14]. We developed two methods of PGC demodulation, differential cross multiplication (DCM), and arc tangent algorithm (Arctan). The flow chart of the PGC demodulation technology is shown in Fig. 4. We made program development and optimization by LabView software and completed the demodulation algorithm performance test and calibration. The demodulation frequency range was 10 Hz – 10 kHz. Its wavelength demodulation resolution rate was less than 10^{-6} pm/Hz$^{1/2}$ (between 10 Hz and 10 kHz), and the linearity was larger than 0.99. Parameters comparison of two methods is shown in Table 1. The linearity of the arc tangent algorithm is better. The total harmonic distortion (THD) of the differential cross multiplication is less. The signal to noise and distortion (SINAD) and wavelength demodulation resolution are almost the same.

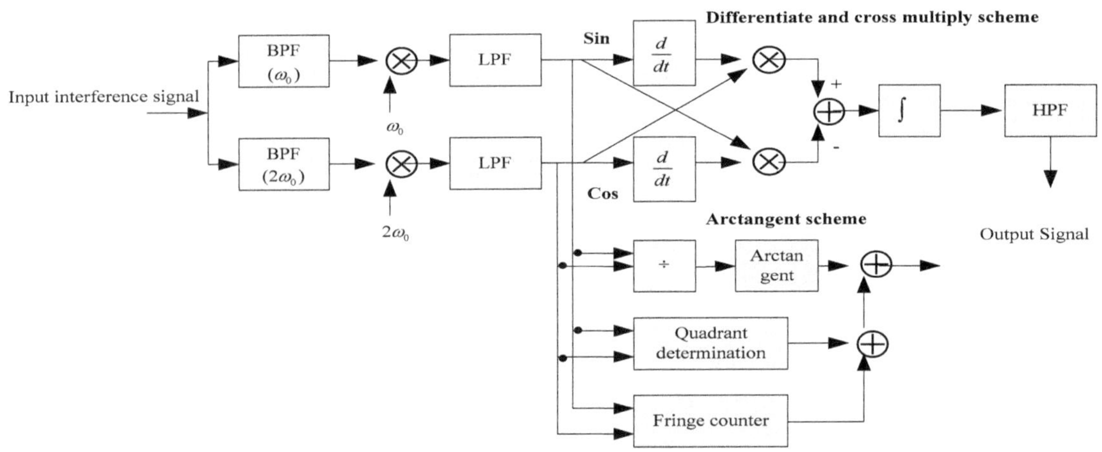

Fig. 4 Flow chart of the PGC demodulation technology.

Table 1 Parameters comparison of two demodulation methods.

Parameters	PGC-DCM	PGC-Arctan
Linearity	99.40%	99.97%
SINAD	23.32 dB	23.21 dB
THD	1.35%	3.44%
Wavelength resolution	<10^{-6} pm/Hz$^{0.5}$ (at 1 kHz)	<10^{-6} pm/Hz$^{0.5}$ (at 1 kHz)

5. Design of eight-element multiplexed fiber laser acoustic pressure hydrophone array system

In the wet end of system, the wavelengths of the eight DFB fiber laser hydrophones were 1528.9 nm, 1530.1 nm, 1536.6 nm, 1539 nm, 1541.3 nm,

1544.6 nm, 1546.1 nm, and 1552.5 nm, respectively. In the dry end of the system, the length difference of Michelson interferometric arm was 30 m. In the case of the same phase resolution, the longer the arm length is, the higher the wavelength shift demodulation resolution is. The signal acquisition card was PXI6368 provided by NI Company. The pump laser source was a 1480-nm semiconductor laser of 300 mW. The system diagram is shown in Fig. 5.

By the test in the relatively quiet environment, the background noise of the fiber laser hydrophone system was about 10^{-5} pm in the frequency range of 10 Hz – 1 kHz, and the noise spectral density was $2.6\times10^{-7}\ \mathrm{pm/Hz^{1/2}}$ ($32.5\ \mathrm{dB/Hz^{1/2}}$), as shown in Fig. 6. The laser frequency acoustic pressure sensitivity was about 115 dB re Hz/Pa (between 10 Hz and 10 kHz), and the dynamic range was more than 120 dB. The equivalent noise acoustic pressure was $57.8\ \mu\mathrm{Pa/Hz^{1/2}}$ (–84.7 dB re $\mathrm{Pa/Hz^{1/2}}$) at 1 kHz, because the deep sea state zero noise pressure (DSS0) at 1 kHz was about $100\ \mu\mathrm{Pa/Hz^{1/2}}$, and the equivalent noise acoustic pressure was less than DSS0.

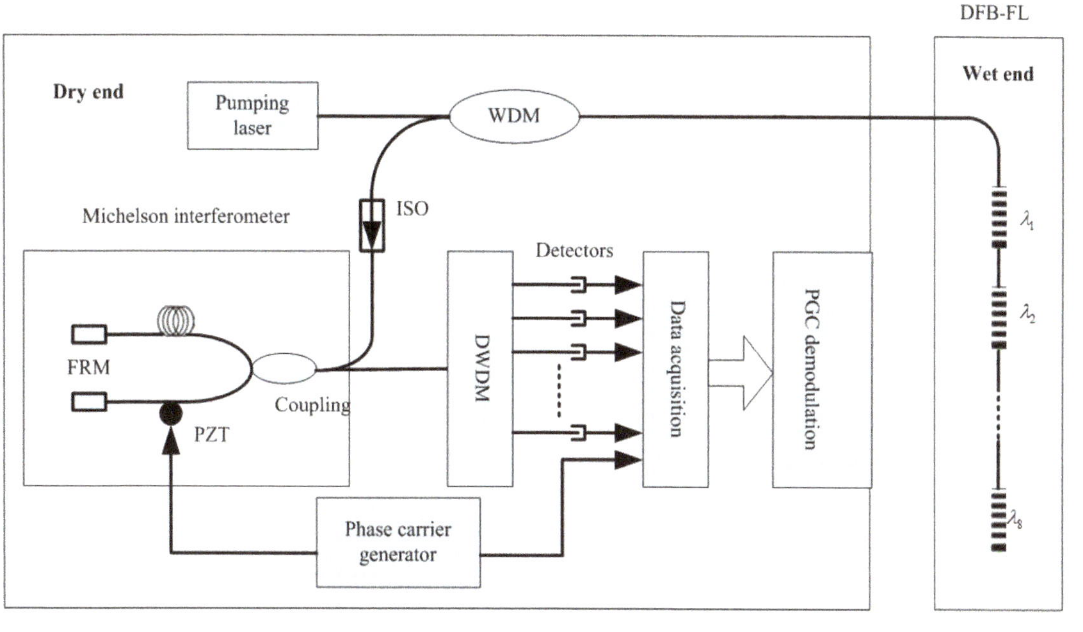

Fig. 5 Fiber laser hydrophone system schematic diagram.

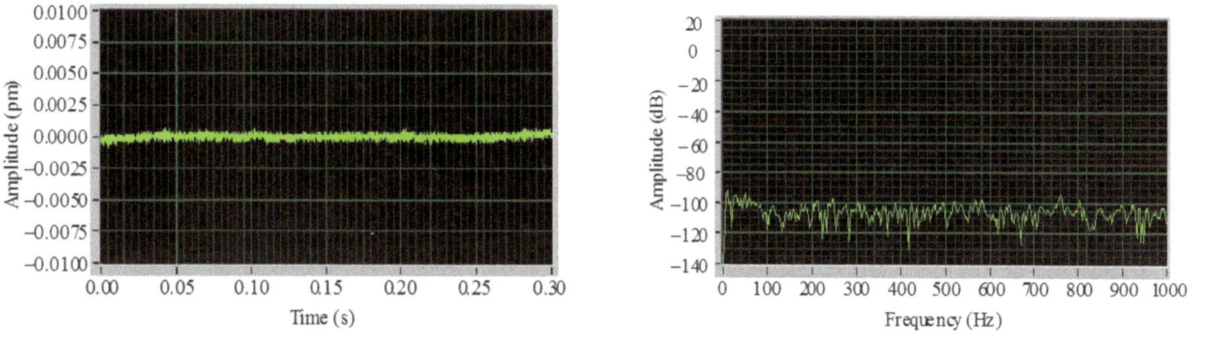

Fig. 6 Background noise of the fiber laser hydrophone system.

6. Conclusions

The fiber laser hydrophone has the advantages of small volume, high sensitivity, large dynamic range, strong anti-electromagnetic interference, and so on. It has wide application prospects and important research value in military and civil fields. Abroad, the fiber laser hydrophone array has been practically applied. In this paper, we develop a wide frequency response, large dynamic range, and high sensitivity eight-element multiplexed fiber laser hydrophones array system. The low-noise distributed feedback fiber laser is used as the core component of the hydrophone, and the polyurethane is used for the sensitization package of the DFB-FL. Two different PGC technologies are developed to demodulate the signal. The frequency response range of the system is 10 Hz – 10 kHz, the laser frequency acoustic pressure sensitivity reaches 115 dB re Hz/Pa, the dynamic range is more than 120 dB, and the equivalent noise acoustic pressure is less than $60\mu\text{Pa}/\text{Hz}^{1/2}$, which is lower than DSS0. The work promotes the domestic development of more than eight-element multiplexed fiber laser hydrophones array system practical process.

Acknowledgment

This work was financially supported by the National Natural Science Foundation of China (No. 60317101), Key Research and Development Plan of Shandong Province (No. 2015GSF11500), and Natural Science Foundation of Shandong Province (No. ZR2013FL029).

References

[1] Y. L'eguillon, K. H. Tow, P. Besnard, A. Mugnier, D. Pureur, and M. Doisy, "First demonstration of a 12 DFB fiber laser array on a 100 GHz ITU grid for underwater acoustic sensing application," in *Proceeding of SPIE Conference on Optical Sensing and Detection II*, Brussels, Belgium, 2012, pp. 17.

[2] S. Foster, A. Tikhomirov, J. Harrison, and J. V. Velzen, "Field demonstration of an 8-element fibre laser hydrophone array," in *Proceeding of 23rd International Conference on Optical Fibre Sensors*, Santander, Spain, 2014, pp. 91570H–1 –91570H–4.

[3] F. X. Launay, R. Bouffaron, R. Lardat, G. Roux, M. Doisy, and C. Bergogne, "Acoustic antenna based on fiber laser hydrophones," in *Proceeding of 23rd International Conference on Optical Fibre Sensors*, Santander, Spain, 2014, pp. 91570Y–1 –91570Y–4.

[4] Q. Zhu, X. B. Chen, J. P. Chen, and G. D. Peng, "DFB fiber laser fabrication by moving phase mask," *Optical Fiber & Electric Cable & Their Applications*, 2006, 1(1): 17–20.

[5] B. Tan and J. B. Huang, "Design about a novel encapsulation structure of DFB fiber laser hydrophones," in *Proceeding of International Symposium on Photonics and Optoelectronics*, Suzhou, China, 2014, pp. 92330X–1–92330X–7.

[6] L. N. Ma, "Fiber laser hydrophone," Ph.D. dissertation, National University of Defense Technology, Changsha, 2010.

[7] W. T. Zhang, Y. L. Liu, F. Li, and H. Xiao, "Fiber laser hydrophone based on double diaphragms: theory and experiment," *Journal of Lightwave Technology*, 2008, 26(10): 1349–1352.

[8] W. T. Zhang, F. Li, and Y. L. Liu, "Field test of an eight-element fiber laser hydrophone array," in *Proceeding of 22nd International Conference on Optical Fiber Sensors*, Beijing, China, 2012, pp. 8421AX–1–8421AX–4.

[9] H. C. Gu, J. B. Huang, H. B. Yu, R. Z. Li, B. Tang, and J. Wu, "A 4-element fiber laser hydrophone towed linear array," in *Proceeding of International Symposium on Optoelectronic Technology and Application (IPTA)-Laser and Optical Measurement Technology and Fiber Optic Sensors*, Beijing, China, 2014, pp. 92972Y–1 –92972Y–6.

[10] S. W. Lovseth and E. Ronnekleiv, "Fundamental and higher order mode thresholds of DFB fiber lasers," *Journal of Lightwave Technology*, 2002, 20(3): 494–501.

[11] D. Sacastru, S. Miclos, and I. Lancranjan, "Theoptical analysis of the output noise of a FDB fiber laser sensor," *International Journal of Mathematics & Computers in Simulation*, 2010, 4(4): 116–123.

[12] Z. Q. Song, H. F. Qi, C. Wang, Y. Shang, and W. Chang, "Narrow line-width fiber laser and application in sensing," *Acta Optica Sinica*, 2015, 35(s2): s214005.

[13] H. Xu, C. Wang, G. D. Peng, X. L. Zhang, Y. Y. Ying, Z. H. Sun, *et al.*, "A distributed feedback fiber laser hydrophone with high sensitivity and flat frenquency response," *Shandong Science*, 2013, 26(4): 46–50.

[14] Y. Chen, J. Zhang, M. Zhang, and L. W. Wang, "Calibration of phase-shifted sensitivity of optical fiber hydrophone in the frequency range 20 Hz to 10 kHz," *Acta Optica Sinica*, 2011, 40(11): 1686–1691.

16

Optofluidic Refractive Index Sensor Based on Partial Reflection

Lei ZHANG[1*], Zhang ZHANG[1], Yichuan WANG[1], Meiying YE[2], Wei FANG[1*], and Limin TONG[1]

[1]*State Key Laboratory of Modern Optical Instrumentation, College of Optical Science and Engineering, Zhejiang University, Hangzhou, 310027, China*

[2]*College of Material, Chemistry and Chemical Engineering, Hangzhou Normal University, Hangzhou, 310036, China*

[*]Corresponding author: Lei ZHANG and Wei FANG E-mail: zhang_lei@zju.edu.cn and wfang08@zju.edu.cn

Abstract: We demonstrate a novel optofluidic refractive index (RI) sensor with high sensitivity and wide dynamic range based on partial reflection. Benefited from the divergent incident light and the output fibers with different tilting angles, we have achieved highly sensitive RI sensing in a wide range from 1.33 to 1.37. To investigate the effectiveness of this sensor, we perform a measurement of RI with a resolution of *ca.* 5.0×10^{-5} refractive index unit (RIU) for ethylene glycol solutions. Also, we have measured a series of liquid solutions by using different output fibers, achieving a resolution of *ca.* 0.52 mg/mL for cane surge. The optofluidic RI sensor takes advantage of the high sensitivity, wide dynamic range, small footprint, and low sample consumption, as well as the efficient fluidic sample delivery, making it useful for applications in the food industry.

Keywords: Optofluidic sensor; refractive index; partial reflection

1. Introduction

Miniaturized optofluidic devices offer great potential for realizing more functional and more compact optical systems through the fusion of microfluidics and optics [1, 2]. The past decade has witnessed the realizations of optofluidic devices including tunable waveguides [3, 4], lenses [5, 6], switches [7, 8], apertures [9, 10], reconfigurable lasers [11, 12], interferometers [13, 14], gratings [15, 16], and sensitive sensors with fast response and low sample consumption [17–20], as well as demonstrations of their applications in chemical and biological analysis [21], energy [22], and photonics [23]. For many biological and chemical applications, the measurement of refractive index (RI) is extremely useful for detecting compounds that are: nonionic, transparent in the UV/vis range, or with no fluorescence, moreover, it has great potential for none invasive and label-free biosensing [24–26].

To date, a variety of optofluidic RI sensors have been reported by using integrated interferometers [27, 28], Fabry-Pérot cavities [29, 30], microring resonators [31, 32] gratings [33, 34], microstructured optical fibers [35, 36], and surface plasmon resonance [37, 38]. Although the sensitivity of the aforementioned sensors can be as high as 10^{-6} RIU by measuring the wavelength shift, these sensors require expensive instruments, e.g., optical spectrum analyzers. Alternatively, one can design a highly sensitive RI sensor based on partial reflection or refraction by measuring the intensity of the reflected or transmitted light or both [39, 40]. It is well known that the reflectivity of liquid-solid interface can be approximated by combining Snell's law with the Fresnel equations of reflection. When the RI of the liquid increases to a value where total internal reflection (TIR) at the liquid-solid interface is no longer satisfied, the reflectivity will decrease sharply, leading to a strong increase of the transmitted light. This phenomenon provides an attractive feature for assembling tunable optofluidic device and RI sensors. For example, Lapsley and co-authors [39] reported a variable optical attenuator where the light attenuation was achieved by adjusting the RI of the liquid in the microfluidic channel and thus altering the reflectivity of the light at the sidewall of the microchannel. Recently, Weber and Vellekoop [40] reported an optofluidic sensor, which was consisted of one input fiber and two out fibers with fixed angles for guiding the incident light, reflected light, and refracted light, respectively. Also, integrated microlenses were used for collimating the divergent light. Note that an inherent disadvantage of this sensor is the narrow dynamic range, typically, 0.01 refractive index unit (RIU) for a fixed incident angle. Thus, one has to prepare many chips with different incident angles to determine samples within a wide range of RI.

In this work, we report an optofluidic RI sensor with one straight detection channel, two input fibers, and ten output fibers for RI sensing with high sensitivity and wide dynamic range. Compared with the existing counterparts that require microlenses for light collimation, our design takes advantage of the divergent incident light and the optimized position of the output fibers to realize high sensitive sensing in a dynamic range from 1.33 to 1.37. In this case, we can measure the RI of most aqueous samples with high sensitivity by choosing an output fiber with a specific tilting angle. Since microlenses are removed from our design, the fabrication of the sensor becomes much simpler and cost-effective. To investigate the effectiveness of the RI sensor, we perform a measurement of ethylene glycol (EG) solutions with a RI range from 1.33 to 1.37, achieving a resolution of *ca.* 5.0×10^{-5} RIU. Furthermore, we present measurements of liquid concentrations with a resolution of *ca.* 0.52 mg/mL for cane surge. The optofluidic RI sensor shown here may provide a compact and versatile sensing platform for sensitive and fast detection of low-volume samples.

2. Experimental section

2.1 Reagents and instruments

All reagents and standards were of analytical grade and purchased from Sinopharm Chemical Reagent (Shanghai, China) unless otherwise stated. SU-8 photoresist and SU-8 developer were purchased from MicroChem Corp. (Newton, MA, USA). Polydimethylsiloxane (PDMS) (Sylgard 184) was purchased from Dow Corning (Midland, MI, USA). Ultrapure water (Siemens Labostar2-UV) was used throughout. Sample solutions were prepared before use. Optical micrographs were obtained using a charge coupled device (CCD) camera (DS-Fi1, Nikon, Japan) mounted on a microscope (Eclipse 90i, Nikon, Japan).

2.2 Fabrication of the optofluidic device

Figure 1 shows an optical image of a PDMS microfluidic chip for multi-fiber RI sensing, in which it contains one detection channel, two fiber channels with a tilting angle of 71°, and ten output fibers channels with tilting angles range from 67° to 76°, respectively. The tilting angle of 71° was calculated by using Snell's law [40], and the tilting angles of the output fibers were defined according to the divergent incident light. The microfluidic chip was fabricated by using standard soft lithography process [41]. Briefly, uncured PDMS was poured onto a SU-8 master, followed by curing at 80 ℃ for 20 min. The cured PDMS slab was then peeled from the SU-8 master, and punched holes at the ends of the detection channel and the PDMS channels to introduce sample and uncured PDMS, respectively. The PDMS slab was bonded with a clean glass slide by using a plasma cleaner (PDC-32G-2, Harrick, USA). Typically, the detection channel was 2 cm in length, 125 μm in width, and 150 μm in depth with a rectangular cross section. In this case, optical fibers could be inserted into the fiber channels and well aligned by fiber channels. To remove the scattering light, the gap between the fiber and fiber channel was sealed by introducing uncured PDMS from the PDMS inlets, followed by curing at 80 ℃ for 20 min.

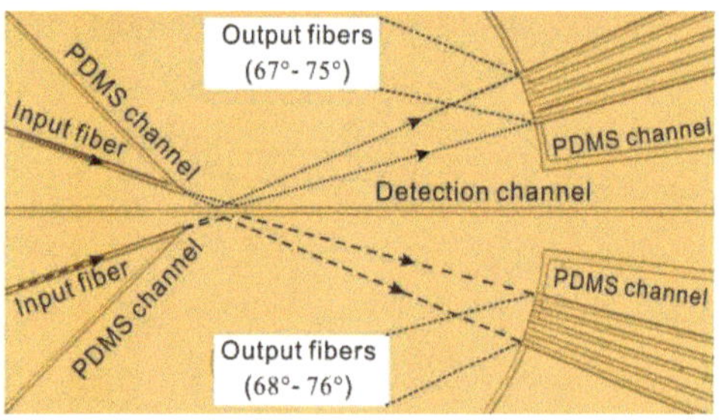

Fig. 1 Optical image of a microfluidic chip for multi-fiber RI sensing, which contains one detection channel, two input fibers channels, four PDMS channels, and ten output fiber channels. The dot and dash lines indicate the divergent incident light and reflected light, respectively.

2.3 Procedures

In the experiment, 5 μL of sample solution was added into the sample inlet and was introduced into the detection channel by negative pressure generated by a syringe. After each measurement, the channel was flushed with ultrapure water. A 473 nm laser and a broadband light from a tungsten halide lamp (PHILIPS 7748XHP) were coupled into input fibers and served as incident light for visualization and RI measurements, respectively. The reflected light was collected by one of the ten output fibers and recorded by a spectrometer (Maya2000 Pro, Ocean optics, Dunedin, FL, USA).

3. Working principle

Figure 2(a) schematically shows a typical structure for illustrating the operating mechanism of the RI sensor, where α and β denote the tilting angles of the input and output fibers, respectively, and d denotes the distance between the end face of an output fiber and the reflection zone.

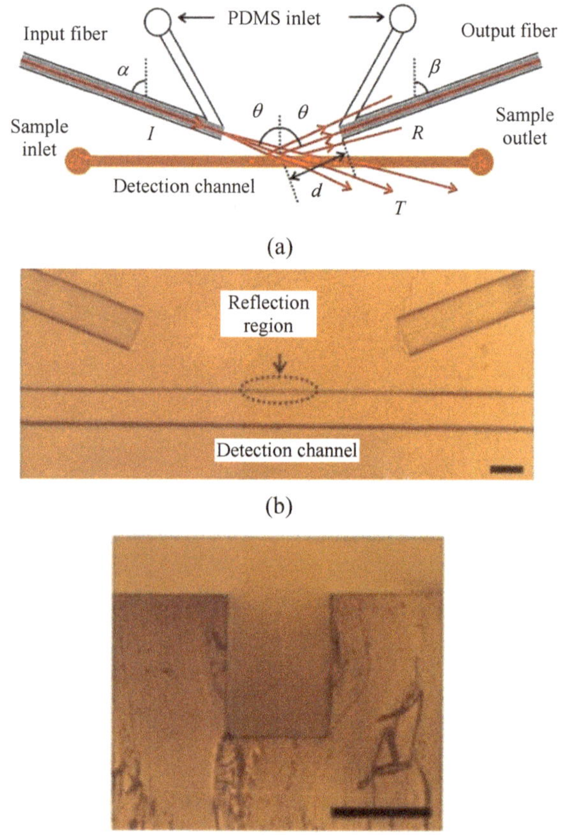

(c)

Fig. 2 Illustration of a microfluidic RI sensor: (a) a schematic illustration of a microfluidic RI sensor with one pair of fibers for optimizing the tilting angles of the input fiber and the distance between the end face of the output fiber to the reflection zone and (b) optical micrograph of a microfluidic RI sensor with α=β=71°, and d=0.6 mm, and (c) cross section of the detection channel. Scale bars: 125 μm.

Figure 2(b) shows an optical micrograph of an optofluidic RI sensor with α=β=71° and d=0.6 mm. As shown in Fig. 2(c), the vertical sidewalls of the detection channel serve as reflective interfaces. When the liquid sample is introduced into the detection channel, a RI mismatch between PDMS and the liquid sample in the detection channel will cause the incident optical beam (I) to be reflected and refracted at the PDMS-liquid interface, resulting in a reflected beam (R) and a transmitted beam (T). The angle of $R(\theta)$ is the same as the angle of I, allowing the reflected beam to be captured by the output optical fiber. The intensity of the reflected light is a function of RI of the liquid sample, and thus, we can determine the concentration of solutes by measuring the intensity of the reflected light. Note that we did not fabricate microlenses to collimate incident light, it offers a possibility to arrange several output fibers with different tilting angles to record the divergent reflected beam. In this case, each output fiber corresponds to an angle of reflection, providing an effective method to achieve a wide dynamic range. On the other hand, the insertion loss of the device is relatively greater than its counterparts with microlenses, and approximately 5% incident light could be collected by the output fiber for the microfluidic chip as shown in Fig. 2(b).

4. Results and discussion

4.1 Visualization of the reflected and refracted beam

To visualize the reflected and refracted beams, a 473-nm-wavelength laser was used as incident light and 0.01 mM fluorescein solutions with different RIs were used to visualize the refracted light. Figure 3 shows four typical images of an optofluidic RI sensor (α=β=71°) with different fluids in the detection channel. When air (n=1) was in the detection channel, TIR occurred at the PDMS-air interface, and the divergent reflected light could be clearly seen in Fig. 3(a). When the tilting angle of the input fiber was 71°, the estimated angle of reflection was from 67° to 76°. When fluorescein solutions were injected into the detection channel, the fluorescence intensity increased obviously with an increase in RI from 1.33, 1.36, to 1.42 [Figs. 3(b)–(d)], indicating an increase in the refracted light intensity and a decrease in the reflected one. When the fluorescein solution with a RI of 1.42 was in the detection channel, the sidewall of the detection channel could hardly be identified, and the incident beam transmitted through the detection channel without obvious refraction and reflection [Fig. 3(d)].

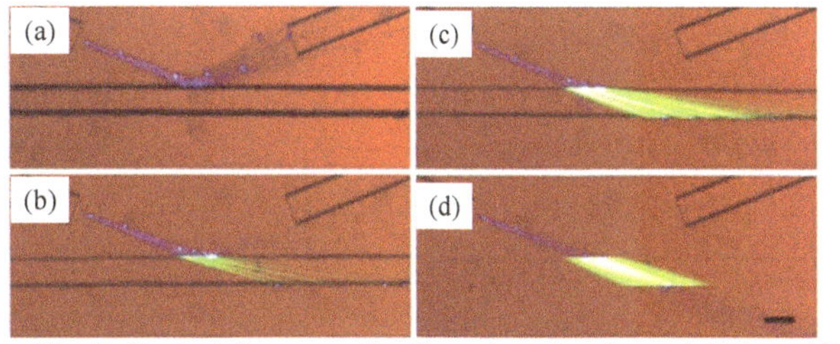

Fig. 3 Optical images of reflected and refracted light with different fluids within the detection channel: (a) air (n=1) and (b)–(d) 0.01 mM fluorescein was introduced into the detection channel as an indicator of the refracted light. The RI of the solutions were 1.33, 1.36, and 1.42, respectively. Scale bar: 125 μm.

4.2 Theoretical calculation

To achieve a fundamental understanding of the sensor's RI response, theoretical analysis was conducted using Fresnel equations. We calculated the reflection coefficients r_s and r_p for s- and p-polarizations by (1) and (2), respectively, with the total reflectivity calculated by (3)

$$r_s = \frac{n_1 \cos\theta - n_2 \sqrt{1-\left(\frac{n_1}{n_2}\sin\theta\right)^2}}{n_1 \cos\theta + n_2 \sqrt{1-\left(\frac{n_1}{n_2}\sin\theta\right)^2}} \tag{1}$$

$$r_p = \frac{n_2 \cos\theta - n_1 \sqrt{1-\left(\frac{n_1}{n_2}\sin\theta\right)^2}}{n_2 \cos\theta + n_1 \sqrt{1-\left(\frac{n_1}{n_2}\sin\theta\right)^2}} \tag{2}$$

$$\bar{R} = \frac{|r_s|^2 + |r_p|^2}{2} \tag{3}$$

where n_1=1.41 is the RI of PDMS, and n_2 is the RI of the liquid in the detection channel. Thus, the reflectivity is a function of n_2. Figure 4(a) shows a theoretical analysis of the reflected light intensity of an RI sensor with different angles of reflection. For a fixed angle θ, we found that the intensity kept as unity for small n_2 because of the TIR at the PDMS-liquid interface. When n_2 increased to a value where TIR was no longer satisfied, a very sharp decrease in reflectivity was observed with very small change in n_2. Based on theoretical prediction, the slope can be infinity right at the critical point, thus a very sensitive RI sensor could be realized. For example, when n_2 changed 0.001

RIU, the maximum change in the intensity could be 30%, however, the dynamic range of the sensor was *ca.* 0.01 RIU, which was relatively narrow for real application. As shown in Fig. 4(a), when θ increases from 69° to 74°, the critical RI of n_2 shifts from 1.33 to 1.37. Thus, we could achieve a wide dynamic range by choosing an output fiber with different θ.

3.3 Effect of *d* on the measurement of RI

To optimize *d* of the RI sensor, we fabricated a series of microfluidic chips with one input fiber and one output fiber for characterization. Owing to the large aperture and core diameter, a multi-mode fiber could collect light from a wide angle, and thus smearing out the sharp decrease of the light coming from a single collection angle as indicated by the theoretical calculations [Fig. 4(a)]. In this work, we adopted SMFs with a core diameter around 10 μm as input and output optical fibers, thus, the reflection light within a narrow range of angle could be collected by the output fiber. We found that *d* had a great impact on the performance of the sensor. Figures 4(b) and 4(c) show normalized intensity of the light collected by the output fiber with different tilting angles as a function of RI for *d* of 0.6 mm, and 1.5 mm, respectively. The results did not match with the theoretical analysis [Fig. 4(a)], especially, the response curves for β=69° and β=71° can be hardly resolved in the RI range of 1.33–1.34. This mismatch could be attributed to the output fiber collecting the light with different angles of reflection. Increasing *d* is an effective method to narrow the range of the angle of reflection. When we increased *d* to 5 mm, the flare angle of the reflected light that could be collected by a SMF with a core diameter of 10 μm decreased dramatically. To obtain the estimated flare angle of the reflected light, we considered the reflection area as a point, thus, the sides of the isosceles triangle were 5 mm and 10 μm, respectively, resulting in a flare angle of approximately 0.1°. As shown in Fig. 4(d), all of the four response curves for the different tilting angles are well separated, and the slopes after the critical points become steeper, which matched well with the theoretical calculations [Fig. 4(a)]. It is worth mentioning that we can arrange more output fiber on one side the detection to collect reflected light with different angles when *d* is 5 mm.

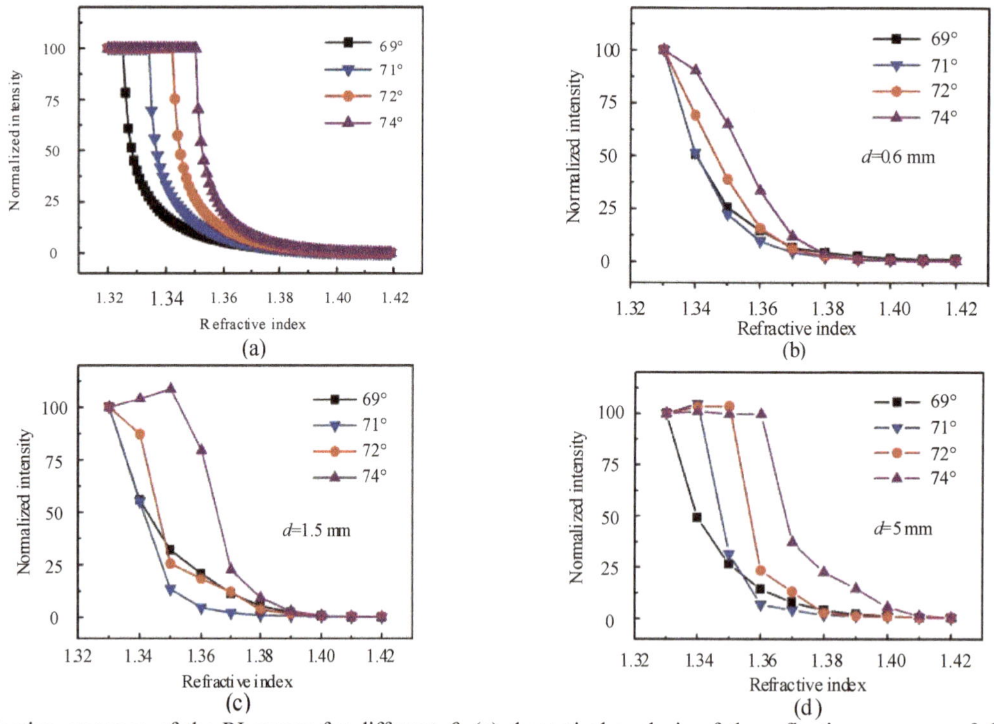

Fig. 4 Reflection response of the RI sensor for different θ: (a) theoretical analysis of the reflection response of the RI sensor for different θ and (b)–(d) normalized intensity of the reflected light collected by four output fibers with reflected angles of 69°, 71°, 72°, and 74°, respectively, except for the variations to the distance between the endface of an output fiber and the reflection zone as indicated in the figures.

4.4 Performance of the multi-fiber RI sensor

Figure 5 shows four typical curves recorded by the output fibers with a tilting angle of 69°, 71°, 72°, and 74°, respectively, which covers a RI range from 1.33 to 1.37. The result confirmed that the dynamic range could be extended by using different output fibers for sensing. When RI increased from 1.35 to 1.36, the normalized intensity for the output fiber with a tilting angle of 72° decreased from 103 to 23. Because the stability of the light source was measured with an outstanding reproducibility of 0.13% RSD, the RI resolution could be estimated as 5.0×10^{-5} RIU based on 3 times the standard deviation of the light source. Note that the signals collected by the output fibers with a tilting angle less than 68° or greater than 74° were too weak for sensing owing to the Gaussian intensity distribution of the divergent incident beam.

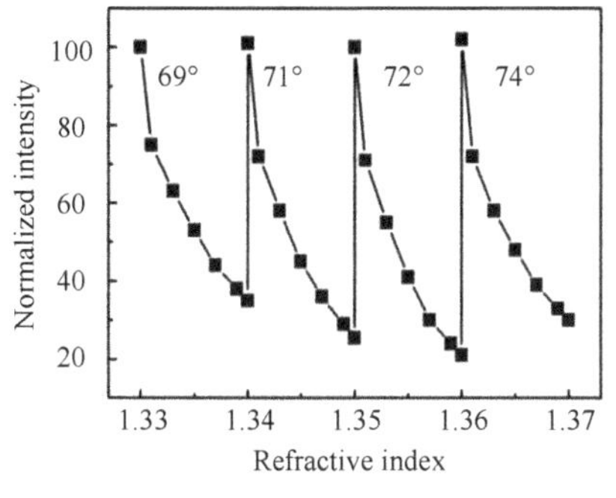

Fig. 5 Normalized intensity of the reflected light collected by four output fibers with tilting angles of 69°, 71°, 72°, and 74°, respectively.

Table 1 Measurement results of the concentration of cane surge obtained by using three output fibers with different tilting angles.

Referenced concentration (g/mL)	Referenced RI	Experimental value (g/mL)	Tilting Angle	RE (%)
0.100	1.345 9	0.105	71°	5.0
0.150	1.352 5	0.154	72°	2.7
0.200	1.358 6	0.192	72°	−4.0
0.250	1.364 2	0.246	74°	−1.6
0.300	1.369 2	0.307	74°	2.3

To further evaluate the performance of the RI sensor, aqueous cane surge solutions with different concentrations (0.1 g/mL–0.3 g/mL) were prepared and analyzed. Because the RI of the cane surge solution is in a range from 1.34 to 1.37, we adopted three output fibers with tilting angle of 71°, 72°, and 74° for collecting the signals, respectively. Table 1 gives obtained values and the relative errors (RE) for the sensor, showing a wide dynamic range and high accuracy with an estimated resolution of 0.52 mg/mL.

5. Conclusions

In summary, we have developed a sensitive, robust, and inexpensive optofluidic RI sensor. We have realized highly sensitive RI sensing in a wide dynamic range of 1.33–1.37 with a RI resolution of 5.0×10^{-5} RIU by choosing an output fiber with different tilting angles for signal recording. Also, we have measured a series of liquid solutions by using different output fibers, achieving a resolution of *ca.* 0.52 mg/mL for cane surge. To further enhance the sensing performance, one might use a higher resolution lithography instrument to fabricate a SU-8 master with a smoother sidewall. We believe this optofluidic RI sensor can be used for real-time, low-cost, and multifunctional measurements in a wide range of applications.

Acknowledgment

This work has been supported in part by National Basic Research Program of China (Nos. 2013CB328703 and 2014CB921303), and National Natural Science Foundation of China (61275217, 21407039).

References

[1] D. Psaltis, S. R. Quake, and C. H. Yang, "Insight review: developing optofluidic technology through the fusion of microfluidics and optics," *Nature*, 2006, 442(7101): 381–386.

[2] C. Monat, P. Domachuk, and B. J. Eggleton, "Integrated optofluidics: a new river of light," *Nature Photonics*, 2007, 1(2): 106–114.

[3] D. B. Wolfe, R. S. Conroy, P. Garstecki, B. T. Mayers, M. A. Fischbach, K. E. Paul, *et al.*, "Dynamic control of liquid-core/liquid-cladding optical waveguides," *Proceedings of the National Academy of Sciences*, 2004, 101(34): 12434–12438.

[4] B. T. Mayers, D. V. Vezenov, V. I. Vullev, and G. M. Whitesides, "Arrays and cascades of fluorescent liquid-liquid waveguides: broadband light sources for spectroscopy in microchannels," *Analytical Chemistry*, 2005, 77(5): 1310–1316.

[5] X. L. Mao, J. R. Waldeisen, B. K. Juluri, and T. J. Huang, "Hydrodynamically tunable optofluidic cylindrical microlens," *Lab on a Chip*, 2007, 7(10): 1303–1308.

[6] P. Fei, Z. He, C. Zheng, T. Chen, Y. Men, and Y. Huang, "Discretely tunable optofluidic compound microlenses," *Lab on a Chip*, 2011, 11(17): 2835–2841.

[7] A. Groisman, S. Zamek, K. Campbell, L. Pang, U. Levy, and Y. Fainman, "Optofluidic 1×4 switch," *Optics Express*, 2008, 16(18): 13499–13508.

[8] W. Z. Song and D. Psaltis, "Electrically tunable optofluidic light switch for reconfigurable solar lighting," *Lab on a Chip*, 2013, 13(14): 2708–2713.

[9] Y. Hongbin, Z. Guangya, C. F. Siong, and L. Feiwen, "Optofluidic variable aperture," *Optics Letters*, 2008, 33(6): 548–550.

[10] C. L. Song, N. T. Nguyen, A. K. Asundi, and C. L. N. Low, "Tunable optofluidic aperture configured by a liquid-core/liquid-cladding structure," *Optics Letters*, 2011, 36(10): 1767–1769.

[11] Z. Li, Z. Zhang, T. Emery, A. Scherer, and D. Psaltis, "Single mode optofluidic distributed feedback dye laser," *Optics Express*, 2006, 14(2): 696–701.

[12] M. Aas, A. Jonas, A. Kiraz, O. Brzobohaty, J. Jezek, Z. Pilat, and P. Zemanek, "Spectral tuning of lasing emission from optofluidic droplet microlasers using optical stretching," *Optics Express*, 2013, 21(18): 21380–21394.

[13] C. Grillet, P. Domachuk, V. Ta'eed, E. Mägi, J. Bolger, B. Eggleton, *et al.*, "Compact tunable microfluidic interferometer," *Optics Express*, 2004, 12(22): 5440–5447.

[14] Y. Zou, Z. Hen, X. Hen, Z. Di, and X. Chen, "An integrated tunable interferometer controlled by liquid diffusion in polydimethylsiloxane," *Optics Express*, 2012, 20(17): 18931–18936.

[15] M. Hashimoto, B. Mayers, P. Garstecki, and G. M. Whitesides, "Flowing lattices of bubbles as tunable, self-assembled diffraction gratings," *Small*, 2006, 2(11): 1292–1298.

[16] J. Q. Yu, Y. Yang, A. Q. Liu, L. K. Chin, and X. M. Zhang, "Microfluidic droplet grating for reconfigurable optical diffraction," *Optics Letters*, 2010, 35(11): 1890–1892.

[17] P. Liu, H. Huang, T. Cao, X. Liu, Z. Qi, Z. Tang, *et al.*, "An ultra-low detection-limit optofluidic biosensor with integrated dual-channel Fabry-Pérot cavity," *Applied Physics Letters*, 2013, 102(16): 163701-1–163701-4.

[18] A. A. P. Trichet, J. Foster, N. E. Omori, D. James, P. R. Dolan, G. M. Hughes, *et al.*, "Open-access optical microcavities for lab-on-a-chip refractive index sensing," *Lab on a Chip*, 2014, 14(21): 4244–4249.

[19] Z. Xu, K. Han, I. Khan, X. Wang, and G. L. Liu, "Liquid refractive index sensing independent of opacity using an optofluidic diffraction sensor," *Optics Letters*, 2014, 39(20): 6082–6085.

[20] C. Wu, M. L. V. Tse, Z. Liu, B. O. Guan, A. P. Zhang, C. Lu, *et al.*, "In-line microfluidic integration of photonic crystal fibers as a highly sensitive refractometer," *Analyst*, 2014, 139(21): 5422–5429.

[21] X. D. Fan and I. M. White, "Optofluidic microsystems for chemical and biological analysis," *Nature Photonics*, 2011, 5(10): 591–297.

[22] D. Erickson, D. Sinton, and D. Psaltis, "Optofluidics for energy applications," *Nature Photonics*, 2011, 5(10): 583–590.

[23] L. Pang, H. M. Chen, L. M. Freeman, and Y. Fainman, "Optofluidic devices and applications in photonics, sensing and imaging," *Lab on a Chip*, 2012, 12(19): 3543–3551.

[24] P. J. Viskari and J. P. Landers, "Unconventional detection methods for microfluidic devices," *Electrophoresis*, 2006, 27(9): 1797–1810.

[25] Y. Wang, S. Meng, Y. Liang, L. Li, and W. Peng, "Fiber-optic surface plasmon resonance sensor with multi-alternating metal layers for biological measurement," *Photonic Sensors*, 2013, 3(3): 202–207.

[26] J. Zhu, L. Qin, S. Song, J. Zhong, and S. Lin, "Design of a surface plasmon resonance sensor based on grating connection," *Photonic Sensors*, 2015, 5(2): 159–165.

[27] L. K. Chin, A. Q. Liu, Y. C. Soh, C. S. Lim, and C. L. Lin, "A reconfigurable optofluidic Michelson interferometer using tunable droplet grating," *Lab on a Chip*, 2010, 10(8): 1072–1078.

[28] M. I. Lapsley, I. K. Chiang, Y. B. Zheng, X. Y. Ding, X. L. Mao, and T. J. Huang, "A single-layer, planar, optofluidic Mach-Zehnder interferometer for label-free detection," *Lab on a Chip*, 2011, 11(10): 1795–1800.

[29] W. Z. Song, X. M. Zhang, A. Q. Liu, C. S. Lim, P. H. Yap, and H. M. M. Hosseini, "Refractive index

measurement of single living cells using on-chip Fabry-Pérot cavity," *Applied Physics Letters*, 2006, 89(20): 203901-1–203901-3.

[30] Y. Guo, H. Li, K. Reddy, H. S. Shelar, V. R. Nittoor, and X. Fan, "Optofluidic Fabry-Pérot cavity biosensor with integrated flow-through micro-/nanochannels," *Applied Physics Letters*, 2011, 98(4): 041104-1–041104-3.

[31] L. Q. Ren, X. Wu, M. Li, X. W. Zhang, L. Y. Liu, and L. Xu, "Ultrasensitive label-free coupled optofluidic ring laser sensor," *Optics Letters*, 2012, 37(18): 3873–3875.

[32] M. Li, X. Wu, L. Y. Liu, X. D. Fan, and L. Xu, "Self-Referencing optofluidic ring resonator sensor for highly sensitive biomolecular detection," *Analytical Chemistry*, 2013, 85(19): 9328–9332.

[33] W. Liang, Y. Y. Huang, Y. Xu, R. K. Lee, and A. Yariv, "Highly sensitive fiber Bragg grating refractive index sensors," *Applied Physics Letters*, 2005, 86(15): 151122-1–151122-3.

[34] F. Xu, G. Brambilla, and Y. Q. Lu, "A microfluidic refractometric sensor based on gratings in optical fibre microwires," *Optics Express*, 2009, 17(23): 20866–20871.

[35] J. Wu, D. Day, and M. Gu, "A microfluidic refractive index sensor based on an integrated three-dimensional photonic crystal," *Applied Physics Letters*, 2008, 92(7): 071108-1–071107-3.

[36] A. P. Zhang, G. Yan, S. Gao, S. He, B. Kim, J. Im, *et al.*, "Microfluidic refractive-index sensors based on small-hole microstructured optical fiber Bragg gratings," *Applied Physics Letters*, 2011, 98(22): 221109-1–221109-3.

[37] Y. Wang, S. Meng, Y. Liang, L. Li, and W. Peng, "Fiber-optic surface plasmon resonance sensor with multi-alternating metal layers for biological measurement," *Photonic Sensors*, 2013, 3(3): 202–207.

[38] J. Zhu, L. Qin, S. Song, J. Zhong, and S. Lin, "Design of a surface plasmon resonance sensor based on grating connection," *Photonic Sensors*, 2015, 5(2): 159–165.

[39] T. J. Huang, M. I. Lapsley, S. C. S. Lin, and X. L. Mao, "An in-plane, variable optical attenuator using a fluid-based tunable reflective interface," *Applied Physics Letters*, 2009, 95(8): 083507-1–083507-3.

[40] E. Weber and M. J. Vellekoop, "Optofluidic micro-sensors for the determination of liquid concentrations," *Lab on a Chip*, 2012, 12(19): 3754–3759.

[41] D. Qin, Y. Xia, and G. M. Whitesides, "Soft lithography for micro- and nanoscale patterning," *Nature Protocols*, 2010, 5(3): 491–502.

Alcohol Sensing Over O+E+S+C+L+U Transmission Band Based on Porous Cored Octagonal Photonic Crystal Fiber

Bikash Kumar PAUL[1, 2], Md. Shadidul ISLAM[1], Kawsar AHMED[1,2*], and Sayed ASADUZZAMAN[1,2,3]

[1]*Department of Information and Communication Technology, Mawlana Bhashani Science and Technology University Santosh, Tangail-1902, Bangladesh*

[2]*Group of Bio-Photomatriχ, Tangail-1902, Bangladesh*

[3]*Department of Software Engineering, Daffodil International University, Sukrabad, Dhaka, Bangladesh*

[*]Corresponding author: Kawsar AHMED E-mail: kawsar.ict@mbstu.ac.bd and k.ahmed.bd@ieee.org

Abstract: A micro structure porous cored octagonal photonic crystal fiber (P-OPCF) has been proposed to sense aqueous analysts (alcohol series) over a wavelength range of 0.80 μm to 2.0 μm. By implementing a full vectorial finite element method (FEM), the numerical simulation on the proposed O-PCF has been analyzed. Numerical investigation shows that high sensitivity can be gained by changing the structural parameters. The obtained result shows the sensitivities of 66.78%, 67.66%, 68.34%, 68.72%, and 69.09%, and the confinement losses of 2.42×10^{-10} dB/m, 3.28×10^{-11} dB/m, 1.21×10^{-6} dB/m, 4.79×10^{-10} dB/m, and 4.99×10^{-9} dB/m at the 1.33 μm wavelength for methanol, ethanol, propanol, butanol, and pentanol, respectively can satisfy the condition of much legibility to install an optical system. The effects of the varying core and cladding diameters, pitch distance, operating wavelength, and effective refractive index are also reported here. It reflects that a significant sensitivity and low confinement loss can be achieved by the proposed P-OPCF. The proposed P-OPCF also covers the wavelength band (O+E+S+C+L+U). The investigation also exhibits that the sensitivity increases when the wavelength increases like $S_{\text{O-band}} < S_{\text{E-band}} < S_{\text{S-band}} < S_{\text{C-band}} < S_{\text{L-band}} < S_{\text{U-band}}$. This research observation has much pellucidity which has remarkable impact on the field of optical fiber sensor.

Keywords: Porous cored OPCF; alcohol sensor; sensitivity; confinement loss; transmission band

1. Introduction

A massive technological change emerges through the innovative invention of researchers. As an optical medium, photonic crystal has some identical properties [1–11]. The photonic crystal fibers are enormously used in the optical fiber technology due to its idiosyncratic properties. In 1996, Knight *et al.* firstly proposed the photonic crystal fiber for its novel competence [11]. The photonic crystal fiber is a fiber in which artificial frequent capillaries occur. Optical characteristics are mainly controlled by the number and size or magnitude of microstructure capillaries of air holes [12] and various structural diversities. Photonic crystal fibers (PCF) are becoming more and more popular for its better guiding properties, robustness, and flexibility.

Some identical properties are seen like as nonlinearity [1, 2], high birefringence [3, 4], high sensitivity [5, 6], low confinement loss [7, 8], improved effective area, and endlessly single mode [9–11]. For these reasons, PCF has drawn a great attraction among researchers. At the beginning time, PCF has limited applications areas. Now PCF is used in fields of THz telecommunications [13], nonlinear optics [14], bio sensing [15], chemical sensing [16, 17], spectroscopy [18], optical chromatography [19], and fuel adulteration detection [20]. In medical science, it is becoming more and more popular for bio-photonics [21] and neuro-photonics [22]. It is also used in the fields of cancer cell detection, HIV viral load detection [23], protein detection [24], diabetics detection [25], etc. Different types of PCF are used for numerous applications for their superb guiding capability.

As a novel method, optical indexed guiding PCF is well rebounded [26]. Depending on the light guiding mechanism process, a crystal fiber PCF can be categorized into two classes. One is the photonic band gap fiber which uses the photonic band gap (PBG) effect for guiding light through that fiber thus it is called PBG-PCF. The other one uses common optical properties total internal reflection (TIR) thus it is called TIR-PCF. PCFs are formed by the crystal which is made of silica glass. The main reason for choosing silica glass as a background material is its paramount property which provides momentous amenities. Recently, graphene [27], chalconide glass with polymer [28], telluride, and topus [29] are also used to fabricate PCF.

The first proposal of PCF came from Yeh *et al.* in 1978 [30]. But a photonic crystal fiber was designed using two-dimensional (2D) photonic crystals by Russel [31]. This fiber consisted of single air core, which was reported in Optical Fiber Conference at San Jose, CA in 1996 [32]. As a consequence in 1997, endlessly single mode fiber was invented. Highly birefringent PCF (HB-PCF) opened a new horizon in 2000. Later Bragg fiber, PCF laser, and PCF with ultra-flatted dispersion were invented. Recently, gas and liquid filled PCF created a new diversity in different application areas. In 2014, Ademgil proposed an octagonal PCF sensor for chemical sensing [26]. At operating the 1.00 μm wavelength, it shows approximately 47 % sensitivity for aqueous analytes. In 2015, Ahmed *et al.* designed an O-PCF which gives the indication of relatively high sensitivity [33].

In this paper, we propound a design and numerical analysis of microstructure circular porous core PCF with high sensitivity and low confinement for chemical sensing applications covering the O+S+C+L+U wavelength band. A five-layer octagonal air hole lattice array which acts as a cladding encompasses the core region. For electromagnetic computation, an anisotropic perfectly match layer (PML) plays a vital role in sensing performance. PML is also used in the proposed PCF to attenuate unwanted reflection of light wave [33]. Here PML acts as an absorbing layer. Variations of geometrical properties (parameters) are also examined to analyze the effects of guiding properties on P-OPCF.

2. Geometries of the proposed PCF

The transverse cross sectional view of the proposed P-OPCF is shown in Fig. 1(a). There are five rings in the designed structure in the cladding part, and each ring gives a view of an octagonal that's why it can be called as OPCF. For the first ring (the first layer) of the cladding, it contains 8 air holes. The number of air holes gets increased geometrically where the nearest next level number ring has 8 air holes more than the previous level of the ring. For the five rings, it can be represented as $8n$ where n is the number of levels, and the first ring is n=1 and will increase by 1 for each level until it touches the ring number five where total number of air hole is $8n=8\times5=40$. Each air hole has the same diameter where d=1.75 μm. The distance between two adjacent air holes of two conjugative rings is called pitch and denoted by Λ_1.

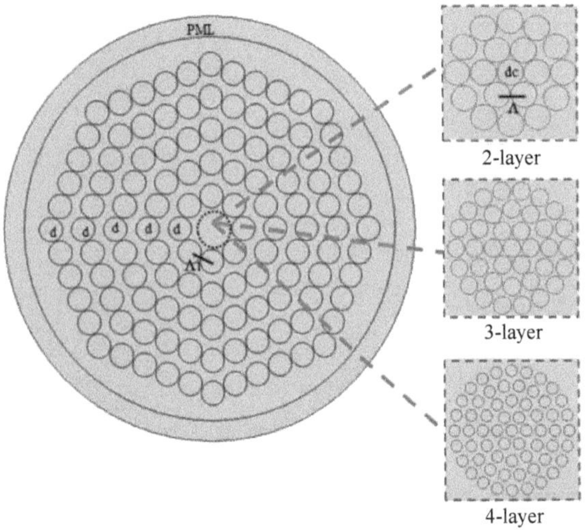

Fig. 1 Transverse cross-sectional view of the proposed P-OPCF.

The pivotal part of the proposed PCF contains a small area with two circular rings in porous shape which is known as core. Six holes filled with aqueous analysts are required to design the inner circle of the core where the outer circle has twelve holes filled. The core diameter d_c processes a very small area where d_c= 0.57 μm and each hole is filled with aqueous analysts that has the same diameter. The pitch of the core (Λ_c) is equal to Λ_c= 0.61 μm, and the two adjacent air holes are filled with aqueous analysts both in the same circle and between two circles. For two-layer circular porous core, the first layer has 6 holes in total which makes an angle of 60°, and the second layer has 12 holes which makes an angle of 30°. According to the sequence for three layers having 6, 12, and 18 holes, they make angles of 60°, 30°, and 20°, respectively. The four-layer porous core makes angles of 60°, 30°, 20°, and 15° for consequence layers, respectively.

The core cladding geometric structure is surrounded by anisotropic PML. Hence, we use a 10% of the PML size additive with an overall cladding size. The key component of the back ground, core, and cladding is pure silica where Sellmeier equation [35] is followed to determine the refractive index for the defined wavelength range.

3. Numerical analysis

A finite element method (FEM) is used to simulate our proposed photonic crystal fiber. The FEM sub-divides a larger complicated design into a smaller and simpler one which produces a more approximate solution. To design the issue of a PCF sensor, two fundamental parameters are directly associated.

The background material of the proposed P-OPCF is silica. As for the theoretical and experimental aspects, the refractive index (RI) is varying with wavelength λ for different materials. There is a relationship between the wavelength and RI of different transparent materials and liquids. Based on the relationship, an equation is derived which is highly used to determine RI of different crystals and liquids. This relationship was first found by Willhem Sellmeier in 1871. The equation is as follows:

$$n(\lambda)=\sqrt{1+\frac{B_1\lambda^2}{\lambda^2-C_1}+\frac{B_2\lambda^2}{\lambda^2-C_2}+\frac{B_3\lambda^2}{\lambda^2-C_3}} \quad (1)$$

where n and λ represent the refractive index and wavelength, respectively, and B_1, B_2, B_3, C_1, C_2, and C_3 are the coefficients of silica.

The effective refractive index can be calculated by

$$n_{\text{eff}}=B/k_0 \quad (2)$$

where B is the propagation constant, and $k_0 = 2\pi/\lambda$ is the free space wave number.

The sensitivity can be calculated by (3) which was reported in [33]:

$$r=\frac{n_r}{n_{\text{eff}}}f \quad (3)$$

where n_r is the RI of using materials within the air holes, and n_{eff} is the modal effective index. Here, f represents the percentage ratio between air hole power and total power which can be calculated by

$$f=\frac{\int_{\text{sample}}\text{Re}(E_xH_y-E_yH_x)dxdy}{\int_{\text{total}}\text{Re}(E_xH_y-E_yH_x)dxdy}\times 100 \quad (4)$$

where E_x and E_y are transverse and longitudinal electric fields, respectively, and H_x and H_y are transverse and longitudinal magnetic fields, respectively.

Light in the PCF should be confined into the core region. Due to the material impurity or structure combined, some light passes the outer of the core region, and the losses of light energy are treated as confinement losses or leakage losses. The loss can be calculated by using

$$L_c = 8.68k_0 \operatorname{Im}[n_{\text{eff}}] \tag{5}$$

where $\operatorname{Im}[n_{\text{eff}}]$ is the imaginary part of the effective refractive index. Using the discussed equation above and FEM, a numerical result analysis is attained.

4. Result analysis

To construct a prognostic computational model for real world continuity numerous software use FEM. The accuracy of the computational model is directly involved with a numerical analysis. Different well-known numerical methods such as FEM, boundary element method (BEM), finite difference method (FDM), finite volume method (FVM), and mesh-less method (MLM) are used in the numerical computation. Among these numerical methods, FEM provides the most accurate computation. Along with FEM, the finer mesh analysis is used to gain more accurate solutions. By using the finer mesh analysis, we get 42,294 elements (DOF degree of Freedom). COMSOL Multiphysics 4.2 has been used to investigate for the mesh and numerical analysis. According to the simulation result, Fig. 2 shows wavelength verses effective RI of our proposed optimized P-OPCF. It shows that the curves go downward due to an increase in the wavelength. This means that if we decrease the wavelength we get much higher effective RI and vice versa. This figure also represents the confinement of more light due to *X*-polarization and *Y*-Polarization at wavelength 1.33 μm for ethanol (n = 1.354).

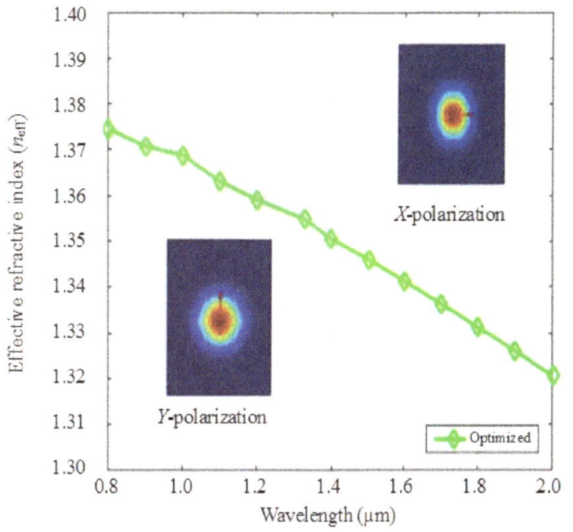

Fig. 2 Effective refractive index verses wavelength in *X*-polarization fundamental mode for optimized P-OPCF and n = 1.354.

Our desired goal is to getting much sensitivity and lower confinement loss, so we have investigated the core region by setting two, three, and four layers. Figure 3 shows the relative sensitivity verses wavelength for two-, three-, and four-layer porous core PCFs where two-layer core PCF is shown too much higher relative sensitivity than others. In addition, the confinement loss of 3.28×10^{-11} dB/m, 1.32×10^{-10} dB/m, and 6.96×10^{-11} dB/m for two, three, and four layers core respectively at the 1.33 μm wavelength for ethanol. Here the relative sensitivity of two-layer core is superior to others three- and four-layer core. Based on the above argument, we choose two-layer porous core O-PCF for the further investigation.

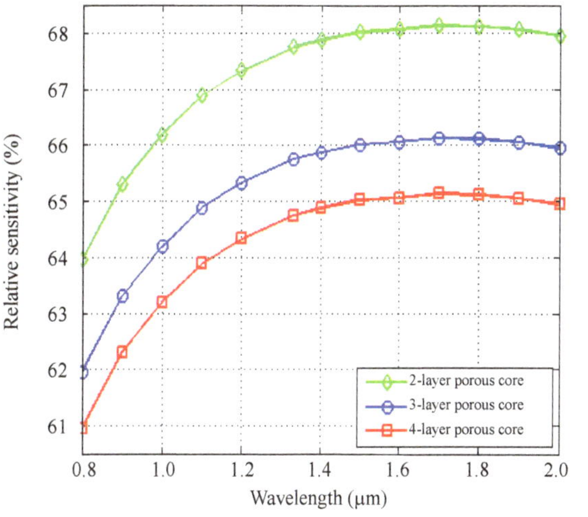

Fig. 3 Comparison of the relative sensitivity with core layer variations for n=1.354.

A fundamental factor confinement loss (CL) is a vital issue which should be remembered in mind at the fabrication time of PCF. To reduce the confinement loss of micro-structured P-OPCF is always desirable. For the confinement loss, some light energy encroaches into cladding area which must cause the main hindrance to the flow of informative light energy. Relative sensitivities of methanol, ethanol, propanol, butanol, and pentanolare are 66.78%, 67.66%, 68.34%, 68.72%, and 69.09 %, respectively at the 1.33 μm operating wavelength. Figure 4 depicts the relative sensitivity performance for alcohol analytes as methanol< ethanol< propanol< butanol< pentanol, respectively. It also exhibits that the sensitivity goes upward with an increase in the wavelength. Figure 5 shows the confinement losses for aqueous analysts of alcohol series methanol, ethanol, propanol, butanol, and pentanol, which are 2.42×10^{-10} dB/m, 3.28×10^{-11} dB/m, 1.21×10^{-6} dB/m, 4.79×10^{-10} dB/m, and 4.99×10^{-9} dB/m, respectively at the same wavelength.

Diameter/pitch (d/Λ), known as the air filling ratio, is tuned by modifying the core hole diameter and core pitch distance. Figure 6 shows that relative sensitivity curves are upward where in core region air filling ratio is high. So the air filling ratio has strong positive bonding with the relative sensitivity. We have investigated the relative sensitivity and

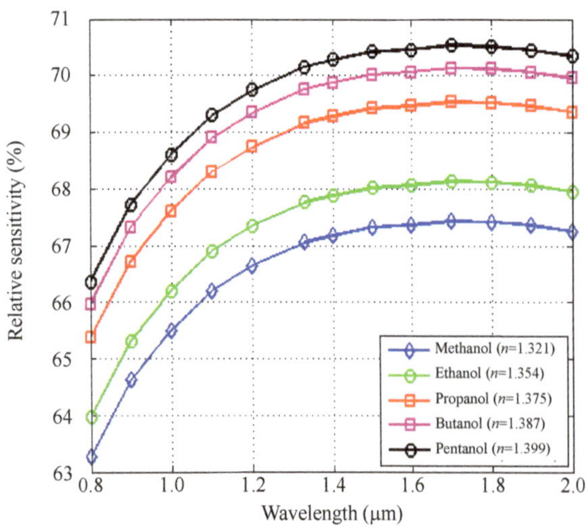

Fig. 4 Comparison of relative sensitivity with alcohol series for two-layer core.

confinement loss by varying the cladding air holes diameter which is shown in Figs. 7 and 8, respectively. It recounts comparatively higher sensitivity at d/Λ_1=0.931 and relatively lower sensitivity at d/Λ_1= 0.808 over the wavelength range 0.2 μm to 2.0 μm. To avoid fabrication tolerance, d/Λ_1= 0.931 is selected because that the maximum air filling ratio 0.94 is considerable for fabrication without any collapse [36].

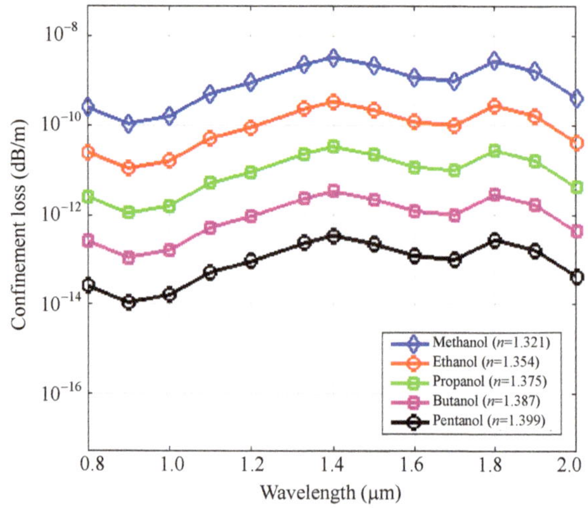

Fig. 5 Comparison of the confinement loss with alcohol series for two-layer core.

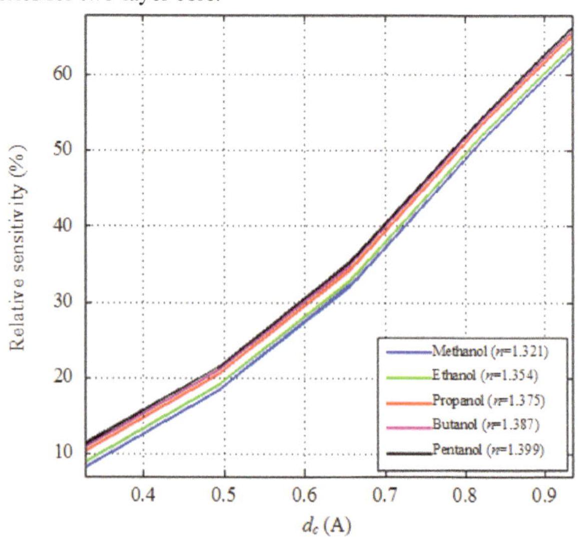

Fig. 6 Comparison of relative sensitivity with core air filling ratio (d_c/Λ) variations for alcohol series.

Fabrication is an important issue because expected characteristics of the proposed P-OPCF greatly depend on the fabrication accuracy. After the fabrication process, teeny change in parameters may occur. For this unexpected error, we have changed different parameters of the proposed P-OPCF. At the time of fabrication, by using the standard draw there may occur ±1% variations on their overall parameters of PCF [38]. For this reason, we inspect the behaviors of the proposed P-OPCF by tuning the structural parameters in range from –2% to + 2%.

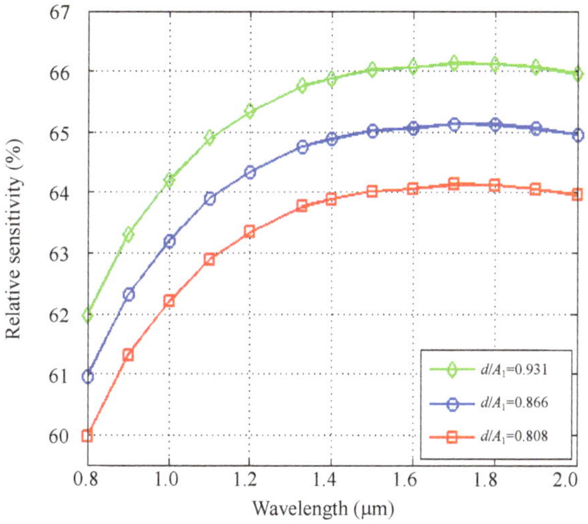

Fig. 7 Comparison of relative sensitivity with cladding air filling ratio (d/Λ_1) variations for n=1.354.

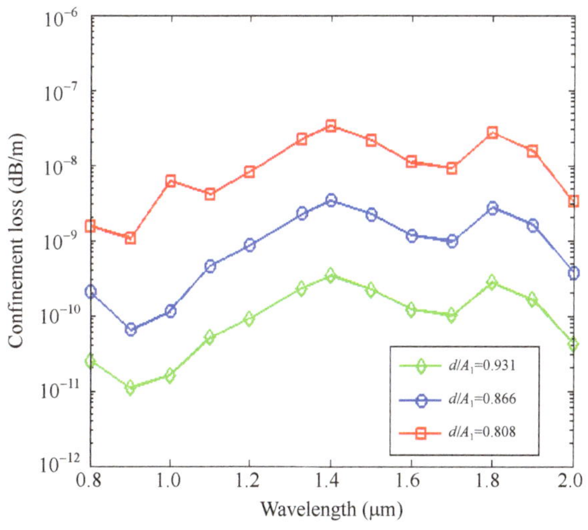

Fig.8 Comparison of confinement loss with cladding air filling ratio (d/Λ_1) variations for n = 1.354.

Figure 9 shows the inspection of wavelength verses the relative sensitivity, making –2%, –1%, +1%, and +2% variations with the optimum parameters. The simulation process goes down for a wider wavelength limit over 0.8 μm to 2.0 μm which does not make a great difference between the proposed P-OPCFs. By changing +2%, +1%, –1%, and –2% of overall parameters, the relative sensitivity as well as confinement loss is not affected greatly.

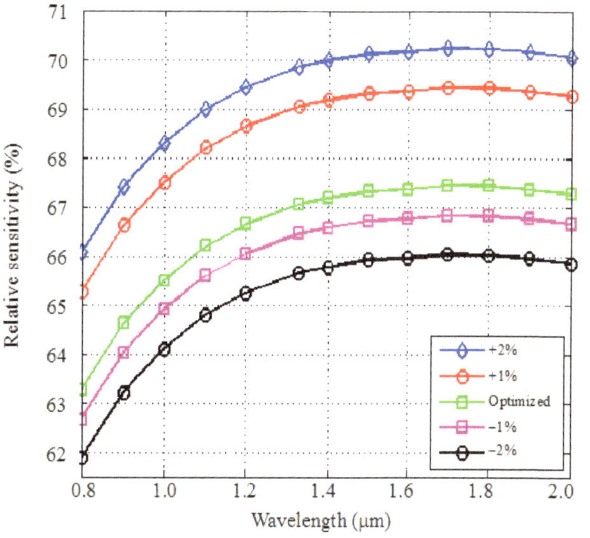

Fig. 9 Comparison of relative sensitivity with optimum structural parameters variations for n = 1.354.

On the other hand, we have examined that wavelength (λ) verses relative sensitivity increases in the wide range of wavelength and vice versa, which is more convenience for the fiber sensor fabrication. The main obstruction is to fabricate such a complex photonic crystal fiber sensor. But the massive achievement occurs in different technological sectors. By the progression of the technological achievement, numerous fabrication techniques such as stack and draw process [38], extrusion [39], drilling [40], sol gel casting [41], and die cast process [42] have been proposed. The stack and draw process is generally preferable for honey comb and triangular lattices. It has some limitations to construct the circular pattern. Different background materials are recently used to fabricate PCF. Some of these materials are fragile. For material fragility, the drilling technique is not allowed. Extrusion is another well-known fabrication technique for soft glass nevertheless wastage of material. For design flexibility and adaptive large scale manufacturing capability, the sol gel technique helps fabricate different micro-structure PCFs. So with the help of sol gel technique, the proposed P-OPCF can be fabricated easily without any damages.

5. Conclusions

A two-layer microstructure porous cored with octagonal cladding structure inclusion with perfectly match layer (PML) has been simulated using the finite element method. In the field of electromagnetic study, the finite element method has a leading impact on analyzing and calculating the complex geometry. This study of octagonal microstructure two-layer porous core has been proposed in favor of chemical sensing. The relative sensitivity and confinement loss have been analyzed by tuning different geometrical parameters of the proposed P-OPCF structure. The simulation is also employed to calculated sensitivity for two-, three-, and four-layer porous core. It shows that the proposed micro-cored two-layer structures reflect a higher level of sensitivity 67.66 % and a low confinement loss 3.28×10^{-11} dB/m at the 1.33 μm for ethanol (RI=1.354). In fiber optics communication, wavelength band O+S+C+L is extensively used. Bearing with the wavelength band, the proposed fiber optic sensor can be applied in the wide range of wavelength (0.80 μm to 2.0 μm). Hence, it is vivid, and we confide that our designed P-OPCF will be very efficient for sensing application in the area of optical systems.

Acknowledgment

The authors are grateful to all of the subjects who participated in this research.

References

[1] X. Sang, P. L. Chu, and C. Yu, "Applications of nonlinear effects in highly nonlinear photonic crystal fiber to optical communications," *Optical and Quantum Electronics*, 2005, 37(10): 965–994.

[2] K. P. Hansen, "Introduction to nonlinear photonic crystal fibers," *Journal of Optical and Fiber Communications Reports*, 2005, 2(3): 226–254.

[3] A. Ortigosa-Blanch, J. C. Knight, W. J. Wadsworth, J. Arriaga, B. J. Mangan, T. A. Birks, *et al.*, "Highly birefringent photonic crystal fibers," *Optics Letters*, 2000, 25(18): 1325–1327.

[4] T. P. Hansen, J. Broeng, S. E. Libori, E. Knudsen, A. Bjarklev, J. R. Jensen, *et al.*, "Highly birefringent index-guiding photonic crystal fibers," *IEEE Photonics Technology Letters*, 2001, 13(6): 588–590.

[5] G. An, S. Li, X. Yan, X. Zhang, Z. Yuan, and Y. Zhang, "High-sensitivity and tunable refractive index sensor based on dual-core photonic crystal fiber," *Journal of the Optical Society of America B*, 2016, 33(7): 1330–1334.

[6] W. Qianet, C. L. Zhao, S. He, X. Dong, S. Zhang, Z. Zhang, *et al.*, "High-sensitivity temperature sensor based on an alcohol-filled photonic crystal fiber loop mirror," *Optics Letters*, 2011, 36(9): 1548–1550.

[7] S. Olyaee and F. Taghipour, "Ultra-flattened dispersion hexagonal photonic crystal fiber with low confinement loss and large effective area," *IET Optoelectronics*, 2012, 6(2): 82–87.

[8] Aihan-Yin and L. Xiong, "Highly nonlinear with low confinement losses square photonic crystal fiber based on a four-hole unit," *Infrared Physics & Technology*, 2014, 66(9): 29–33.

[9] K. Kishor, R. K. Sinha, and A. D. Varshney, "Experimental verification of improved effective index method for endlessly single mode photonic crystal fiber," *Optics and Lasers in Engineering*, 2012, 50(2): 182–186.

[10] H. Ademgil and S. Haxha, "Endlessly single mode photonic crystal fiber with improved effective mode area," *Optics Communications*, 2012, 285(6): 1514–1518.

[11] T. A. Birks, J. C. Knight, and P. S. J. Russell, "Endlessly single-mode photonic crystal fiber," *Optics Letters*, 1997, 22(13): 961–963.

[12] M. Morshed, M. H. Imran, T. K. Roy, M. S. Uddin, and S. M. A. Razzak, "Microstructure core photonic crystal fiber for gas sensing applications," *Applied Optics*, 2015, 54(29): 8637–8643.

[13] M. Tonouchi, "Cutting-edge terahertz technology," *Nature Photonics*, 2007, 1(2): 97–105.

[14] J. M. Dudley and J. R. Taylor, "Ten years of nonlinear optics in photonic crystal fiber," *Nature Photonics*, 2009, 3(2): 85–90.

[15] N. Skivesen, A. Têtu, M. Kristensen, J. Kjems, L. H. Frandsen, and P. I. Borel, "Photonic-crystal waveguide biosensor," *Optics Express*, 2007, 15(6): 3169–3176.

[16] Y. L. Hoo, W. Jin, C. Shi, H. L. Ho, D. N. Wang, and S. C. Ruan, "Design and modeling of a photonic crystal fiber gas sensor," *Applied Optics*, 2003, 42(18): 3509–3515.

[17] A. M. Cubillaset, S. Unterkofler, T. G. Euser, B. J. Etzold, A. C. Jones, P. J. Sadler, *et al.*, "Photonic crystal fibers for chemical sensing and photochemistry," *Chemical Society Reviews*, 2013, 42(22): 8629–8648.

[18] S. Okaba, T. Takano, F. Benabid, T. Bradley, L. Vincetti, Z. Maizelis, *et al.*, "Lamb-Dicke spectroscopy of atoms in a hollow-core photonic crystal fiber," *Nature Communications*, 2014, 5(5): 4096–4105.

[19] P. C. Ashok, R. F. Marchington, P. Mthunzi, T. F. Krauss, and K. Dholakia, "Optical chromatography using a photonic crystal fiber with on-chip fluorescence excitation," *Optics Express*, 2010, 18(6): 6396–6407.

[20] S. Roy, "Fiber optic sensor for determining adulteration of petrol and diesel by kerosene," *Sensors and Actuators B: Chemical*, 1999, 55(2–3): 212–216.

[21] P. N. Prasad, *Introduction to biophotonics*. New York, United States: Wiley-Interscience, 2003.

[22] L. V. Doronina-Amitonova, V. F Il'ya, O. I. Ivashkina, M. A. Zots, A. B. Fedotov, K. V. Anokhin, *et al.*, "Photonic-crystal-fiber platform for multicolor multilabel neurophotonic studies," *Applied Physics Letters*, 2011, 98(25): 253706-1–253706-3.

[23] H. Shafiee, E. A. Lidstone, M. Jahangir, F. Inci, E. Hanhauser, T. J. Henrich, *et al.*, "Nanostructured optical photonic crystal biosensor for HIV viral load measurement," *Scientific Reports*, 2014, 4(6174): 1032–1035.

[24] S. C. Buswell, V. A. Wright, J. M. Buriak, V. Van, and S. Evoy, "Specific detection of proteins using photonic crystal waveguides," *Optics Express*, 2008, 16(20): 15949–15957.

[25] P. Sharma and P. Sharan, "Design of photonic crystal-based biosensor for detection of glucose concentration in urine," *IEEE Sensors Journal*, 2015, 15(2): 1035–1042.

[26] H. Ademgil, "Highly sensitive octagonal photonic crystal fiber based sensor," *Optik–International Journal for Light and Electron Optics*, 2014, 125(20): 6274–6278.

[27] J. N. Dash and R. Jha, "Graphene-based birefringent photonic crystal fiber sensor using surface plasmon resonance," *IEEE Photonics Technology Letters*, 2014, 26(11): 1092–1095.

[28] B. Temelkuran, S. D. Hart, G. Benoit, J. D. Joannopoulos, and Y. Fink, "Wavelength-scalable hollow optical fibers with large photonic bandgaps for CO_2 laser transmission," *Nature*, 2002, 420(6916): 650–653.

[29] S. Atakaramians, K. Cook, H. Ebendorff-Heidepriem, J. Canning, D. Abbott, and T. M. Monro, "Cleaving of extremely porous polymer fibers," *IEEE Photonics Journal*, 2009, 1(6): 286–292.

[30] P. Yeh, A. Yariv, and E. Marom, "Theory of Bragg fiber," *Journal of the Optical Society of America*, 1978, 68(9): 1196–1201.

[31] P. Russell, "Photonic crystal fibers," *Science*, 2003, 299(5605): 358–362.

[32] J. C. Knight, T. A. Birks, P. S. J. Russell, and D. M. Atkin, "All-silica single-mode optical fiber with photonic crystal cladding," *Optics Letters*, 1996, 21(19): 1547–1549.

[33] K. Ahmed and M. Morshed, "Design and numerical analysis of microstructured-core octagonal photonic crystal fiber for sensing applications," *Sensing and Bio-Sensing Research*, 2016, 7: 1–6.

[34] T. Rylander and J. M. Jin, "Perfectly matched layer for the time domain finite element method," *Journal of Computational Physics*, 2004, 200(1): 238–250.

[35] G. Ghosh, "Sellmeier coefficients and dispersion of thermo-optic coefficients for some optical glasses," *Applied Optics*, 1997, 36(7): 1540–1546.

[36] S. Asaduzzaman, K. Ahmed, M. F. H. Arif, and M. Morshed, "Application of microarray-core based modified photonic crystal fiber in chemical sensing," in *International Conference on Electrical & Electronic Engineering*, Bangladesh, Nov. 4–6, 2015.

[37] M. S. Habib, M. S. Habib, S. M. A. Razzak, and M. A. Hossain, "Proposal for highly birefringent broadband dispersion compensating octagonal photonic crystal fiber," *Optical Fiber Technology*, 2013, 19(5): 461–467.

[38] D. Pysz, I. Kujawa, R. Stępień, M. Klimczak, A. Filipkowski, M. Franczyk, *et al.*, "Stack and draw fabrication of soft glass microstructured fiber optics," *Bulletin of the Polish Academy of Sciences Technical Sciences*, 2014, 62(4): 667–682.

[39] H. Ebendorff-Heidepriem and T. M. Monro, "Extrusion of complex performs for microstructured optical fibers," *Optics Express*, 2007, 15(23): 15086–15092.

[40] S. Liu, L. Jin, W. Jin, D. Wang, C. Liao, and Y. Wang, "Structural long period gratings made by drilling micro-holes in photonic crystal fibers with a femtosecond infrared laser," *Optics Express*, 2010, 18(6): 5496–5503.

[41] H. H. El, Y. Ouerdane, L. Bigot, G. Bouwmans, B. Capoen, A. Boukenter, *et al.*, "Sol-gel derived ionic copper-doped microstructured optical fiber: a potential selective ultraviolet radiation dosimeter," *Optics Express*, 2012, 20(28): 29751–29760.

[42] G. Y. Zhou, Z. Y. Hou, S. G. Li, and L. T. Hou, "Fabrication of glass photonic crystal fibers with a die-cast process," *Applied Optics*, 2006, 45(18): 4433–4436.

18

Electromagnetic Resonant Properties of Metal-Dielectric-Metal (MDM) Cylindrical Microcavities

Hang HENG[1*] and Rong WANG[2,3]

[1]*Center for Analysis and Testing, Nanjing Normal University, Nanjing, 210097, China*

[2]*Department of Neurosurgery, Nanjing Drum Tower Hospital, Nanjing, 210008, China*

[3]*The Affiliated Hospital of Nanjing University Medical School, Nanjing, 210008, China*

[*]Corresponding author: Hang HENG E-mail: 3079576@qq.com

Abstract: Optical metamaterials can concentrate light into extremely tiny volumes to enhance their interaction with quantum objects. In this paper, a cylindrical microcavity based on the Au-dielectric-Au sandwiched structure is proposed. Numerical study shows that the cylindrical microcavity has the strong ability of localizing light and confining 10^3- – 10^4-fold enhancement of the electromagnetic energy density, which contains the most energy of the incoming light. The enhancement factor of energy density G inside the cavity shows the regularities as the change in the thickness of the dielectric slab, dielectric constant, and the radius of gold disk. At the normal incidence of electromagnetic radiation, the obtained reflection spectra operate in the range from 4.8 μm to 6 μm and with the absorption efficiency C (C=1–$R_{\min}$), which can reach 99% by optimizing the structure's geometry parameters, and the dielectric constant. Due to the symmetry of the cylindrical microcavities, this structure is insensitive to the polarization of the incident wave. The proposed optical metamaterials will have potential applications in the surface enhanced spectroscopy, new plasmonic detectors, bio-sensing, solar cells, etc.

Keywords: Microcavity; metal-semiconductor-metal; metamaterial

1. Introduction

Recently, surface plasmon polaritons (SPPs) and localized surface plasmon polaritons (LSPs) are widely researched in the basis theory of optical metamaterials [1, 2]. Metal-dielectric photonic band gap structures [3, 4] are usually used to excite SPPs and LSPs. They have many extraordinary optical properties, such as surface enhanced spectroscopy [5, 6], enhanced transmission [7–9], sensitive bio-sensing [10, 11], negative refractive index [12–14], double negative materials [15], and high resolution imaging [16, 17]. The metallic gratings consisting of narrow slits [18] may become transparent for extremely broad bandwidths under the oblique incidence. In addition, ultrathin, broadband, highly efficient [19, 20], and freely tunable [21] metamaterial-based polarization converters have been realized. In the visible and infrared range, as the wavelength increases, the penetration depth into the dielectric layer increases. At the same time, the ohmic loss of metal layers reduces reversely [22]. In particular, the ohmic loss is very low in the far infrared region, where it has a perfect reflection characteristic. Because of these features, we can break the diffraction limitation and confine light within a sub-wavelength volume [23]. So far, confining light within a sub-wavelength volume has become one of the hot issues in the optical metamaterial research. In the terahertz region, dimensions can be reduced to extreme subwavelength [23, 24]. Furthermore, Fevillet-Palma *et al.* [25] studied the near-field enhancement properties of the antenna in the terahertz region and originally calculated the electromagnetic energy density in the cavities. However, the cuboid microcavities have the lower absorption efficiency, and this cuboids cavity is sensitive to the polarization of the incident wave.

In this work, we first proposed metal-dielectric-metal (MDM) cylindrical microcavities and numerically calculated the electromagnetic energy density in the cavities and the concentration mechanism and the absorption properties of the cylindrical microcavities through the mathematical model by high frequency structure simulator (HFSS, commercial software). Compared with the literature [25], we got a higher absorption efficiency *C*. Because the gold disk is of circular symmetry, the structure was insensitive to the polarization of the incident light. The proposed model and calculation method will play a guidance role in the optimization of substrates for the surface enhanced spectroscopy.

2. Simulation and structures

The designed MDM structure is simulated by HFSS, a finite element method based program. The MDM cylindrical cavities are periodically arranged in a square lattice in the *X*-*Y* plane, as shown in Fig. 1(a). Figure 1(b) is the schematic of a unit cell of the cylindrical cavity array. A dielectric layer is sandwiched between two metallic layers: the top layer is a gold disk, and the bottom layer is a gold ground layer. The deep black area between the gold disk and gold ground layer in Fig. 1(b) is the cylindrical microcavity formed between a gold disk

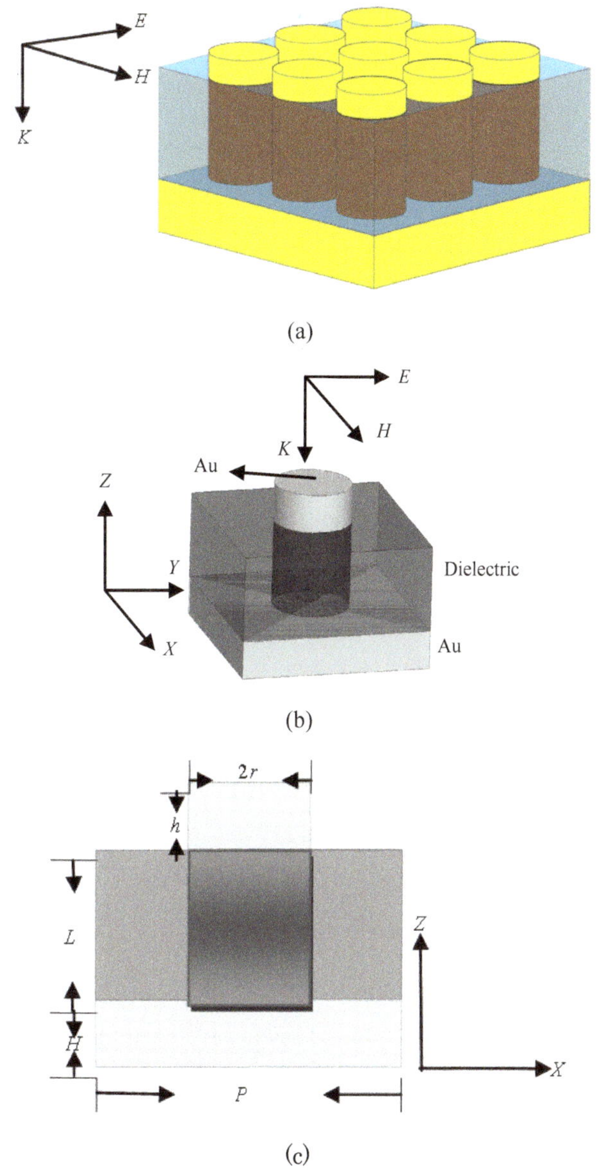

Fig. 1 Schematic of the designed MDM structure: (a) schematic of the cylindrical cavity array, (b) schematic geometry of a unit cell of the cylindrical microcavity at the normal incident TM plane wave (the top is a gold disk, the middle is a dielectric layer, the bottom is a gold plane, and the deep black area between the gold disk and gold ground layer is the cylindrical microcavity), and (c) cross section view of the cylindrical microcavity in the *Y*-*Z* plane (*r*, *h*, *L*, and *P* represent the gold disk radius, thickness of the gold disk, height of the microcavity, and period, respectively).

and a gold ground layer. If the thickness *L* of the dielectric layer is smaller than the wavelength, such a structure supports a TM_{100} mode [25]. Figure 1(c) shows the cross section of the cavity in the *Y*-*Z* plane, and the structural parameters are shown in the figure. The thickness of the dielectric layer is chosen as *L*, and *h* and *r* are defined as the thickness and radius of the gold disk, respectively. In this paper, the thickness of the gold ground layer is chosen as 0.45 μm. The gold ground layer has two significant functions. One is to serve as an optical mirror. As the thickness is greater than the penetration depth of the incident light, the transmittance is close to zero. The other is to couple with the gold disk and gold ground layer to create electric and magnetic dipoles which extremely concentrate electromagnetic energy into the microcavities. Period P is 3 μm. The orientations of the electric field (E), the magnetic field (H), and the wave vector (K) are shown in Fig. 1(b). The incident angle is 0. The boundary conditions are set as the perfect electric conductor and perfect magnetic conductor on two-pair faces. The dielectric permittivity of gold is modeled by using the Drude model [26]

$$\varepsilon = 1 - \frac{\omega_P^2}{\omega^2 + i\omega\gamma} \quad (1)$$

with the plasma frequency ω_P=1.37×10^{16} rad/s and scattering frequency γ =4.08×10^{13} rad/s.

3. Results and discussion

In order to study the resonant characteristics of the cylindrical microcavity, we calculate the reflectivity spectra $R(\omega)=|S_{11}(\omega)|^2$ at the normal incident TM plane wave. As shown in Fig. 2, the resonant cavity mode appears as a reflection dip in the spectra. Moreover, the grating period is kept subwavelength, so that the only contribution to the reflected signal is the specular reflection (0th diffraction order) [25]. We also calculate the transmission of the structure shown in Fig. 1 and obtain near zero transmission for all geometry parameters (the transmission spectra are not given here). According to the equation

$$T + R + C = 1 \quad (2)$$

where T, R, and C are the transmission, reflection, and absorption, respectively. Due to T = 0, the minimum of reflection appears as the maximum of absorption. If we define C=1–R_{min}, where R_{min} is the minimum reflectivity at the resonant frequency, a larger C indicates a stronger absorption efficiency.

3.1 Absorption of the configuration

Figure 2(a) is the reflectivity spectra of the cylindrical microcavity in the MDM structure at the normal incident with variable dielectric constant ε, here L=1 μm, h=0.45 μm, H=0.45 μm, and P=3 μm. When ε changes from 10 to 14 (with a step of 1), the resonant frequency is around 60.3 THz, 57.5 THz, 54.95 THz, 52.9 THz, and 51 THz, and C is 85%, 88%, 97%, 99%, and 97%, respectively. It can be seen that the reflection peaks show red-shift with an increase in the dielectric constant, and this means that the greater the dielectric constant is, the greater the near-field coupling is, which reaches an agreement with that in [27]. This is because the change in the effective dielectric permittivity [ε, (ω)] of the system results in a shift of the electric resonance frequency. And the rate of red-shift is proportional to the dielectric constant. Figure 2(b) shows the reflectivity spectra as a function of L. As shown in Fig. 2(b), the resonant frequency mainly locates in the range of (53.7 ± 0.75) THz, whose dependence on L is relatively weak [27]. In contrast, the dielectric constant ε has much more greater impacts on the electric dipole resonances due to the excitement of the external electric field than L. At L = 0.98 μm, C is 96%. When L increases to 0.99 μm, C reaches the maximum 99%. A further increase in L will reduce C. At L=1.00 μm, C declines to 98%. When L changes from 1.01 μm to 1.03 μm, the absorption efficiency inside the cavity continues to fall. The results show that the absorption efficiency inside the cavity is closely related to L. Figure 2(c) presents the reflectivity spectra as a function of r, here L=1 μm, h=0.45 μm, H=0.45 μm, and P=3 μm. When r increases from 0.4 μm to 0.9 μm (with a step of 0.1 μm), the resonant frequency locates at 56.255 THz, 54.255 THz, 53.75 THz, 53.355 THz, 53.25 THz, and 53.15 THz, respectively and C equals to 71%, 89%, 99%, 88%, 69%, and 50%, respectively. As shown in Fig. 2(c), the resonant peaks show red-shift with an increase in the radius, which has a similar trend as Fig. 2(a), and this near-field coupling phenomenon was reported in [25]. The results also show that the absorption efficiency inside the cavity is sensitive to the radius of the gold disk.

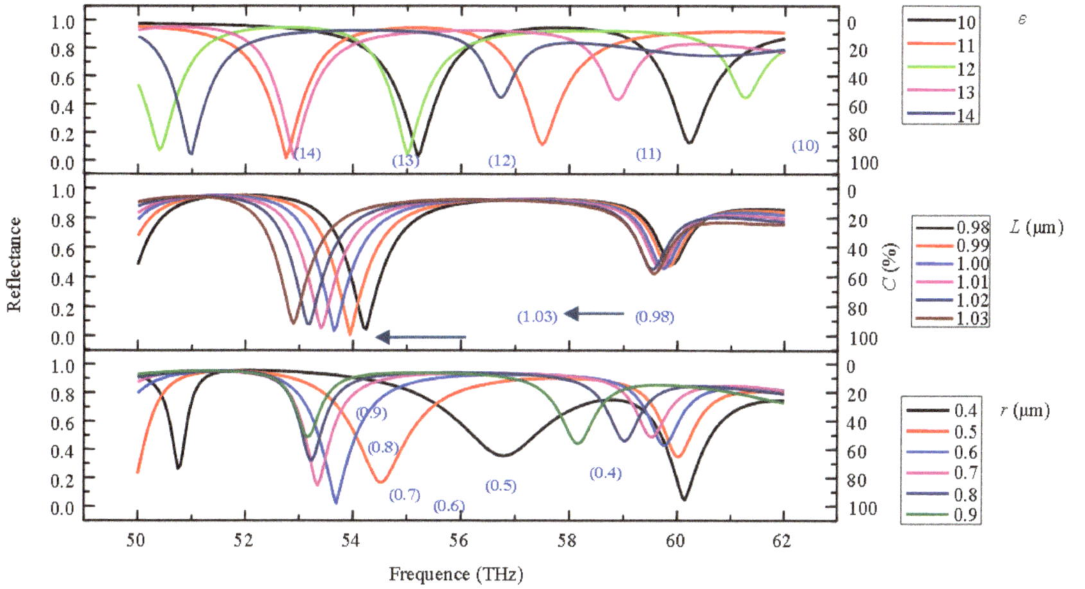

Fig. 2 Reflection spectra of the cylindrical microcavity in the MDM structure at the normal incident TM plane wave: (a) reflectivity spectra with the change in the dielectric constant ε, L=1 μm, h=0.45 μm, H=0.45 μm, and P=3 μm, (b) reflection spectra with the change in L, r=0.6 μm, H=0.45 μm, and P=3 μm, and (c) reflectivity spectra with the change in r, L=1 μm, h=0.45 μm, H=0.45 μm, and P=3 μm.

3.2 Localizing features

As reported in [25], the energy of the incidence wave is dissipated by both the metallic and the dielectric losses, and the large part of the energy can been localized in the dielectric microcavity. In order to study the regularity of the energy density in the condition of different parameters, we calculate the energy density of the electromagnetic field inside the cylindrical cavity with the change in the dielectric constant, the thickness of the dielectric slab, and the radius of the gold disk. Because of the energy conservation, the peak electrical energy is equal to the average total energy inside the cavity, and the solution is achieved by integrating $E\cdot E^*$ within the volume

$$E_d = \int_V \frac{E\cdot E^*}{V} dv \tag{3}$$

where V is the volume of the cavity, E_d is the energy density inside the cavity, and E represents the electric field vector inside the cavity. Firstly, we calculate the energy density E_{d1} inside the dielectric cavity in the MDM structure. Then, we calculate the energy density E_{d2} inside the vacuum cavity without the MDM structure, whose shape and size are the same as the dielectric cavity. We define $G= E_{d1} / E_{d2}$ as the enhancement factor of the energy density. A cavity with too big or small volume will affect the electromagnetic (EM) energy absorption and confinement inside the cavity [23]. Only when the parameters of the material and geometry are optimized, the impedance of the structure matching to the impedance of vacuum will the extreme confinement of the EM field and strong absorption inside the cavities appear. Figure 3(a) shows the enhancement factor of the energy density G as a function of ε. It can be seen that the enhancement factor G increases with an increase in the dielectric constant ε. Because the near-field coupling between the top and bottom metal layers become stronger [27], the ability of localizing the light inside the dielectric cavities is strengthened as shown in Fig. 3(a). Figure 3(b) shows the energy density ratio G as a function of the thickness of the dielectric slab L.

As shown in Fig. 3(b), when L changes from 0.98 μm to 0.99 μm, G shows the trend of growth. When L increases to 0.99 μm, G reaches the maximum 4180. A further increase in L will reduce G. At L=1.03 μm, G declines to 2945. Figure 3(b) reveals very important field spreading in the plane of the metallic strip [23] when L changes from 0.99 μm to 0.98 μm. So the enhancement factor G is sensitive to L, and the most optimal G should also be related to the effective dielectric permittivity [$\varepsilon(\omega)$] of the system. Figures 4(a) and 4(b) present the energy density ratio G and absorption efficiency C as a function of the radius r of the gold disk, respectively. From Fig. 4(a), we can see that G shows the trend of increase first and when r = 0.5 μm, G reaches the maximum 3314. Afterwards, G begins to decrease. As shown in Fig. 4(b), their variation trend is the same as that shown in Fig. 4(a). Their difference is that C reaches the maximum 97% when r = 0.6 μm. It can be seen that the maximum of C does not indicate the maximum of the energy density ratio. To explain this phenomenon, we use the following model (4):

$$\frac{\varepsilon|E_{Z0}|^2}{|E_{\text{in}}|^2}=\frac{\lambda_{\text{res}}\Sigma}{2\pi V}Q_{\text{ohm}}C\cos\theta \tag{4}$$

where ε is the dielectric constant of the middle layer, $V=\pi r^2L$ is the volume of the cylindrical resonator, E_{in} is the electric field amplitude of the incoming wave, E_{Z0} is the amplitude of the mode resonantly excited in the microcavity, and θ is the incident angle. For the data reported here, we have $\cos\theta = 1$, $\Sigma = P^2$, and Q_{ohm} is the quality factor Q of the Ohmic loss. When r = 0.6 μm, C = 0.97, G = 3232, r= 0.5 μm, and R = 3314, C is 0.85. Because Q_{ohm} approximately equals in the bands studied here, we can get $\frac{\varepsilon|E_{Z0}|^2}{|E_{\text{in}}|^2}=\lambda_{\text{res}}C/V$, indicating that the energy density ratio inside the cavity is decided by $\lambda_{\text{res}}C/V$. At r = 0.6 μm, its $\lambda_{\text{res}}C$ is greater than that at r = 0.5 μm. As the volume plays the most important role in R, R reaches the maximum (3314) at r = 0.5 μm.

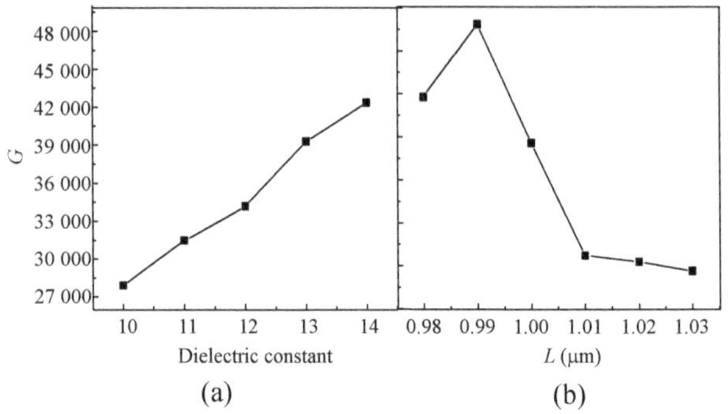

Fig. 3 Energy density ratio G as the dielectric constant ε and the thickness L of the dielectric slab: (a) energy density ratio G as a function of the dielectric constant ε and (b) energy density ratio G as a function of the thickness L of the dielectric slab.

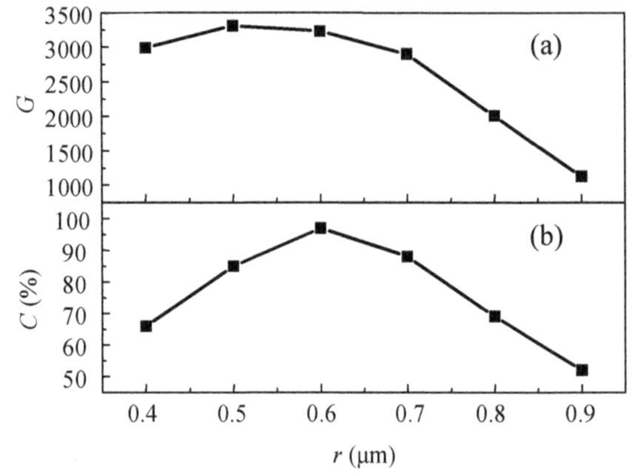

Fig. 4 Energy density ratio G and absorption efficiency C as a function of the radius r of the gold disk: (a) energy density ratio G as a function of the radius r of the gold disk and (b) absorption efficiency C as a function of the radius r of the gold disk.

3.3 Physical mechanism

To explore the physical mechanism of the extreme confinement and perfect absorption inside the cavity, the distributions of the EM field intensity and current at the resonant mode (f = 53.7 THz) are simulated, as shown in Fig. 5. Figure 5(a) shows the amplitude of the electric field in the middle of the dielectric layer in the X-Y plane. The gray circle represents the cylindrical microcavity. A standing wave pattern is clearly shown in Fig. 5(a). The amplitude of the electric field can be approximated by the expression

$$|E|\sim\cos\frac{\pi X}{\Delta x}(0\le X<r). \tag{5}$$

At r = 0, $|E|$ reaches its maximum, that is, the energy is strongly localized around the center of the cylindrical microcavity. Figure 5(b) presents the amplitude of the electric field in the middle of the dielectric layer in the Y-Z plane. In this figure, the white area represents the metal, the middle layer is dielectric, and the EM field is highly localized between the two metal layers, which can be approximated as a standing wave too. The confining mechanism, which arises from the impedance mismatch between the metal-metal and single metal regions [24], is responsible for the formation of the standing wave patterns. The perfect absorbing

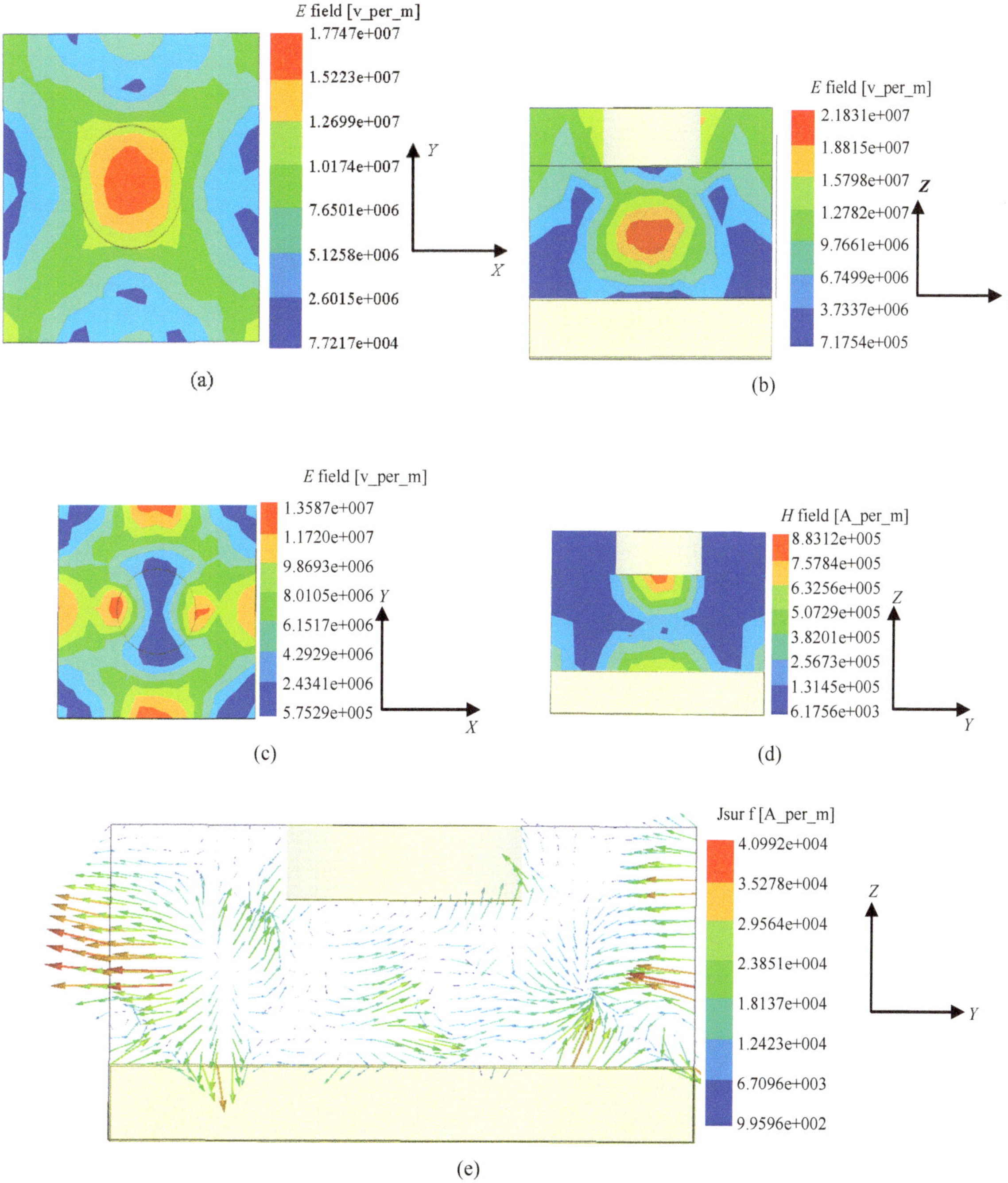

Fig. 5 Distributions of the EM field intensity and current within cylindrical cavities at the normal incident TM plane wave: (a) amplitude of the electric field in the middle of the dielectric layer in the X-Y plane at f = 53.7 THz, (b) amplitude of the electric field in the middle of the dielectric layer in the Y-Z plane at f=53.7 THz, (c) amplitude of the electric field on the top surface of the dielectric layer at f = 53.7 THz, (d) amplitude of magnetic field in the middle of the dielectric layer in the Y-Z plane at f = 53.7 THz, and (e) distribution of the current in the middle of the dielectric layer in the Y-Z plane at f = 53.7 THz.

characteristics may be explained by Figs. 5(c)–5(e). Figure 5(c) shows the amplitude of the electric field on the top surface of the dielectric layer, and here the gray circle represents the position of the gold disk. As indicated in Fig. 5(c), the distribution of the electric field mainly concentrates around the gold disk, forming a square area. However, the amplitude of the electric field in the center of the gold disk is weaker than that in the surrounding. Meanwhile, it can be seen clearly that six "hot spots" symmetrically locate in the four edges of the square area and both sides of the gold disk in the *Y*-axis direction. In this case, the gold disk can be regarded as the electric dipole resonator by the external electric field excitation. A change in the disk diameter will result in a shift of the electric resonance frequency and hence result in a change in the effective dielectric permittivity [$\varepsilon(\omega)$] of the system. Figure 5(d) shows the distribution of the current in the middle of the dielectric layer in the *Y-Z* plane. The white areas in this figure are the two metal layers. The magnetic field distribution shown in Fig. 5(d) clearly indicates the localization of the magnetic field between two metal layers, and the formation of the magnetic dipole resonance. A change in the dielectric layer thickness mainly affects the magnetic resonance and the effective magnetic permeability [$\mu(\omega)$] of the system. By optimizing the parameters of the top gold layer and dielectric layer thickness, we can tune ω of the ε and μ resonances. When we achieve $\varepsilon=\mu$ $\left(Z=\sqrt{\frac{\varepsilon(\omega)}{\mu(\omega)}}=1\right)$, an impedance near the free space value is formed. With strongly coupling the incident radiation to the resonant structure and perfect absorbance being theoretically possible, it can only occur when the metal-metal layer is impedance-matched to the free space [28]. The distribution of the current in the middle of the dielectric layer in the *Y-Z* plane is shown in Fig. 5(e), here the arrows in the dielectric cavity represent the magnitude and direction of the current. Figure 5(e) clearly indicates that the currents on the disk and the ground plane are opposite, forming a circulating current known as a magnetic resonance, which results in an artificial magnetic moment that interacts strongly with the magnetic field of the incident light. If the electric and magnetic dipole resonances occur at the same frequency, then there will be a strong localization of the electromagnetic energy inside the cylindrical microcavity. To realize the high confinement and perfect absorption, the optical metamaterials are usually engineered by simultaneously minimizing the reflectance with perfect impedance matching and eliminating the transmittance by maximizing material losses.

4. Conclusions

We have numerically studied the electromagnetic energy density and absorption efficiency inside the MDM cylindrical microcavities in the mid-infrared range. As the change in the thickness of the dielectric slab, dielectric constant, and the radius of gold disk, *G* and *C* present some important regularities for us to modulate the resonance frequency and strengthen electromagnetic energy density. The enhancement factor *G* increases with an increase in the dielectric constant ε. When *L* changes from 0.98 μm to 1.03 μm, *G* increases first and then decrease. When L = 0.99 μm, *G* reaches the maximum 4180. At L = 1 μm, H = 0.45 μm, h = 0.45 μm, and r = 0.6 μm, *C* is 0.97, and *R* is 3232. When r = 0.5 μm with other conditions unchanging, *C* is 0.85, and *R* is 3314. It can be seen that the maximum of *C* does not indicate the maximum of the energy density ratio. This kind of strong absorption and high localization makes the proposed structures have broad applications in the areas such as the surface enhanced spectroscopy, new plasmonic detectors, bio-sensing, and solar cells.

Acknowledgment

The authors gratefully acknowledge the financial

support provided to this study by the Program of Natural Science Research of Jiangsu Higher Education Institutions of China (Grant No. 14KJB140005).

References

[1] G. Q. Liu, F. L. Tang, L. Li, L. X. Gong, and Z. Q. Ye, "Concentration detection of quantum dots in the visible and near-infrared range based on surface plasmon resonance sensor," *Materials Letters*, 2011, 65(12): 1998–2000.

[2] Y. Y. Yang, Y. L. Zhang, F. Jin, X. Z. Dong, and X. M. Duan, "Steering the optical response with asymmetric bowtie 2-color controllers in the visible and near infrared range," *Optics Communications*, 2011, 284(13): 3474–3478.

[3] D. Owens, C. Fuentes-Hernandez, and B. Kippelen, "Optical properties of one-dimensional metal-dielectric photonic band-gap structures with low index dielectrics," *Thin Solid Films*, 2009, 517(8): 2736–2741.

[4] J. Y. Seo, S. Cho, H. Lim, and S. Lee, "Optical and structural properties of metal-dielectric photonic band gap structures," *Current Applied Physics*, 2006, 6(6): 62–66.

[5] P. Z. El-Khoury, E. J. Bylaska, and W. P. Hess, "Time domain simulations of chemical bonding effects in surface-enhanced spectroscopy," *Journal of Chemical Physics*, 2013, 139(17): 174303-1–174303-5.

[6] C. L. Du, C. J. Du, Y. M. You, Y. Zhu, S. L. Jin, C. J. He, *et al.*, "Numerically investigating the enhanced Raman scattering performance of individual Ag nanowire tips," *Applied Optics*, 2011, 50(25): 4922–4926.

[7] L. Zhu, L. Dong, F. Y. Meng, J. H. Fu, and Q. Wu, "Influence of symmetry breaking in a planar metamaterial on transparency effect and sensing application," *Applied Optics*, 2012, 51(32): 7794–7799.

[8] T. W. Ebbesen, H. J. Lezec, H. F. Ghaemi, T. Thio, and P. A. Wolff, "Extraordinary optical transmission through sub-wavelength hole arrays," *Nature*, 2001, 86(6): 1114–1117.

[9] H. F. Ghaemi, T. Thio, and D. E. Grupp, "Surface plasmons enhance optical transmission through subwavelength holes," *Physical Review B*, 1998, 58(11): 357–368.

[10] A. J. Haes, S. L. Zou, G. C. Schatz, and R. P. Van Duyne, "Nanoscale optical biosensor: short range distance dependence of the localized surface plasmon resonance of noble metal nanoparticles," *Journal of Physical Chemistry B*, 2003, 108(22): 6961–6968.

[11] A. V. Kabashin, P. Sergiy, and A. N. Grigorenko, "Phase and amplitude sensitivities in surface plasmon resonance bio and chemical sensing," *Optics Express*, 2009, 17(23): 21191–21204.

[12] R. A. Shelby, D. R. Smith, and S. Schultz, "Experimental verification of a negative index of refraction," *Science*, 2001, 292(292): 77–79.

[13] J. Wu, B. Ng, S. P. Turaga, M. B. H. Breese, S. A. Maier, M. Hong, *et al.*, "Free-standing terahertz chiral meta-foils exhibiting strong optical activity and negative refractive index," *Applied Physics Letters*, 2013, 103(14): 141106-1–141106-4.

[14] G. Dolling, M. Wegener, C. M. Soukoulis, and S. Linden, "Negative-index metamaterial at 780 nm wavelength," *Optics Letters*, 2007, 32(1): 53–55.

[15] F. L. Zhao, B. A. Kamil, C. Evrim, and O. Ekmel, "Complementary chiral metamaterials with giant optical activity and negative refractive index," *Applied Physics Letters*, 2011, 98(16): 161907-1–161907-3.

[16] R. Tomer, L. Ye, B. Hsueh, and K. Deisseroth, "Advanced clarity for rapid and high-resolution imaging of intact tissues," *Nature Protocols*, 2014, 9(7): 1682–1697.

[17] T. Taubner, D. Korobkin, Y. Urzhumov, G. Shvets, and R. Hillenbrand, "Near-field microscopy through a SiC superlens," *Science*, 2006, 313(5793): 1595–1595.

[18] F. I. Baida, M. Boutria, R. Oussaid, and D. Van Labeke, "Enhanced-transmission metamaterials as anisotropic plates," *Physical Review B*, 2011, 82(3): 2109–2119.

[19] X. R. Huang, R. W. Peng, and R. H. Fan, "Making metals transparent for white light by spoof surface plasmons," *Physical Review Letters*, 2010, 105(24): 119–127.

[20] R. H. Fan, R.W. Peng, X. R. Huang, J. Li, Y. Liu, Q. Hu, *et al.*, "Transparent metalsfor ultrabroadband electromagnetic waves," *Advanced Materials*, 2012, 24(15): 1980–1986.

[21] N. K. Grady, J. E. Heyes, D. R. Chowdhury, Y. Zeng, M. T. Reiten, A. K. Azad, *et al.*, "Terahertz metamaterials for linear polarization conversion and anomalous refraction," *Science*, 2013, 340(6138): 1304–1307.

[22] I. Kogelbauer, E. Heine, C. D'Amboise, C. Müllebner, W. Sokol, and W. Loiskandl, "Adaptation of soil physical measurement techniques for the delineation of mud and lakebed sediments at neusiedler see," *Sensors*, 2013, 13(12): 17067–17083.

[23] C. Fevillet-Palma, Y. Todorov, R. Steed, A. Vasanelli, G. Biasiol, L. Sorba, *et al.*, "Extremely sub-wavelength THz metal-dielectric wire

microcavities," *Optics Express*, 2012, 20(27): 29121–29130.

[24] Y. Todorov, L. Tosetto, J. Teissier, A. M. Andrews, P. Klang, R. Colombelli, *et al.*, "Optical properties of metal-dielectric-metal microcavities in the THz frequency range," *Optics Express*, 2010, 18(13): 13886–13907.

[25] C. Fevillet-Palma, Y. Todorow, A. Vasanelli, and C. Sirtori, "Strong near field enhancement in THz nano-antenna arrays," *Sientific Reports*, 2013, 3(1): 299–308.

[26] X. D. Wang, Y. H. Ye, C. Zhang, Y. Qin, and T. J. Cui, "Tunable figure of merit for a negative-index metamaterial with a sandwich configuration," *Optics Letters*, 2009, 34(22): 3568–3570.

[27] K. Chen, Q. Y. Wen, and H. B. Znang, "Study on the broadband terahertz metamaterial absorber," *Electronic Components and Materials*, 2011, 30(7): 56–59.

[28] N. I. Landy, S. Sajuyigbe, J. J. Mock, D. R. Smith, and W. J. Padilla, "Perfect metamaterial absorber," *Physical Review Letters*, 2008, 100(20): 1586–1594.

19

Design and Optimization of the CFRP Mirror Components

Lei WEI[1,2], Lei ZHANG[2,3*], and Xiaoxue GONG[1,2]

[1]*University of Chinese Academy of Sciences, Beijing, 100039, China*

[2]*Changchun Institute of Optics, Fine Mechanics and Physics, Chinese Academy of Sciences, Changchun, 130033, China*

[3]*Chang Guang Satellite Technology LTD.Co, Changchun, 130000, China*

[*]Corresponding author: Lei ZHANG E-mail: 18686344285@163.com

Abstract: As carbon fiber reinforced polymer (CFRP) material has been developed and demonstrated as an effective material in lightweight telescope reflector manufacturing recently, the authors of this article have extended to apply this material on the lightweight space camera mirror design and fabrication. By CFRP composite laminate design and optimization using finite element method (FEM) analysis, a spherical mirror with φ316 mm diameter whose core cell reinforcement is an isogrid configuration is fabricated. Compared with traditional ways of applying ultra-low-expansion glass (ULE) on the CFRP mirror surface, the method of nickel electroplating on the surface effectively reduces the processing cost and difficulty of the CFRP mirror. Through the FEM analysis, the first order resonance frequency of the CFRP mirror components reaches up to 652.3 Hz. Under gravity affection coupling with +5℃ temperature rising, the mirror surface shape root-mean-square values (RMS) at the optical axis horizontal state is 5.74 nm, which meets mechanical and optical requirements of the mirror components on space camera.

Keywords: Pure-CFRP mirror; lay-up design; bipod; integrated optimization

1. Introduction

With the development of space technology, high resolution, wide field of view, and ultra lightweight are heavily required for space camera nowadays. But the mass of the primary mirror used for the space camera is increased as a sacrifice to meet the requirement. Therefore, the lightweight research on primary mirror with high optical performance has become an important issue worldwide [1].

Carbon fiber composite materials (CFRP) have the characteristics of high stiffness, low thermal expansion coefficient, and changeable material properties, which make it a possible excellent material for the space camera reflector fabrication. Due to the development of the non-contact ion beam optical fabrication method, CFRP mirror processing is no longer a challenge. Researchers from Europe and the United States have already studied on the lightweight of the CFRP mirror for many years [2–9]. Composite Optics, Inc. (COI) designed and manufactured a 2-m caliber CFRP primary mirror which works at a far-infrared wave band. Chen *et al.* proposed the process method using space-flight-qualified graphite-fiber-reinforced cyanate ester composite materials to print mirror surface through at the mould surface and made a tremendous breakthrough in achieving the replicas with areal density 2 kg/m at 42-cm aperture by vacuum drying and ion milling [10]. Romeo *et al.* developed an ULTRA telescope with a 1-m caliber primary mirror [11–13].

Researches on CFRP reflector in China started relatively later than several other countries, but they are making a rapid progress. There have been several related institutions working on the CFRP reflector for several decades. For example, Peng *et al.* from The Shanghai Institute of Technical Physics of the Chinese Academy of Sciences have studied in mirror made of carbon fiber honeycomb sandwich panel [14]. Zhao *et al.* from Xi'an Institute of Optics and Precision Mechanics applied actuators to change the curvature of the mirror surface and produced the CFRP surface-changeable reflector [15].

The CFRP reflect mirror mainly adopts two structure forms now. One form applies a sandwich structure with carbon fiber material as the top and bottom panel and honeycomb material as the intermediate support [14, 16, 17]. The other form uses all carbon fiber not only as the top and bottom but also as the intermediate support [11–13, 18]. Compared with the 1th form, the 2nd form has better thermal suitability and higher stiffness. Therefore, this structure form satisfies space application demands better. However, great limit on the application of pure carbon fiber mirror appears because how to smooth the rough CFRP material surface is still a challenge worldwide. Due to the limitation of current technology, it's impossible to process the CFRP mirror surface at a precise optical standard. A traditional solution to this problem is to cover ultra-low-expansion glass (ULE) chip on the surface of the CFRP mirror body, but the processing of the ULE chip is difficult and costly. This article comes up with a new solution to this problem-electroplating nickel on the surface of CFRP mirror body instead of covering the ULE chip, which is also a useful way to solve the problem. Besides, an optimal design is also given in this paper.

2. Optimized design of the reflector

The basic requirements for optical material are low thermal coefficient of expansion, isotropic property, and high specific stiffness. However, carbon fiber represents a complicated anisotropic property, leading to variation in the elastic modulus and thermal coefficient of expansion as ply-orientation changes. In order to apply carbon fiber in optical mirrors, coupling the ULE mirror surface on honeycomb sandwich structure mirror substrate is proposed as a common method [14]. The CFRP mirror manufactured in this way can be described in Fig. 1. However, this paper adopts a different pure carbon fiber grid support reinforcement structure form instead of honeycomb sandwich structure to manufacture the CFRP mirror. The preparation processes of pure carbon fiber mirror which also take the optical material demands into account are shown in Fig. 2. The reflect mirror studied is a φ316-mm caliber spherical CFRP mirror with 1580-mm spherical radius.

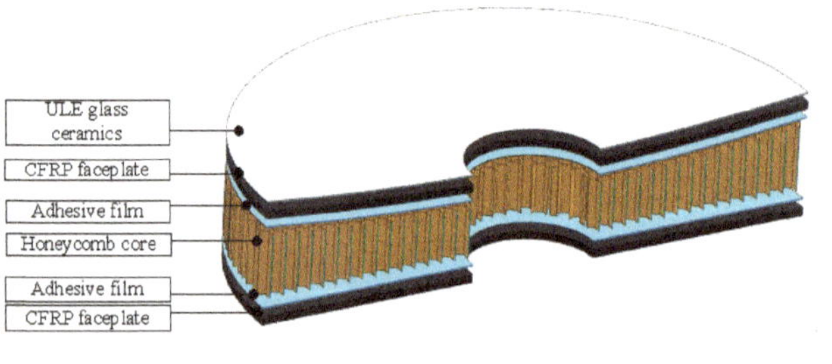

Fig. 1 Component of the CFRP reflector.

2.1 Mirror design

To meet with the high-specific-stiffness, isotropic, and low-thermal-expansion-coefficient characteristics of optical material, the plies used to form CFRP mirror surface should be specially laid up with high modulus fiber to ensure the quasic-isotropic property, and the resin which is used to adhere plies should maintain low thermal coefficient of expansion. Besides, low shrinkage rate, low solidified rate, and process stability and affinity to carbon fiber should also be taken into consideration for the resin. Above all, taking existing resources into account, this paper selects M40 fiber and epoxy resin as the CFRP mirror processing material.

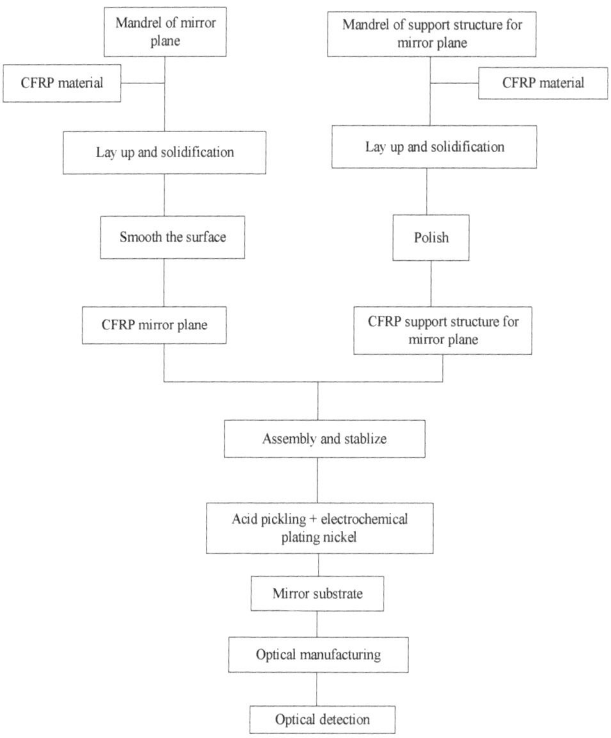

Fig. 2 Preparation processes of the CFRP reflector.

2.1.1 Lay-up design

The lay-up design of the mirror surface part is the key work of the whole mirror design. The quality of lay-up determines the quality of the mirror. A lay-up usually consists of several plies. The ply-orientations, the adhesive order of plies with different orientations, proportion of each orientation, and the whole ply number affect the quality of lay-up. The theoretical formulas which are always used to design the plies are described as follows:

Longitudinal tensile elastic modulus:

$$E_1 = E_t V + E_m(1-V_t) \tag{1}$$

lateral tensile elastic modulus:

$$E_2 = \frac{E_f E_m (V_f + \eta_2 V_m)}{E_m V_f + E_f V_m \eta_2} \tag{2}$$

longitudinal Poisson's ratio:

$$v_1 = v_f V_f + V_m(1-V_f) \tag{3}$$

lateral Poisson's ratio:

$$v_2 = v_1 \frac{E_1}{E_2} \tag{4}$$

shear modulus:

$$G_{12} = \frac{G_f G_m (V_f + \eta_2 V_m)}{G_m V_f + G_f V_m \eta_2} \tag{5}$$

where E_f and E_m are the elastic moduluses of carbon fiber and resin, respectively, v_f and v_m are the corresponding Poisson's ratios, G_f and G_m are the corresponding shear elastic moduluses, V_f is the fiber volume percentage, $V_m = 1 - V_f$ is the resin volume percentage, and η_2 and η_{12} are given by experiments and equal to 0.97 for plies in this paper.

Lay-up of mirror surface part generally has the following principles:

(1) Plies should be symmetrical with the mid of the mirror surface part to prevent warping.

(2) Proportion of each ply-orientation should be no less than 6%–10%.

(3) Ply-orientation should not be too complicated. Orientations at 45° and 60° are mostly chosen to ensure quasic-isotropic property of the surface part.

According to Chen *et al.* [19], the volume of a fiber/resin ratio should be 67% with 24 plies. Each plies is 0.1-mm thick.

Above all, the author works on the optimization research of the ply-orientation design to find the minimal root-mean-square values (RMS) along the optical axis of mirror surface under the affection of gravity. Hypermesh is applied to the FEM analysis. Due to the principle of permutation and combination, total 15 plies orientations are selected to conduct the analysis. The analysis results are given in Table 1.

Due to the results given in Table 1, [0°/60°/–60°] section performs the minimum RMS and is chosen to manufacture the mirror surface. Microstructure of this section is described in Fig. 3, and the surface deformation is described in Fig. 4.

Table 1 Mirror surface deformation under the different ply-orientations combined.

Ply fiber orientation	PV (nm)	RMS (nm)
$[0°/60°/-60°]_8$	6.6	372
$[0°/-60°/60°]_8$	6.59	372
$[60°/0°/-60°]_8$	10.71	349
$[60°/-60°/0°]_8$	16.4	352
$[-60°/60°/0°]_8$	16.4	352
$[-60°/0°/60°]_8$	7.91	356
$[-45°/0°/45°/90°]_6$	6.88	377
$[-45°/90°/45°/0°]_6$	16.6	351
$[0°/45°/90°/-45°]_6$	10.36	349
$[0°/-45°/90°/45°]_6$	10.36	349
$[45°/0°/-45°/90°]_6$	6.9	377
$[45°/90°/-45°/0°]_6$	16.6	351
$[90°/45°/0°/-45°]_6$	10.1	365
$[90°/-45°/0°/45°]_6$	10.1	364

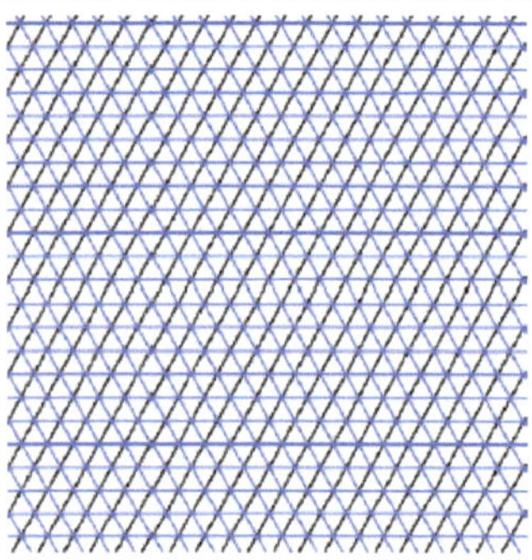

Fig. 3 Microstructure of the surface with different ply-orientations combined.

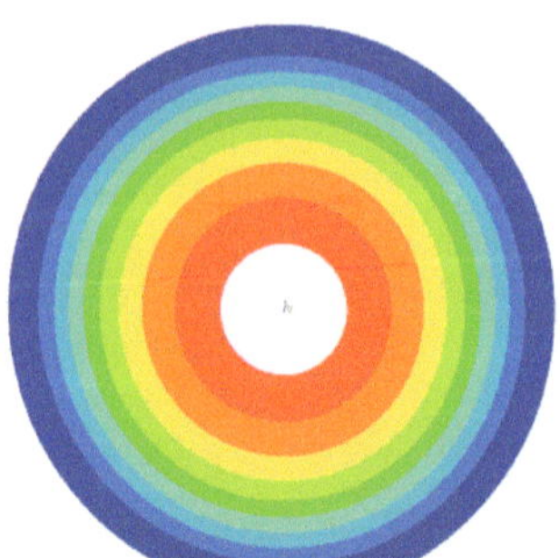

Fig. 4 Surface deformation of the reflector under loads.

2.1.2 Mirror surface fabrication

Carbon fiber mirror fabrication, which is also called optical surface transformation (OST), requires a pair of forming dies which are composed of convex die and concave die. In addition, the forming surface of concave die must be polished into the required shape of the mirror. With forming dies, the carbon fiber material that has been correctly laid up can mold for the mirror plane at the appropriate temperature and pressure conditions. The molding process is shown in Fig. 5.

2.2 Fabrication of mirror body

Different from the honeycomb structure which is adopted in the traditional reflector as mirror body, the carbon fiber grid structure mirror body has higher support stiffness and better thermal matching with the rest part. Unlike the mirror surface part ([0°/60°/-60°] style), the [0°/45°/90°/-45°] style is applied in the body fabrication because better mechanical property and thermal suitability are required for the mirror body. To ensure better thermal stability, the grid structure and its back surface are adhered together as mirror body after the dependent fabrication. To meet with the concave interface between the mirror surface part and mirror body, grid forming die should be manufactured at corresponding curvature, as shown in Fig. 6. The finished mirror body is shown in Fig. 7.

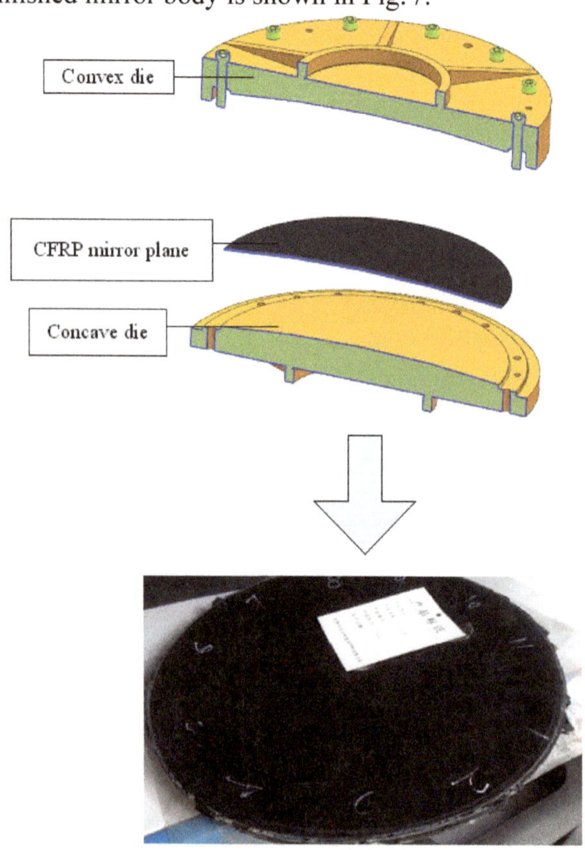

Fig. 5 Forming process of the mirror.

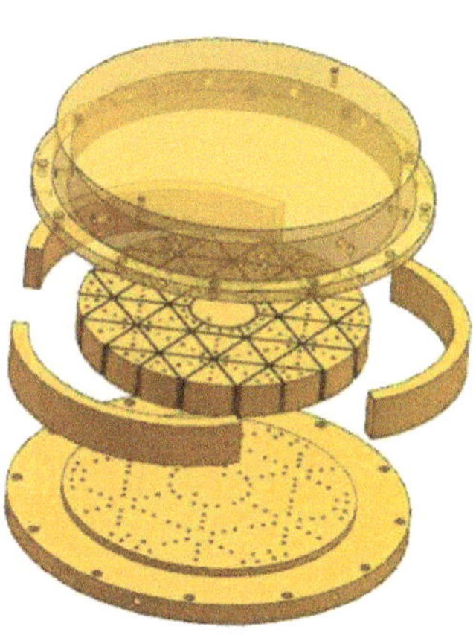

Fig. 6 Grid forming die.

Fig. 7 Mirror body with carbon fiber grid reinforced structure.

2.3 Carbon fiber mirror assembly and optical processing

The carbon fiber mirror assembly processes mainly contain 3 steps as follows:

(1) Splice the mirror surface part and mirror body with epoxy resin and assembly tooling.

(2) Expose the conductive carbon fiber mirror surface by acid pickling of the epoxy resin which is left on the mirror surface during Step 1.

(3) Electroplate 0.2 mm thick nickel on the mirror surface.

Now that a carbon fiber mirror substrate has been manufactured, optical processing is on the next to ensure a precise mirror face. Single diamond turning of the nickel layer with ultraprecise diamond lathe is applied in mirror surface rough finishing, this method is time-saving, and a mirror surface wavefront accuracy of 0.2λ (λ=632 nm) can be achieved after this process. At last ion milling is applied during mirror surface fine finishing, and this non-contact finishing method can guarantee the mirror surface unstressed during the whole fine finishing process.

3. Support design of the carbon fiber mirror

In order to meet with the primary mirror surface deformation requirement under gravity affection coupling with +5℃ temperature rise at optical axis horizontal state and take mirror stiffness requirement also into consideration, 3-point peripheral bipod support as shown in Fig. 8 is adopted to support the CFRP reflector. The detail size parameters of the bipod flexure hinge are shown in Fig. 9.

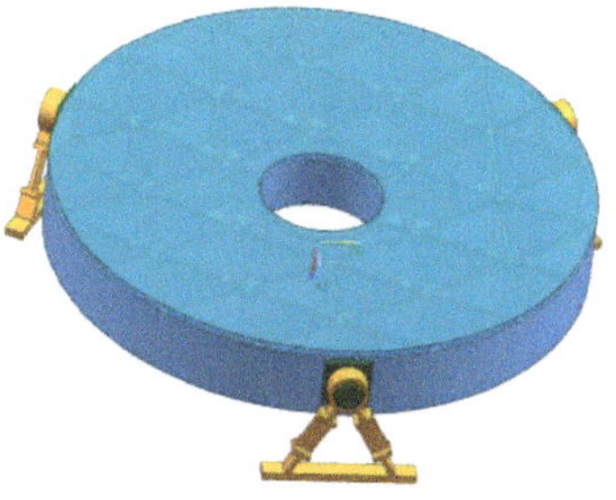

Fig. 8 Three-point peripheral bipod support around the mirror.

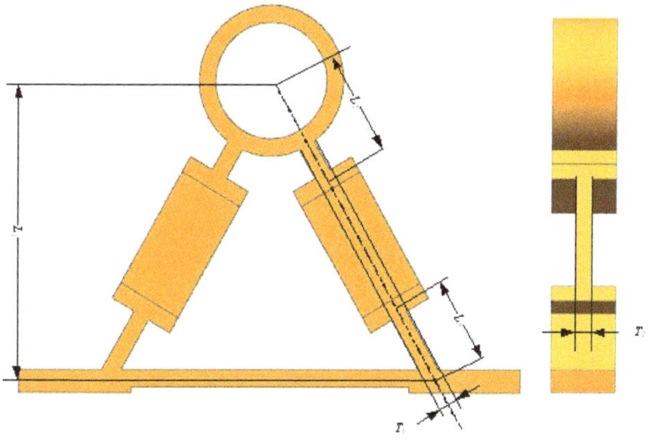

Fig. 9 Each parameter of the flexure hinge.

Set the value of L_1 as limit, and L_2, L_3, T_1, and T_2 as optimization variables, to seek the minimal RMS value under gravity affection coupling with +5℃ temperature rise at the optical axis horizontal state. Several programs contained of Partran, Nastran, UG, Sigfitare are integrated to complete the optimization work. The optimized model goes as follows:

(1) Optimization target

Achieve the minimal RMS value (at axis horizontal state, considering gravity affection coupling with +5℃ temperature rising as well).

(2) Optimization variables

The L_2, L_3, T_1, and T_2 sizes of flexible hinge constraints:

L_1=51 mm;

8 mm≤L_2≤30 mm;

8 mm≤L_3≤30 mm;

1.5 mm≤T_1≤5 mm;

1.5 mm≤T_2≤5 mm.

After 78 iterations, the optimization work is done, and the model converges. Results are shown in Table 2, and Fig. 10 shows the RMS value in each iteration.

Table 2 Optimized size parameters of the bipod hinge type and the surface RMS values.

L_2 (mm)	L_3 (mm)	T_1 (mm)	T_2 (mm)	RMS (nm)
20.096	8.4353	4.52	1.72	4.3

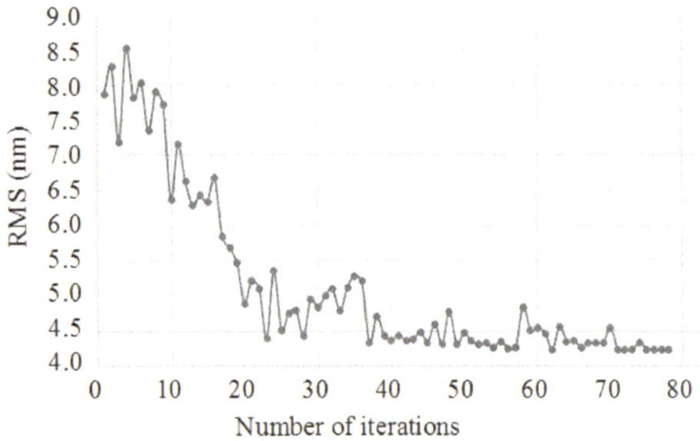

Fig. 10 Iteration history of mirror surface shape.

According to the optimal results above and processing craft, the flexible supporting structure is given in Fig. 11.

Fig. 11 Optimized structure of the bipod support.

4. Analysis and validation

Apply the FEM analysis for CFRP mirror, the FEM meshes are shown in Fig. 12, and the analysis results are shown in Table 3.

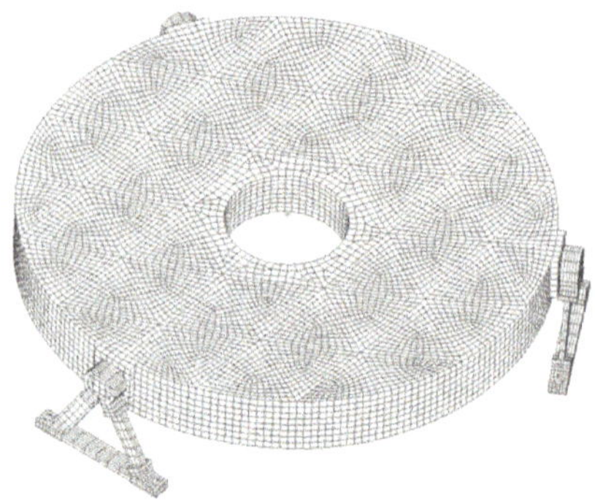

Fig. 12 Finite model of the CFRP mirror.

Table 3 FEM analysis results.

Modal analysis		
Primary frequency	652.3 Hz	
Second frequency	680.4 Hz	
Third frequency	716.9 Hz	
Surface deformation (axis horizontal state)		
Temperature rising	PV (nm)	RMS (nm)
+5℃	86.53	5.74
0℃	84.09	5.65

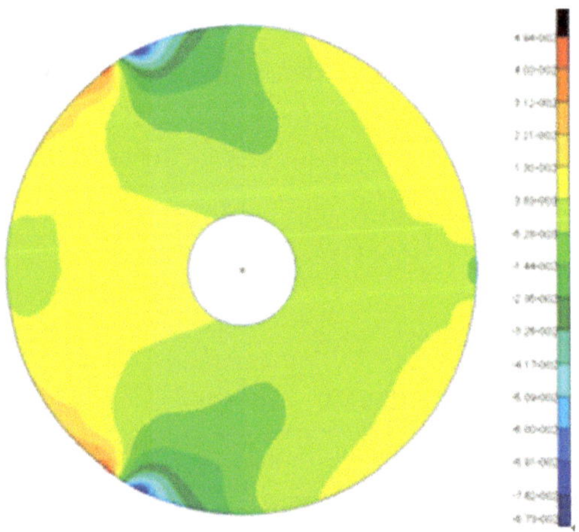

Fig. 13 Mirror surface deformation nephogram under gravity affection coupling with +5°C temperature rise at optical axis horizontal state.

Considering gravity affection coupling with +5 ℃ temperature rising, the mirror surface deformation on the detecting direction (axis horizontal state) is shown in Fig. 13. Based on the analysis results above, the RMS value of designed CFRP mirror components is 5.74 nm, less than $\lambda/50$

(λ=632 nm), which satisfies the space mirror demand at visible wavelengths. The primary frequency (652.3 Hz) of designed CFRP mirror components is far higher than 100 Hz, which satisfies the requirement during rocket launch.

5. Conclusions

The authors have designed a CFRP mirror and its supports for space camera, and the CFRP mirror substrate which is assembled by carbon fiber mirror surface part and carbon fiber mirror body has been fabricated. The carbon fiber mirror surface part applies the FEM method in the ply-orientation optimal design and finds the best result at [$0^o/60^o/-60^o$] style. The mirror body applies carbon fiber grid reinforced structure instead of honeycomb. Three-point peripheral bipod flexure hinge support is chosen to support the mirror. An integrated optimization analysis has revealed the best structure parameters of bipod support. Through the FEM analysis of the whole components, the primary frequency of the designed CFRP mirror components reaches up to 652.3 Hz, and the RMS value of the mirror surface is 5.74 nm under gravity affection coupling with +5℃ temperature rise at optical axis horizontal state, which meet with the mechanical and optical requirements of space camera. Researches in this paper will offer great help to future applications of CFRP mirror on space camera.

Acknowledgement

This work was supported by the National Natural Science Foundation of China (NSFC) (No. 41501383).

References

[1] B. He and Z. L Li, "Design of the strip SiC mirror supporting structure and lightweight," *Optical Technique*, 2008, 4(34): 593–596.

[2] P. B. Willis and D. R. Coulter, "Durability and reliability of lightweight composite mirrors for space optical systems," *SPIE*, 1993, 1993: 127–136.

[3] M. E. L. Jungwirth, C. C. Wilcox, D. V. Wick, J. Robichaud, R. C. Romeo, R. N. Martin, *et al.*, "Large-aperture active optical carbon fiber reinforced polymer mirror," *SPIE*, 2013, 8725: 1–11.

[4] C. C. Wilcox, M. E. L. Jungwirth, D. V. Wick, R. C. Romeo, and R. N. Martin, "Closed-loop performance of an actuated deformable carbon fiber reinforced polymer mirror," *SPIE*, 2012, 8373: .

[5] S. Kendrew and P. Doel, "Development of a carbon fibler composite active mirror: design and testing," *Optical Engineering*, 2006, 45(3): 535–545.

[6] P. Doel, S. Kendrew, D. Brooks, C. Dorn, C. Yates, R. D. Martin, *et al.*, "Development of an active carbon fiber composite mirror," *SPIE*, 2004, 5490: 1526–1533.

[7] M. E. L. Jungwirth, M. Baker, D. V. Wick, C. C. Wilcox, R. C. Romeo, R. D. Martin, *et al.*, "Actuation for carbon fiber reinforced polymer active optical mirrors," in *Proc. of 2012 IEEE Aerospace Conference*, Big Sky, Montana, 2012, pp. 1–9.

[8] Y. Arao, J. Koyanagi, S. Utsunomiya, S. I. Takeda, and H. Kawada, "Analysis of time-dependent deformation of a CFRP mirror under hot and humid conditions," *Mechanics of Time-Dependent Materials*, 2009, 13(2): 183–197.

[9] S. Tanaka, T. Ikeda, and A. Senba, "Sensitivity analysis of thermal deformation of CFRP laminate reflector due to fiber orientation error," *Journal of Mechanical Science & Technology*, 2016, 30(10): 4423–4426.

[10] P. C. Chen, T. T. Saha, A. M. Smith, and R. Romeo, "Progress in very lightweight optics using graphite fiber composite materials," *Optical Engineering*, 1998, 37(2): 666–676.

[11] R. C. Romeo and R. D. Martin, "Final assembly of the ULTRA 1-m carbon fiber optical telescope," *SPIE*, 2007, 6665: 1–10.

[12] R. C. Romeo and R. D. Martin, "Progress in 1m-class lightweight CFRP composite mirrors for the ULTRA telescope," *SPIE*, 2006, 6273: 1–12.

[13] R. D. Martin and R. C. Romeo, "Lightweight optical telescope structures fabricated from CFRP composites," *SPIE*, 2007, 6665: 1–7.

[14] W. C. Peng, S. L. Sun, G. L. Chen, L. W. Sun, and P. J. Zhang, "Carbon fiber reinforced composites for lightweight optical mirrors," *Optical Technique*, 2006, 8 (32): 138–142.

[15] H. Zhao, X. W. Fan, Z. H. Pang, G. T. Ren, W. Wang, Y. J. Xie, *et al.*, "Carbon-fiber-reinforced polymer variable-curvature mirror used for optical zoom imaging: prototype design and experimental demonstration," *Optical Engineering*, 2015, 54(2): 025103.

[16] D. E. Krumweide, G. D. Wonacott, P. M. Woida, R.

Q. Woida, and W. Shih, "Carbon-carbon mirrors for exoatmospheric and space applications," *SPIE*, 2007, 6666: 1–8.

[17] S. Utsunomiya, T. Kamiya, and R. Shimizu, "CFRP composite mirrors for space telescopes and their micro-dimensional stability," *SPIE*, 2010, 7739: 1–7.

[18] G. J. Barber, A. Braem, N. H. Brook, W. Cameron, C. D'Ambrosio, N. Harnew, *et al*., "Development of lightweight carbon-fiber mirrors for the RICH 1 detector of LHCb," *Nuclear Instruments and Methods in Physics Research*, 2008, 593(3): 624–637.

[19] P. C. Chen, C. W. Bowers, D. A. Content, M. Marzouk, and R. C. Romeo, "Advances in very lightweight composite mirror technology," *Optical Engineering,* 2000, 39(9): 2320–2329.

Parametric Analysis of NO_2 Gas Sensor Based on Carbon Nanotubes

Asama N. NAJE*, Russul R. IBRAHEEM, and Fuad T. IBRAHIM

University of Baghdad, Collage of Science, Department of Physics, Baghdad, Iraq

*Corresponding author: Asama N. NAJE E-mail: Naje.as75@gmail.com

Abstract: Two types of carbon nanotubes [single walled nanotubes (SWCNTs) and multi walled carbon nanotubes (MWCNTs)] are deposited on porous silicon by the drop casting technique. Upon exposure to test gas mixing ratio 3% NO_2, the sensitivity response results show that the SWCNTs' sensitivity reaches to 79.8%, where MWCNTs' is 59.6%. The study shows that sensitivity response of the films increases with an increase in the operating temperature up to 200℃ and 150℃ for MWCNTs and SWCNTs. The response and recovery time is about 19 s and 54 s at 200℃ for MWCNTs, respectively, and 20 s and 56 s at 150℃ for SWCNTs.

Keywords: SWCNTs; MWCNTs; NO_2 gas; sensitivity; response time

1. Introduction

Detecting gas molecules is basic to environmental monitoring, control of chemical processes, space mission, agricultural and medical applications. The sensing of NO_2 is important to monitor environmental pollution resulting from combustion or automotive emissions [1]. Nitrogen dioxide (NO_2) is flammable, colorless, and dangerous, even at very low concentration. Moreover, the interaction of NO_2 and CO in sunlight tends to produce O_3, which is believed to be harmful to plants and the respiratory system of human beings and animals because of its strongly oxidizing behavior. Therefore, from the public health and environmental protection viewpoint, the sensitive detection of nitrogen dioxide is of great scientific importance [2].

A gas sensor is a device which detects the presence of different gases in a region, particularly those gases which might be harmful to humans or animals. The development of the gas sensor technology has received considerable attention in recent years for monitoring environmental pollution. It is well known that chemical gas sensor performance features, such as sensitivity, selectivity, time response, stability, durability, reproducibility, and reversibility, are largely influenced by the properties of the sensing materials used [3] .The basic principle behind gas detection is a change in an electrical property of the detecting material upon exposure to the gas. For example, in the case of carbon nanotubes, the resistance changes with exposure to different gases like NO_2 and NH_3. Other electrical properties of nanotubes like thermoelectric power and dielectric properties also change upon gas exposure [4]. Common gas sensors are metal oxide semiconductor such as tin oxide, zinc oxide, titanium oxide, and aluminum oxide. Problems encountered with these sensors are lack of flexibility and poor response time, and it is operated at the elevated temperature [5]. But nowadays, researchers' goal is to develop the sensors, which can operate at room temperature and consume low power. Carbon nanotubes (CNTs) based sensors have been recently studied due to their excellent electrical, mechanical, and sensing properties. Because of the high surface area, CNTs can adsorb large amount of gases, which make them a probable contender for a gas sensor with very high sensitivity and low response time. Particularly, gas adsorption in CNTs is an important issue for both fundamental research and technical applications [6]. There are two types of carbon nanotubes' morphology. Single-walled nanotubes (SWNTs) consist of a honeycomb network of carbon atoms and can be visualized as a cylinder rolled from a graphitic sheet. The other is multi-walled nanotubes (MWNTs) that are a coaxial assembly of graphitic cylinders generally separated by the plane space of graphite [7].

The present study focuses on the synthesis and chemi-resistive characteristics of carbon nanotubes (CNTs) thin film sensor. The material is prepared by a simple chemical solution method and the thin films are fabricated by drop-coating method.

2. Experiment

2.1 Preparation of the samples

In this work, $2\times2\,cm^2$ dimensions primary n-type silicon wafer substrates were thoroughly cleaned to de-contaminate their surface from any available stains and dirt. A porous silicon layer (PS) was prepared via photochemical wet etching. This process was carried out by using ordinary light source. Its main apparatus consisted of a Quartz Tungsten Halogen lamp (250 W), a focusing lens (3.8 cm) with focal length, and the diluted etching HF acid of 50% concentration mixed with ethanol in (1:1) ratio in a Teflon container. To prepare CNT sample, 0.01 g of CNT was dispersed in Dimethylformamide (DMF). A magnetic stirrer was incorporated for this purpose for 15 minutes, followed by 1 hour sonication. The obtained solution was used for film deposition on porous silicon by the drop casting method.

2.2 Gas sensor testing system

The detail of the gas sensor testing unit, which was used in the current tests, was described elsewhere [8]. A steel cylindrical test chamber of diameter 163 mm and of height 200 mm with the bottom base made removable and of O-ring sealed. The effective volume of the chamber was 4173.49 cc, which had an inlet for allowing the test gas to flow in and an air admittance valve allowing atmospheric air after evacuation. Another third port was provided for the vacuum gauge connection. A multi-pin feed through at the base of the chamber allowed for the electrical connections to be established to the sensor and the heater assembly. The heater assembly consisted of a hot plate and a k-type thermocouple inside the chamber in order to control and set the desired operating temperature of the sensor. The thermocouple sensed the temperature at the surface of the film exposed to the analyte gas. The PC-interfaced multi meter, of type UNI-T UT81B, was used to register the variation of the sensor conductance (reciprocal of resistance) exposed to predetermined air – NO_2 gas mixing ratio. The chamber can be evacuated by using a rotary pump to a rough vacuum of 2×10^{-2} bar. A gas mixing manifold was incorporated to control the mixing ratios of the test and carrier gases prior to being injected into the test chamber. The mixing gas manifold was fed by zero air and test gas through a flow meter and needle valve arrangement. This arrangement of mixing scheme was done to ensure that the gas mixture entering the test chamber was premixed thereby giving the real sensitivity.

3. Results and discussion

The response of a sensor upon the introduction of a particular gas species is called the sensitivity (S).

The most general definition of sensitivity applied to solid state chemi-resistive gas sensors is a change in the electrical resistance (or conductance) relative to the initial state upon exposure to a reducing or oxidizing gas component. It is calculated by using the equations below [9]:

$$S = \frac{\Delta R}{R_o} \times 100\% = \left| \frac{R_{gas} - R_{air}}{R_{air}} \right| \times 100\% \qquad (1)$$

or

$$S = \frac{\Delta G}{G_o} \times 100\% = \left| \frac{G_{gas} - G_{air}}{G_{air}} \right| \times 100\% \qquad (2)$$

where R is the electrical resistance, G is the electrical conductance, and the subscript "air" indicates that background is the initial dry air state and the subscript "gas" indicates the analyte gas has been introduced.

Figure 1 shows the scanning electron microscope (SEM) images for the MWCNT and SWCNTs deposited on PS. In these figures, long nanotubes with large agglomerates and closely packed CNTs are shown.

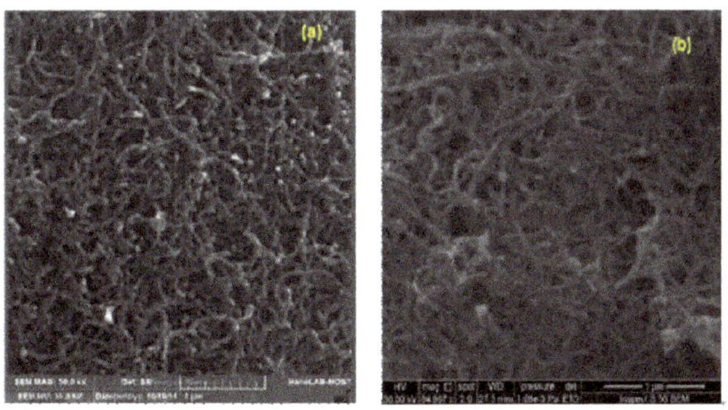

(a) (b)

Fig. 1 SEM images for the CNT on PS: (a) MWCNTs and (b) SWCNTs.

The atomic force microscopic (AFM) of the PSi/CNTs is shown in Fig. 2. Results of surface morphology of the CNTs film had a good uniform surface homegensity and good indication for having nanoporous with a regular distribution of the CNTs. The roughness average (Sa) for this layer of PSi/CNTs was 1.04 nm while the root mean square roughness (Sq) was 1.24 nm and ten point height (Sz) was 5.86 nm.

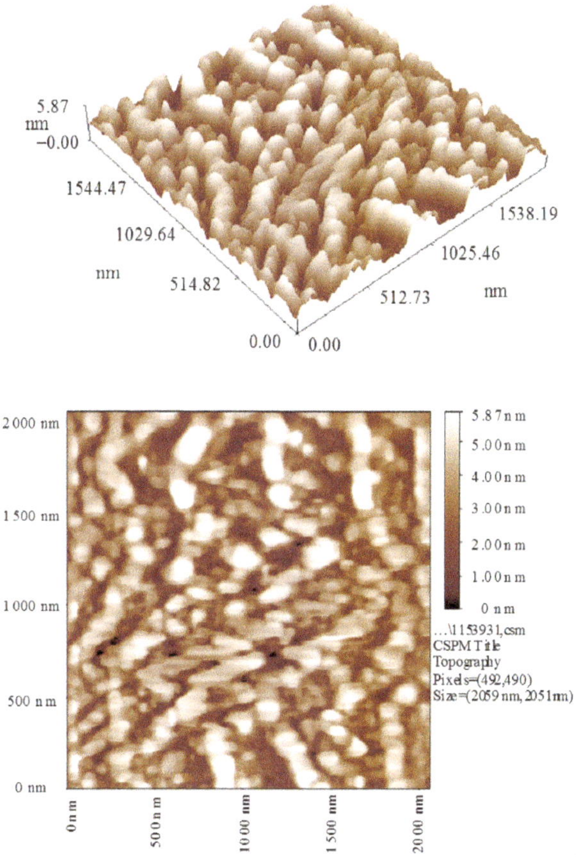

Fig. 2 AFM images for the CNTs deposited on PS.

Figure 3 shows the response of the MWCNTs and SWCNTs thin films upon exposure to NO_2 gas for mixing ratio 3% and at varying operating temperatures. It can be seen that the response of the films increases as the operating temperature increases up to 200℃ and 150℃, for MWCNTs and SWCNTs, respectively, and then decreases. The gas-sensing response increases with temperature in the 50℃ – 200℃ and 50℃ – 150 ℃ ranges for MWCNTs and SWCNTs, respectively, because thermal energy helps the reactions involved overcome their respective activation energy barriers. However, if the operating temperature becomes too high (i.e., >150 ℃ for SWCNTs or 200°C for MWCNTs), the adsorbed oxygen species at the sensing sites on the film surface will be diminished and less available to react with NO_2 molecules, thereby limiting the film's response.

As shown in Fig. 4, the sensitivity $S\%$ increases with increasing the operating temperature T, where the maximum sensitivity of MWCNTs is 59.61%, at 200℃ of the testing temperature after which it begins to drop with increasing *T* and the test is terminated.

While the sensitivity of SWCNTs is 79.81% at 150℃ of testing temperature after which it begins to drop with increasing *T* and the test is terminated.

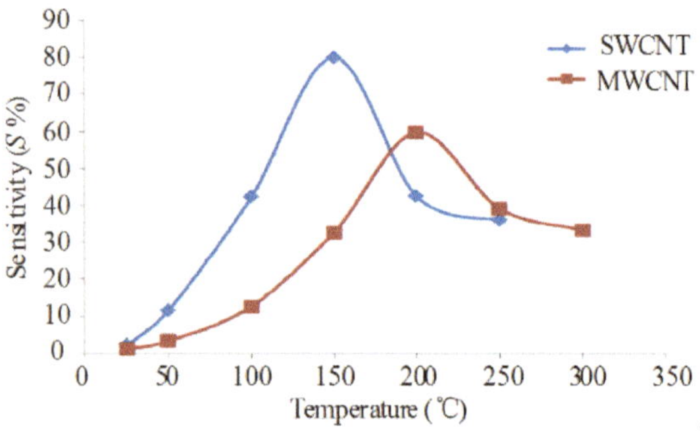

Fig. 3 Sensitivity variation with the operating temperature of the MWCNT and SWCNT gas sensor at 5 V bias voltage.

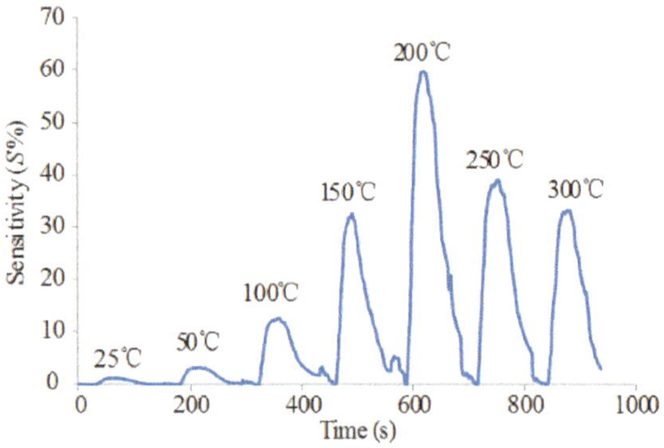

(a)

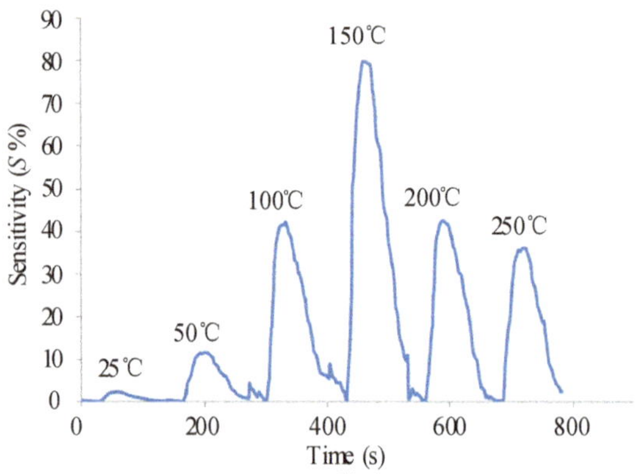

(b)

Fig. 4 Transient response of CNTs thin film at various testing temperatures upon exposure to NO_2 gas, 3% maxing ratio, at 5 V bias voltages for (a) MWCNTs and (b) SWCNTs.

At any operating temperature, the sensor response of the SWCNTs thin film is higher than that of the MWCNTs thin film. The results show that SWCNTs have a better performance than MWCNTs, and these results strongly depend on the structure of CNTs. When the structure of atoms in a carbon nanotube minimizes the collisions between conduction electrons and atoms, a carbon nanotube is highly conductive. The strong bonds between carbon atoms also allow carbon nanotubes to withstand higher electric currents than material. So when the diameter of tubs decreases the current increases. Since SWCNTs have small diameter (1 nm–2 nm), it will have the higher conductivity and better performances [10].

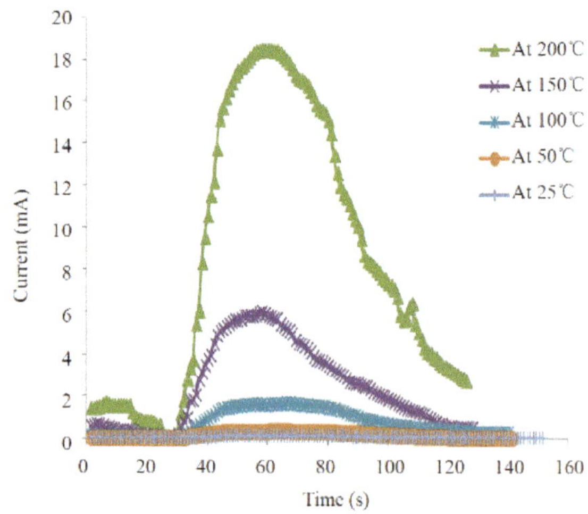

(a)

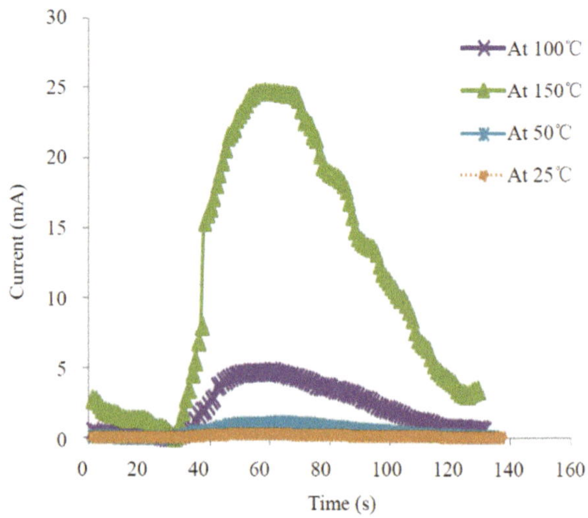

(b)

Fig. 5 Current-time variation of CNTs sensor time at 25˚C, 50˚C, 100˚C, 150˚C and 200˚C testing temperature upon exposure to NO_2 gas and 5 V bias voltage for (a) MWCNTs and (b) SWCNTs.

The observed current appears in Fig. 5 increases (i.e resistance decreases) when exposing the CNT networks to NO_2, which can be due to that the electron transfer occurs from CNTs to NO_2 which has highly oxidizing natures. The variation of device resistance (or current) in the presence of oxidizing gas can be explained as follows: when oxidizing species like NO_2 are adsorbed on the surface of p-type CNTs, the Fermi levels are shifted towards valance band, generating more holes and thus decreasing the electrical resistance.

The response time of the CNTs gas sensor decreases with increasing the operating temperature, and with the shortest response and recovery time being at about 19 s and 54 s at 200℃ for MWCNTs, respectively, and 20 s and 56 s at 150℃ for SWCNTs.

In order to explain the results, the chemical nature of the NO_2 molecule is considered. Recent experimental results show that the electrical conductance of an individual semiconducting carbon nanotube strongly increases upon NO_2 gas exposure and the NO_2 is identified as an electron acceptor. In light of the present work, it is reasonable to propose that this behavior in the nanotube film is also due to adsorbed NO_2 in the tube wall. According to recent theoretical calculations, a possible interpretation of the electrical response of CNT films to NO_2 gas could be explained in terms of the physical absorption of this molecule. NO_2 has an unpaired electron and is known as a strong oxidizer. Upon NO_2 adsorption, a charge transfer is likely to occur from the CNTs to the NO_2 due to the electron-acceptor character of NO_2 molecules. The electrical response of the CNTs indicates that there is a charge transfer between the test gas and sensing element, and hence, the physisorption of gases in the nanotubes is the dominant sensing mechanism.

4. Conclusions

Thin films of two types of carbon nanotubes (SWCNTs and MWCNTs) were prepared by the drop casting method. The maximum variation of resistances to NO_2 was found at an operating temperature of around 200 ℃ and 150 ℃ for MWCNTs and SWCNTs. The response time was 19 s and 20 s for MWCNTs and SWCNTs, respectively.

References

[1] J. Kong, N. R. Franklin, C. Zhou, M. G. Chapline, S. Peng, K. Cho, *et al.*, "Nanotube molecular wires as chemical sensors," *Science*, 2000, 287(5453): 622–625.

[2] X. B. Yan, Z. J. Han, Y. Yang, and B. K. Tay, "NO_2 gas sensing with polyaniline nanofibers synthesized by a facile aqueous/organic interfacial polymerization," *Sensors and Actuators B: Chemical*, 2007, 123(1): 107–113.

[3] B. Ding, M. Wang, J. Yu, and G. Sun, "Gas sensors based on electrospun nanofibers," *Sensors*, 2009, 9(3): 1609–1624.

[4] S. Chopra, K. McGuire, N. Gothard, A. M. Rao and A. Pham, "Selective gas detection using a carbon nanotube sensor," *Applied Physics Letters*, 2003, 83(11): 2280–2283.

[5] M. Y. Faizah, "Room temperature multi gas detection using carbon nanotubes," *European Journal of Scientific Research*, 2009, 35(1): 142–149.

[6] S. Dhall, N. Jaggi, and R. Nathawat, "Functionalized multiwalled carbon nanotubes based Hydrogen gas sensor," *Sensors and Actuators: A Physical*, 2013, 201(10): 321–327.

[7] I. Sayago, E. Terrado, E. Lafuente, M. C. Horrillo, W. K. Maser, A.M. Benito, *et al.*, "Hydrogen sensors based on carbon nanotubes thin films," *Synthetic Metals*, 2005, 148(1): 15–19.

[8] G. Al-zaidi, A. M. Suhail, and W. R. Al-azawi, "Palladium – doped ZnO thin film hydrogen gas sensor," *Applied Physics Research*, 2011, 3(1): 89–99.

[9] L. A. Patil, A. R. Bari, M. D.Shinde, V. V. Deo, and D. P. Amalnerkar, "Synthesis of ZnO nanocrystalline powder from ultrasonic atomization technique, characterization, and its application in gas sensing," *IEEE Sensor Journal*, 2011, 11(4): 939–946.

[10] F. Y. Wu and H. M. Cheng, "Structure and thermal expansion of multi-walled carbon nanotubes before and after high temperature treatment," *Journal of Physics D: Applied Physics*, 2005, 38(24): 4302–4307.

Theoretical and Experimental Investigation of Fiber Bragg Gratings With Different Lengths for Ultrasonic Detection

Zhouzhou YU, Qi JIANG*, Hao ZHANG, and Junjie WANG

School of Control Science and Engineering, Shandong University, Jinan, 250061, China

State Key Lab of Microbial Technology, School of Life Science, Shandong University, Jinan, 250100, China

*Corresponding author: Qi JIANG Email: jiangqi@sdu.edu.cn

Abstract: In this paper, the response of fiber Bragg gratings (FBGs) subjected to the ultrasonic wave has been theoretically and experimentally investigated. Although FBG sensors have been widely used in the ultrasonic detection for practical structural health monitoring, the relationship between the grating length and ultrasonic frequency is not yet to be obtained. To address this problem, an ultrasound detection system based on FBGs is designed and the response sensitivity of different lengths gratings are detected. Experimental results indicate that the grating with 3 mm length has a higher sensitivity when detecting high frequency ultrasonic wave, and the amplitude can be up to 0.6 mV. The 10 mm length grating has better detection sensitivity for low frequency ultrasonic wave and the amplitude is 0.8 mV. The results of this analysis provide useful tools for high sensitivity ultrasound detection in damage detection systems.

Keywords: FBG; nondestructive testing; ultrasonic detection

1. Introduction

The ultrasonic detection based on fiber Bragg gratings (FBGs) is a new nondestructive testing technology being developed in recent years [1]. The basic idea is to use FBG sensors as ultrasonic detectors to analyze the injury condition of the construction. Compared with classical electronic detection sensors, FBGs have unique advantages such as immunity to electromagnetic interference, lightweight, high temperature corrosion resistance, stability, and durability in harsh environment. FBGs-based ultrasonic sensors have been indicated as the ideal candidate for practical structural health monitoring in light of their unique advantages over conventional sensing devices [2, 3].

At present, the use of FBG sensors for the detection of ultrasound has been widely reported in literature. Sorazu presented the results of theoretical and experimental investigations into the interaction between ultrasonic waves and fiber optic sensors, focusing on the birefringence introduced within an optical fiber placed with its axis parallel to the ultrasonic wave front [4]. Lam proposed that fiber Bragg grating sensors could be embedded in the laminate material to detect the signal of ultrasonic Lamb wave, and the damage assessment of the composite material could be found out [5]. Lee designed a new type of FBG ultrasonic receiver based on the embedded system, which detected the impact damage of carbon fiber reinforced plastic laminates by shockwave. They described the design of the structure of the FBG sensors and analyzed characteristics of the ultrasonic [6]. The FBG performance has a great influence on the test results, Liu and Han investigated the affection of the FBG sensors directivity, and their analysis revealed several unique FBG response characteristics [7]. Wild presented results for the spatial performance of fiber Bragg grating sensors to continuous-wave acousto-ultrasonic (AU) signals [8]. However, these researches mainly focused on health monitoring, location defection, and damage degree evaluation of the material structure [9]. There were only a few studies concerning the effect of different grating lengths on the sensitivity. The relationship between the grating length and ultrasonic frequency has not a clear conclusion.

Based on this argument, an ultrasound detection system based on FBGs is designed in this paper, and the response of FBGs to the ultrasonic wave is investigated in terms of gratings length changes. The results of the experiment have a good effect, meanwhile the qualitative relationship between the grating length and ultrasonic frequency is obtained.

2. Theoretical analysis

In this section, the interaction between the Bragg grating length and ultrasonic wave frequency is theoretically analyzed. FBGs consist of a periodic modulation of the refractive index in the core of a single mode fiber. According to the mode coupling theory, the central wavelength of the reflected narrowband light namely λ_B can be described by the Bragg condition [8]:

$$\lambda_B = 2n_{\mathrm{eff}}\Lambda \tag{1}$$

where n_{eff} is the effective refractive index of the optical mode propagating along the fiber, and Λ is the periodicity of the grating. When the Bragg condition is satisfied, the FBG reflects a narrow band of wavelengths, creating a stop band in the transmission spectrum and all of the other wavelengths not meeting the resonant condition are transmitted. The resonance depends on the periodicity of the grating and the effective refractive index of the fiber. In all of the external factors, the strain, temperature, and ultrasonic field are the most common factors that cause the Bragg wavelength shift [10, 11].

In our hypothesis, the ultrasonic wave can be modeled by a longitudinal strain wave, propagating along the axis of the optical fiber. Moreover, the time dependence is assumed to be sinusoidal. Hence, it can be expressed as an equation:

$$\varepsilon(t) = \varepsilon_m \cos(kz - w_s t) \tag{2}$$

where w_s is the ultrasonic wave frequency, ε_m is the amplitude of sound wave, z is the sound path, and $k=2\pi/\lambda s$ is the ultrasonic wave number [12]. The transfer matrix method is utilized to analyze the FBG multilayered structure. When the length of FBG is much smaller than the ultrasonic wavelength, the whole length of the gratings and the refractive index changes caused by the photo elastic effect are uniform [13, 14], and the wavelength shift of the FBG can be given by

$$\Delta\lambda(t) = \lambda_B \varepsilon_m \left[1 - \left(\frac{n_{\mathrm{eff}}^2}{2}\right)[p_{12} - \nu(p_{11} + p_{12})]\right]\cos(w_s t) \tag{3}$$

where ν is Poisson's ratio, and p_{11} and p_{12} represent elastic coefficient of the material, respectively.

In ultrasonic testing, the length of the grating, the reflectivity, and other parameters have a great influence on the detection of signal. In order to obtain a better detection effect, FBG length is less than 1/2 of the ultrasonic wavelength and its reflectivity is more than 0.9 [15, 16]. This can make sure that the FBG can not be affected by the non-uniform stress field of ultrasonic. When the distance between the FBG and the ultrasonic emission source is constant, the relationship between the response amplitude of sensor and the angle is a kind of approximate cosine [10]. The angle is formed by the grating axial and the ultrasonic wave propagation direction. Therefore, the FBG should be arranged in the coaxial direction with the ultrasonic transmission line and avoid the far field.

3. Design and analysis of the experimental test system

3.1 Design of the experimental test system

In order to observe the effect of different length gratings for ultrasonic detection, an ultrasound detection system was set up. The test system consisted of the high frequency signal part, the FBG optical sensing detection part, and the computer processing part. Figure 1 presents the specific experimental platform.

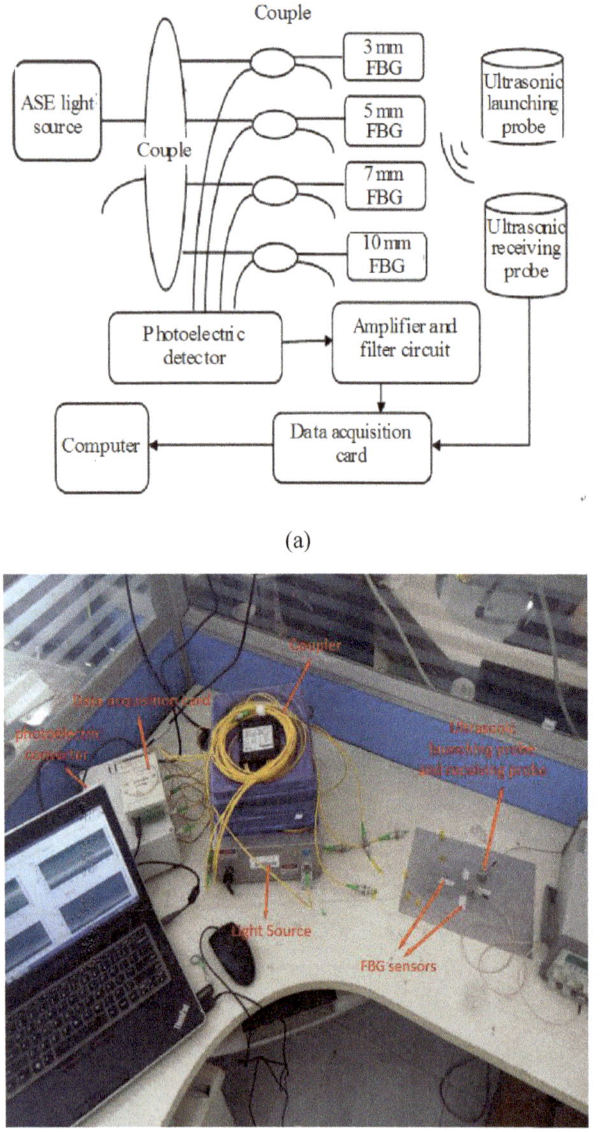

Fig. 1 Experimental setup for FBG sensing system (a) experimental device flow chart and (b) measurement instrument.

In this test system, the FBGs and the ultrasonic transducers were all surface-attached to the aluminum alloy plate. The grating lengths were 3 mm, 5 mm, 7 mm, and 10 mm. The 3 dB bandwidths were 0.524 nm,0.49 nm, 0.254 nm, and 0.235 nm, while the reflectivity were 89.38%, 89.06%, 91.61%, and 95.28%. All gratings were placed at an equal distance (100 mm) from the ultrasonic transducers in order to ensure the same acoustic amplitude. Specific layout position is shown in Fig. 2. The responses of different length gratings excited under the same frequency ultrasonic input wave were detected with their frequency range from 100 kHz to 800 kHz. Furthermore, the resonant piezoelectric sensor was measured at the same time. In order to make sure the experiment goes well, the temperature of the room was monitored and held constantly at 21℃. So the effects of temperature changes were eliminated.

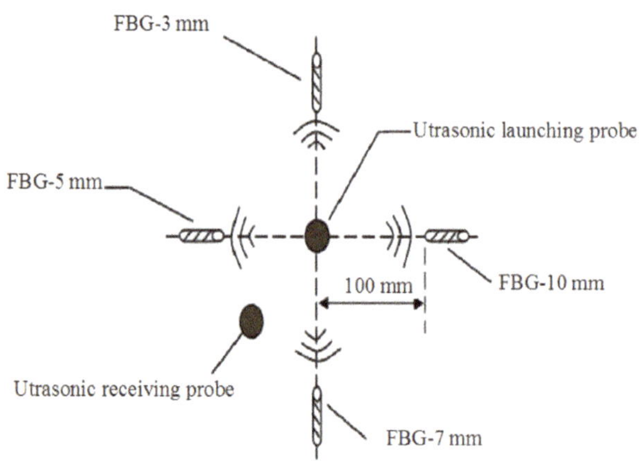

Fig. 2 Position of FBG sensors and the ultrasonic transducers.

3.2 Analysis and results

In the experiment, the ultrasonic frequency range from 100 kHz to 800 kHz and the data were detected at a 100 kHz interval. Under the 400 kHz frequency excitation, the signal responses of five sensors channels (3 mm, 5 mm,, 7 mm, 10 mm, and ultrasonic receiving probe) were obtained. After filtered and Fourier transformed by MATLAB software, the time and frequency domains of response are shown in Fig. 3.

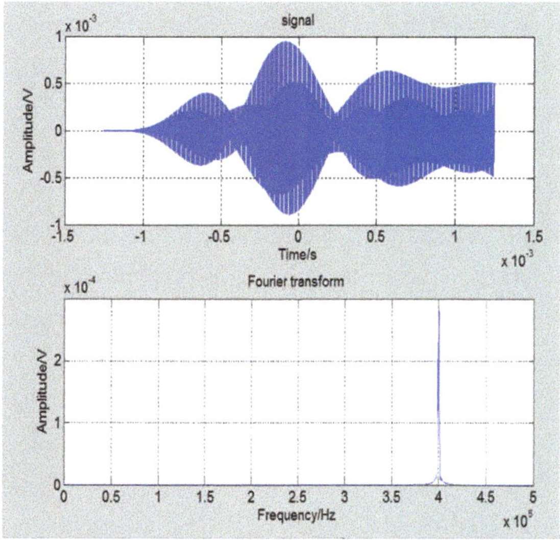

(a)

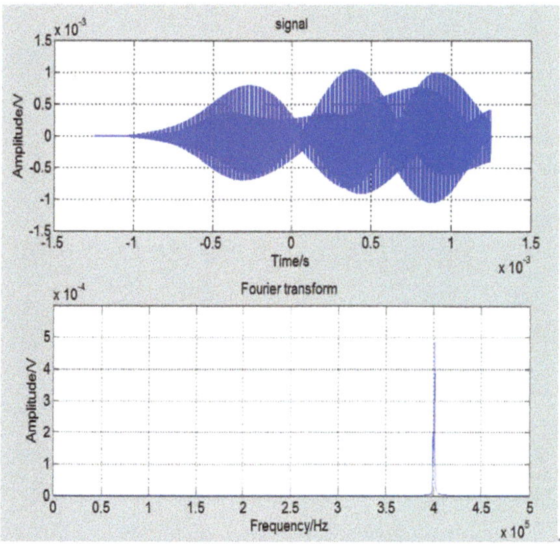

(b)

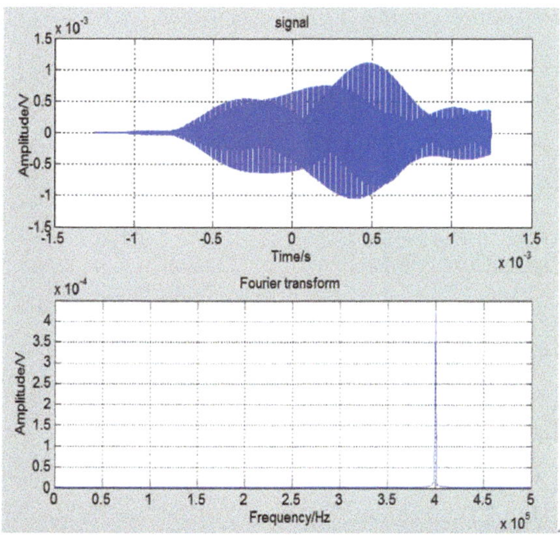

(c)

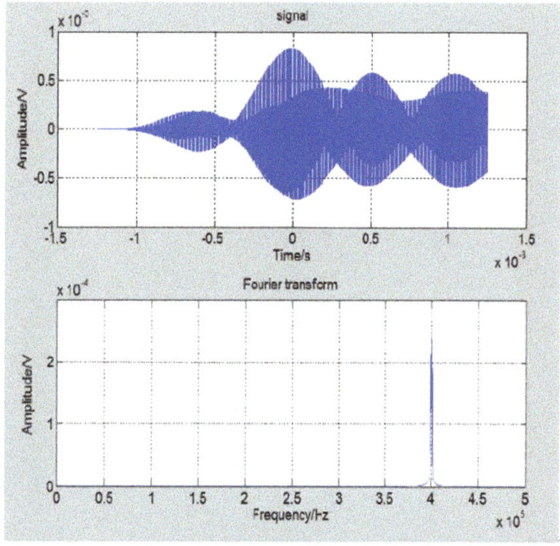

(d)

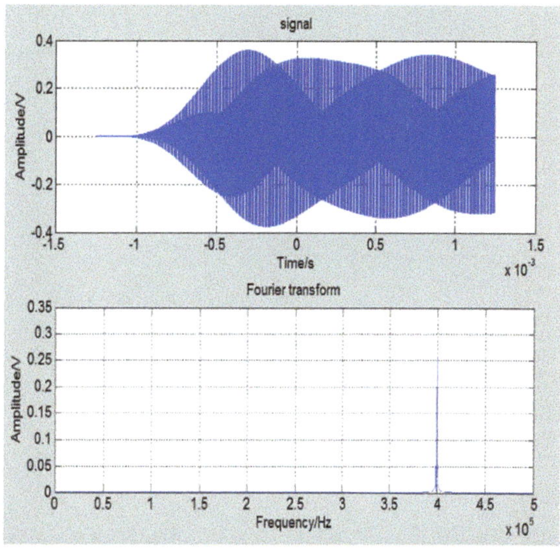

(e)

Fig. 3 Time and frequency domains of signal responses: (a) 3 mm FBG sensor, (b) 5 mm FBG sensor, (c) 7 mm FBG sensor, (d) 10 mm FBG sensor, and (e) ultrasonic receiving probe.

Since the FBG sensors have a smaller contact area than the piezoelectric sensor with the aluminum, the response amplitude detected by ultrasonic receiving probe was much larger than the grating detection. Using the origin to integrate data as shown in Fig. 4, we could see that gratings with 5 mm and 7 mm length had obvious effect under 400 kHz ultrasonic excitation, and the detection amplitude of 7 mm grating could be up to 1.2 mV.

The excitation signal was applied from 100 kHz to 800 kHz in turn, and the eight groups of data are presented in Table 1 and Table 2.

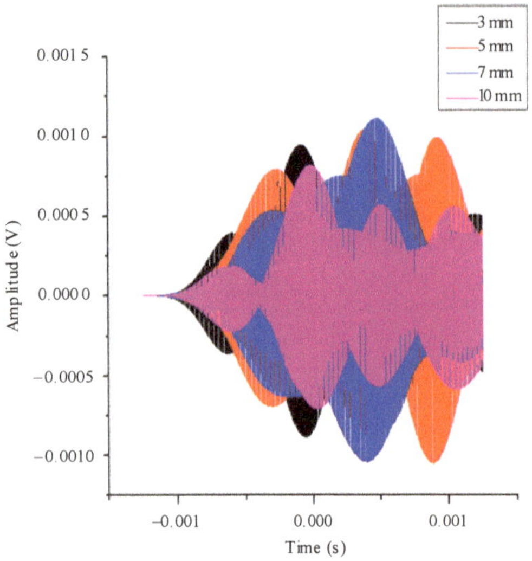

Fig. 4 Integrate data of sensors detection in five channels.

Table 1 Results of ultrasonic signal detection in five channels.

Frequency \ Length of FBG, Amplinude (10^{-4} V)	3 mm	5 mm	7 mm	10 mm	Ultrasonic receiving probe
100 kHz	5	5	11	8	2500
200 kHz	9	9	8	6	4000
300 kHz	4	4	6.5	5	3500
400 kHz	9	9	11	12	6500
500 kHz	2.5	2.5	2.8	6.5	3500
600 kHz	1.5	1.5	1.8	2.6	3500
700 kHz	2	2	4	2.2	1500
800 kHz	6	6	2	3.5	2000

Table 2 Normalization process of the data.

Frequency \ The length of FBG, Amplinude (10^{-2} V)	Data processing	3 mm	5 mm	7 mm	10 mm
100 kHz	2.5	2.5	2.5	2.5	2.5
200 kHz	4	2.5	1.8	1.9	2.2
300 kHz	3.5	2	1.5	1.6	2.5
400 kHz	6.5	4.5	2.5	3.75	2.2
500 kHz	3.5	1.25	0.6	2	1.25
600 kHz	3.5	0.75	0.4	0.8	1.4
700 kHz	1.5	1	0.9	0.69	0.7
800 kHz	2	1.5	0.45	1.1	0.69

The data detected by the ultrasonic probe were compared with FBGs, and the relationship between the normalized FBG response and the ultrasonic frequency is shown in Fig. 5.

The experimental results of four grating sensors and the ultrasonic transducer showed very similar trends in both sensitivity and directivity. Although FBG sensors had a weak response amplitude, they were still showing high sensitivity for ultrasonic detection. In addition, when the range of frequency was from 100 kHz to 800 kHz, the grating with a 3 mm length had the highest sensitivity, and when the detection frequency was 800 kHz, the amplitude could be up to 0.6 mV. The 10 mm length of the grating had the highest sensitivity when the detecting frequency was low, and the amplitude was 0.8 mV as the frequency was 100 kHz. The gratings with the length of 5 mm and 7 mm had a better detection sensitivity for intermediate frequency ultrasonic wave.

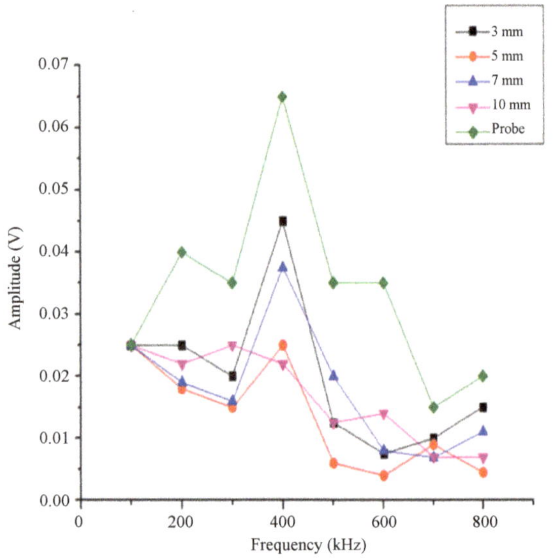

Fig. 5 Normalization of data.

4. Conclusions

In this paper, an ultrasound detection system based on FBGs is designed to analyze the effects of the grating length on the quality of ultrasonic resonance. The detection of different frequency ultrasonic signals in the aluminum plate is further explored, and a clear relationship between the grating length and ultrasonic frequency is obtained. Quantitative analysis has been carried out for uniform gratings, showing a better suitability in using gratings as ultrasound detectors. The theoretical analysis would be useful for sensitivity improvement of the FBG-based ultrasonic and acoustic emission sensing system. Because this experiment is a preliminary study on the response sensitivity of different length gratings to detect different wavelengths ultrasonic, deficiencies cannot be avoided in the experimental plan, which will be optimized in the further research.

Acknowledgment

This work was financially supported by the National Natural Science Foundation of China (No. 61271073 and No. 61473175) and was supported by the Fundamental Research Funds of Shandong University (No. 2015JC040).

References

[1] Y. Tan, L. Meng, and D. Zhang, "Strain sensing characteristic of ultrasonic excitation-fiber Bragg gratings damage detection technique," *Measurement*, 2013, 46(1): 294–304.

[2] Z. LI, M. Xu, J. Tang, Y. Chen, and Y. Zhao, "A study of signal demodulation techniques in a fiber Bragg grating sensing system," *Industrial Instrumentation and Automation*, 2005, (3): 12–16.

[3] V. Italia, A. Cusano, S. Campopiano, A. Cutolo, and M. Giordano, "Analysis of the phase response of fiber Bragg gratings to longitudinal ultrasonic fields in the high frequency regime: towards new interrogation strategies," in *Proceedings of 2005 IEEE/LEOS Workshop on Fibers and Optical Passive Components*, Italy, pp. 389–392, 2005.

[4] B. Sorazu, G. Thursby, and B. Culshaw, "Wavefront integrating fiber sensors for ultrasonic detection," *IEEE Sensors Journal*, 2011, 11(7): 1623–1631.

[5] P. M. Lam, K. T. Lau, H. Y. Ling, Z. Su, and H. Y. Tam, "Acousto-ultrasonic sensing for delaminated GFRP composites using an embedded FBG sensor," *Optics & Lasers in Engineering*, 2009, 47(10): 1049–1055.

[6] J. Lee, H. Tsuda, and N. Toyama, "Impact wave and damage detections using a strain-free fiber Bragg grating ultrasonic receiver," *NDT & E International*, 2007, 40(1): 85–93.

[7] T. Liu and M. Han, "Analysis of fiber Bragg gratings for ultrasonic detection," *IEEE Sensors Journal*, 2012, 12(7): 2368–2373.

[8] G. Wild and S. Hinckley, "Spatial performance of acousto-ultrasonic fiber Bragg grating sensor," *IEEE Sensors Journal*, 2010, 10(4): 805–806.

[9] H. Zhu, J. Yu, and J. Zhang, "A summary review and advantages of vibration-based damage identification methods in structural health monitoring," *Engineering Mechanics*, 2011, 28(2): 1–6.

[10] T. Erdogan, "Fiber grating spectra," *Journal of Lightwave Technology*, 1997, 15(8): 1277–1294.

[11] M. Majumder, T. K. Gangopadhyay, A. K. Chakraborty, K. Dasgupta, and D. K. Bhattacharya, "Fiber Bragg gratings in structural health monitoring—present status and applications," *Sensors and Actuators A: Physical*, 2008, 147(1): 150–164.

[12] J. Song, Q. Jiang, Y. Huang, Y. Li, Y. Jia, X. Rong, *et al*., "Research on pressure tactile sensing technology based on fiber Bragg grating array," *Photonic Sensors*, 2015, 5(3): 1–10.

[13] Q. C. Wang, Y. E. Xian-Feng, Y. L. Cao, and L. Y. Zhan, "Detection of focused ultrasonic fields with fiber Bragg grating," *Journal of Transcluction Technology*, 2005, 18(1): 171–174.

[14] H. Zhang, X. Sun, X. Qi, X. Liu, and H. Dong, "Ultrasonic lamb wave inspection using fiber Bragg gratings," in *IEEE 2008 China-Japan Joint Microwave Conference*, Shanghai, pp. 805–808, 2008.

[15] G. Coppola, A. Minardo, A. Cusano, G. Breglio, L. Zeni, A. Cutolo, *et al*., "Analysis of feasibility on the use of fiber Bragg grating sensors as ultrasound detectors," in *Proc. SPIE*, vol. 4328, pp. 224–232, 2001.

[16] N. E. Fisher, D. J. Webb, C. N. Pannell, D. A. Jackson, L. R. Gavrilov, J. W. Hand, *et al.*, "Ultra-sonic field and temperature sensor based on short in-fiber Bragg gratings," *Electronics Letters*, 1998, 34(11): 1139–1140.

Study of Rayleigh-Backscattering Induced Coherence Collapse in an Asymmetric DFB FL Sensor

Wen LIU[1*], Lina MA[2], Zhengliang HU[2], Ying FENG[1], and Huayong YANG[1]

[1]*College of Optoelectronic Science and Engineering, National University of Defense Technology, Changsha, 410073, China*

[2]*Academy of Ocean Science and Engineering, National University of Defense Technology, Changsha, 410073, China*

[*]Corresponding author: Wen LIU   E-mail: liusenanbei@163.com

Abstract: Rayleigh-back scattering induced coherence collapse of an asymmetric distributed feedback fiber laser (DFB FL) sensor is investigated using a composite cavity model. The coherence collapse threshold condition of the asymmetric DFB FL sensor is measured. The DFB FL sensor shows different dynamic behaviors in different pump configurations. According to the asymmetric behavior to the external optical feedback, a novel method to find the actual phase shift position of the asymmetric DFB FL sensor is presented.

Keywords: Asymmetric DFB fiber laser sensor; phase shift; coherence collapse; Rayleigh backscattering

1. Introduction

Distributed feedback fiber lasers (DFB FLs) have attracted general attention since 1990s, because of the impact structure, low noise figure, and the inherent wavelength-encoded multiplexing capability [1−4]. The laser frequency of a DFB FL is determined by the resonance condition of the laser cavity, which is sensitive to the physical parameters that could disturb the oscillation process. By using the dependence of the laser frequency on the physical parameters, DFB FL can be employed as a sensor [5]. DFB FL sensor arrays are normally arranged through remote pumping and interrogation configuration, which means that a long lead fiber is required to feed the pump light into DFB FLs [6]. In the lead fiber, Rayleigh backscattering may cause excess frequency noise or even unstable output of the DFB FLs, namely, coherence collapse [7], and substantially affects the performance of DFB FL sensors. The maximum multiplexing capacity of DFB FL array remains to be 16 until now [8], and coherence collapse induced by Rayleigh backscattering is considered as one of the main reasons for the limited array scale [9]. To improve the performance of the DFB FL sensor array, it is necessary to study the effect of coherence collapse induced by Rayleigh scattering of the lead fiber on the operation status of the DFB FL. The tolerable length of lead fiber was found to be 135 m−200 m for symmetric DFB FL [6], corresponding to a typical Rayleigh backscattering of −72 dB/m, while the dynamic behavior of asymmetric DFB FL with external optical feedback was rarely mentioned.

In this paper, we study the tolerance of asymmetric DFB FL to Rayleigh backscattering of a long lead fiber. We start to analyze the coherence collapse threshold condition of asymmetric DFB FL in Section 2. The dynamic behaviors of an asymmetric DFB FL sensor to Rayleigh backscattering is evaluated in different pump configurations in Section 3. In Section 4, a novel method to determine the phase shift position of asymmetric DFB FL is proposed based on the coherence-collapse threshold length measurement.

2. Composite cavity model

DFB FL consists of a phase-shift Bragg grating in a short piece of rare earth doped fiber. The symmetric grating structure is designed for DFB FL to obtain the single polarization laser output and reduce the pump threshold [10], but many asymmetric DFB FLs are made on purpose. In a sensor array, DFB FLs with asymmetrical outputs can modify the pump power distribution and improve the multiplexing capacity with fixed pump power budget [11, 12].

In asymmetric DFB FL, each grating segment on either side of the phase shift can be considered as a separate reflector. By simplifying the distributed feedback cavity to be a Fabry-Perot (FP) structure and considering the external optical feedback as the external cavity of the FL [13], a composite cavity model can be established as shown in Fig. 1. r_1 and r_2 are the reflection coefficients of the FP cavity, r_{ext} is the reflection coefficient of the external reflection facet, L is the effective length of the FP cavity, and d is the external path length.

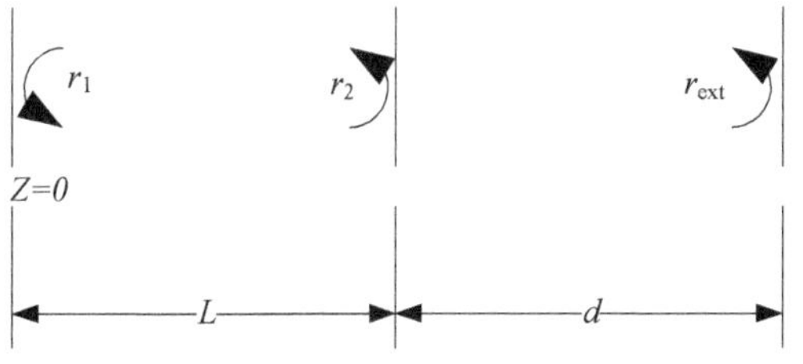

Fig. 1 Schematic diagram of the composite cavity model.

The composite cavity can be equivalent to an FP cavity, and the reflection coefficients of the equivalent FP cavity can be written as

$$\begin{cases} r_L = r_1 \\ r_R = r_2\left[1+\dfrac{T_2 r_{\text{ext}}\cos(2\pi\nu\tau_e)}{r_2}\right] - \mathrm{i}T_2 r_{\text{ext}}\sin(2\pi\nu\tau_e) \end{cases} \tag{3}$$

where T_2 is the transmissivity of right piece of uniform grating segment, and τ_e is the external round trip delay.

The steady-state laser oscillation condition for the equivalent FP cavity can be given as

$$|r_L||r_R|\exp[(g_c-\gamma)L]=1 \tag{2}$$

$$\Delta\varphi = 0 = \frac{\tau_L}{\tau_e}[2\pi\tau_e(\nu-\nu_0)+ \frac{T}{r_2}\frac{\tau_e}{\tau_L}\sqrt{1+\alpha^2}r_{\text{ext}}\sin(2\pi\nu_0\tau_e+2\pi\Delta\nu\tau_e-\theta_\alpha)] \tag{3}$$

where ν_0 is the laser frequency without optical feedback, τ_L is the roundtrip time of the solitary laser, α is the linewidth enhancement factor, and θ_α is the associated phase. The frequency shift $\Delta\nu$ due to the optical feedback and the feedback parameter C can be expressed as

$$\tau_e\Delta\omega = -C\sin(\Delta\omega\tau_e+2\pi\nu_0\tau_e-\theta_\alpha) \tag{4}$$

$$C=\frac{T}{r_2}\frac{\tau_e}{\tau_L}\sqrt{1+\alpha^2}r_{\text{ext}}\,. \tag{5}$$

For very weak feedback with $C<1$, (4) has only one solution. When $C\geq 1$, there are more than one solutions for (4), indicating that the external optical feedback splits the single mode operation into multiple external cavity modes, accompanying with dramatically increased noise level and intensity fluctuation, resulting in coherence collapse. As a result, $C=1$ may be defined as the coherence collapse threshold condition for DFB FL.

For an asymmetric DFB FL, the reflection coefficient and the transmissivity of each grating segment at the Bragg wavelength are given by [13]

$$r_i \approx -\tanh(-\kappa l_i) \tag{6}$$

$$T_i \approx 4\exp(-\kappa l_i) \tag{7}$$

where l_i with i=1, 2 is the length of the grating segment. The effective cavity length of DFB FL can be written as

$$L \approx (r_1+r_2)/2\kappa\,. \tag{8}$$

By substituting (7) and (8) into (5), the feedback parameter of asymmetric DFB FL can be rewritten as

$$C = \frac{8\kappa\sqrt{1+a^2}\, r_{\text{ext}} d \exp(-2\kappa l_2)}{r_2(r_1+r_2)}. \tag{9}$$

For a DFB FL sensor system, the external optical feedback mainly comes from external reflecting facet, splice points, and Rayleigh backscattering [14]. The influence of former two sources can be eliminated to the negligible extent, but the detrimental effect of the last one is never possible to be entirely ignored, which becomes the main factor to limit the performance of the DFB FL sensor. Then the coherence collapse threshold length of lead fiber in the DFB FL system can be given by

$$d = \left(\frac{r_2(r_1+r_2)}{8\kappa\sqrt{1+a^2}\, r_R \exp(-2\kappa l_2)} \right)^{\frac{2}{3}}. \tag{10}$$

When $l_1 = l_2$, (10) can be used to calculate the threshold condition for symmetric DFB FL, which coincides with the conclusion got by E. Rønnekleiv *et al.* [6].

3. Experiments and results

A high resolution frequency shift demodulation system is constructed to evaluate the influence of external optical feedback on the asymmetric DFB FL sensor. The schematic diagram is shown in Fig. 2. The DFB FL sensor is a fiber laser hydrophone with ultrahigh pressure sensitivity [15]. The total length of the laser cavity is 40 mm, and the phase shift section is 4 mm long. The DFB FL sensor is pumped by a 975-nm semiconductor laser at 63 mW through a 980/1550 WDM (wavelength division multiplexer). The output signal of the FL sensor is guided to an unbalanced Michelson interferometer through an isolator. The optical path difference of the interferometer is 15 m. A sinusoidal signal at the frequency of 12.5 kHz is loaded to the piezoelectric ceramic transducer (PZT) in one arm of the interferometer. The interference signal is converted into electric voltage and then acquired by an acquisition card. A phase generated carrier (PGC) scheme is adopted to achieve high resolution interrogation. The whole optical setup is packaged to isolate the environment fluctuation, and the end facet of the pigtail is immersed in the refractive index matched gel to eliminate the back reflection.

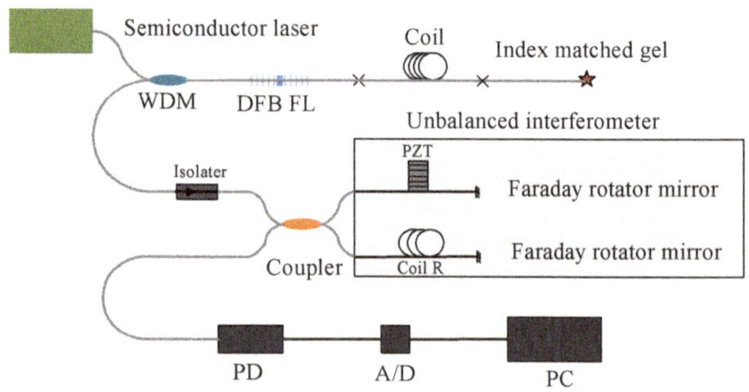

Fig. 2 Schematic diagram of the frequency shift demodulation system.

The distance between the WDM and DFB FL sensor is set to 1 m, while the lead fiber on the other end of the DFB FL sensor is increased by 10 m each time until coherence collapse occurs. The two ends of the asymmetric DFB FL sensor are labeled as Port 1 and Port 2, and the DFB FL sensor is pumped through Port 1 at first. The phase noise floor of the DFB FL sensor with different lengths of lead fiber is shown in Fig. 3, and the demodulated interference signals are shown in Fig. 4.

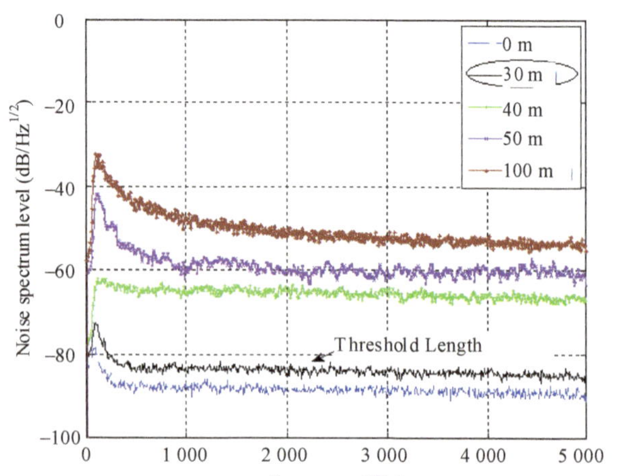

Fig. 3 Phase noise level of DFB FL sensor due to an increase in the length of lead fiber.

It is illustrated that the phase noise level of the DFB FL sensor rises with an increase in the length of the lead fiber. When no extra lead fiber is added to the pigtail, the phase noise of the DFB FL sensor is about –89 dB re. rad/Hz$^{1/2}$ @1kHz. When the lead fiber is 30 m long, the phase noise increases by 5 dB, and the interference signal is still stable as shown in Fig. 4. However, when another 10-m lead fiber increases, the phase noise increases by over 10 dB, and there is self-pulsing in the interference signal, hinting that coherence collapse occurs.

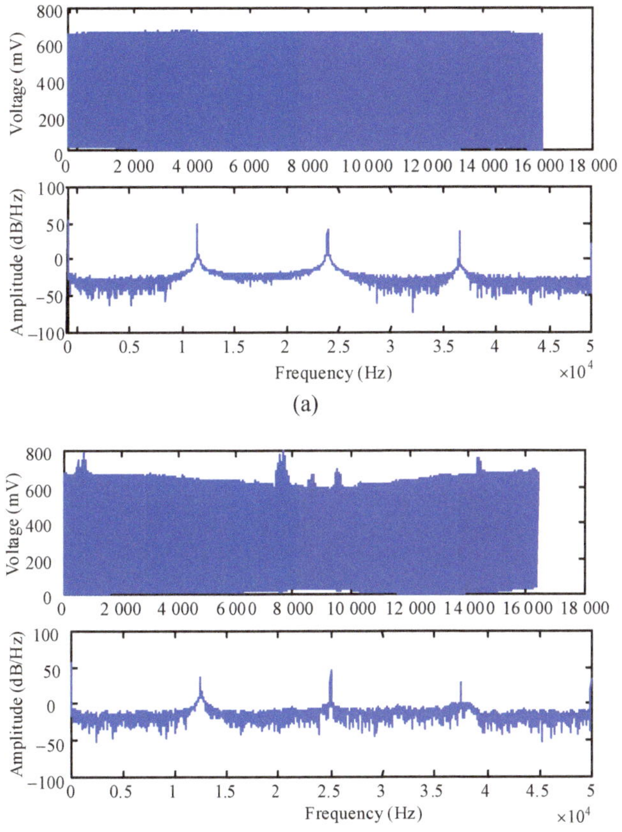

Fig. 4 Demodulated interference signal with different length of lead fiber: (a) d=30 m and (b) d=40 m.

Through a similar approach, the coherence collapse threshold length of the lead fiber of the DFB FL sensor is measured when pumped from Port 2. The threshold length and the laser output power of the DFB FL sensor in different pump configurations are presented in Table 1.

Table 1 Coherence collapse threshold length and laser output power of the asymmetric DFB FL sensor.

Pump port	Port 1	Port 2
Threshold length (m)	14.13	43.89
Output power (μW)	30	100

It is indicated that the asymmetric DFB FL sensor has distinctive threshold lengths of the lead fiber in different pump configurations, and the difference of the threshold length exceeds 50 m. When the DFB FL sensor system is arranged through a remote-pumping and interrogation configuration, the DFB FL sensor should be pumped from Port 2 to extend the total length of the sensing system.

4. Phase shift assessment method based on coherence collapse threshold length

Based on (6) and (10), the coherence collapse threshold length ratio of the asymmetric DFB FL in different pump configurations satisfies

$$\frac{d_1}{d_2}=\left[\frac{\tanh(\kappa l_2)}{\tanh(\kappa l_1)}\exp(2\kappa l_1-2\kappa l_2)\right]^{2/3}. \quad (11)$$

For an asymmetric DFB FL with 40-mm cavity length and 4-mm phase shift section, the threshold length ratio to the grating segment length near Port 1 is shown in Fig. 5, where the estimated grating coupling coefficient is κ=150 m^{-1}.

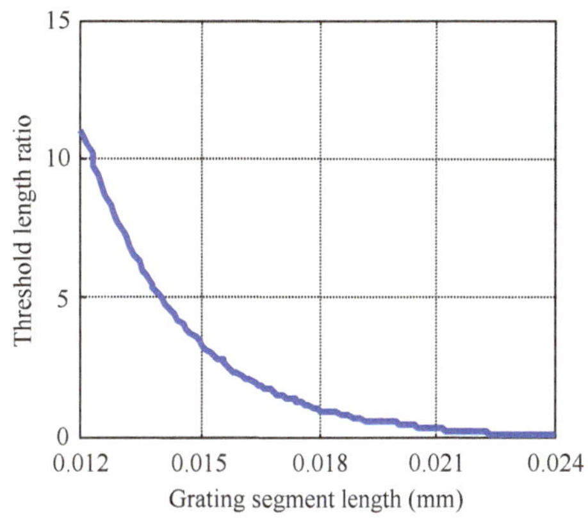

Fig. 5 Phase noise level of the DFB FL sensor due to an increase in the external path length.

As the threshold length ration of the DFB FL sensor is 10/3, the lengths of grating segments beside of the phase shift section can be deduced as 15 mm and 21 mm, respectively.

In the field test, the performance of the DFB FL sensor always differs from the initial DFB FL before mounting, as non-uniform extension or twisting of the laser cavity is introduced by the mounting structure. The method proposed here can be used to understand the real physical state of the FL cavity, and help to optimize the sensor performance.

5. Conclusions

The stability of the asymmetric DFB FL sensor with Rayleigh scattering of the lead fiber is studied. The coherence collapse threshold length of an asymmetric DFB FL sensor is measured based on a high resolution frequency shift demodulation system. The difference in the threshold length of the same FL sensor in different pump configurations suggests a novel method to attain the phase shift position, which is significant to enlarge the total length of the DFB FL sensor system. More details need to be revealed for the asymmetric DFB FL with different phase shifts at different positions, and the coherence collapse characteristics of the asymmetric DFB FL sensor array will be studied in further work.

Acknowledgment

This work was supported by the National Science Foundation of China (NSFC 11304388).

References

[1] J. T. Kringlebotn, J. L. Archambault, L. Reekie, and D. N. Payne, "Er^{3+}:Yb^{3+}-codoped fiber distributedfeedback laser," *Optics Letters*, 1994, 19(24): 2101−2103.

[2] D. J. Hill, P. J. Nash, D. A. Jackson, D. J. Webb, S. F. O'Neill, I. Bennion, *et al.*, "A fiber laser hydrophone array," *Proc. SPIE*, 1999, 3860: 55−66.

[3] D. J. Hill, J. D. Hodder, S. D. Freitas, S. D. Thomas, and L. Hickey, "DFB fibre laser sensor developments," *Proc. SPIE*, 2005, 5855: 904−907.

[4] G. A. Cranch, G. M. H. Flockhart, and C. K. Kirkendall, "Distributed feedback fiber laser strain sensors," *IEEE Sensors Journal*, 2008, 8(7): 1161−1172.

[5] Y. Liu, W. Zhang, T. Xu, J. He, F. Zhang, and F. Li, "Fiber laser sensing system and its applications," *Photonic Sensors*, 2011, 1(1): 43−53.

[6] E. Rønnekleiv, O. Hadeler, and G. Vienne, "Stability of an Er-Yb-doped fiber distributed-feedback laser with external reflections," *Optics Letters*, 1999, 9(24): 617−619.

[7] R. W. Tkach, "Regimes of feedback effects in 1.5μm distributed feedback lasers," *Journal of Lightwave Technology*, 1986, 4(11): 1655−1661.

[8] S. Foster, A. Tikhomirov, M. Englund, H. Inglis, G. Edvell, and M. Milnes, "A 16 channel fibre laser sensor array," in *ACOFT/AOS*, 2006: 40−42.

[9] A. Tikhomirov and S. Foster, "DFB FL sensor multiplexing noise," in *ACOFT/AOS*, 2006: 60−62.

[10] M. Yamada and K. Sakuda, "Analysis of almost-periodic distributed feedback slab waveguides via a fundamental matrix approach," *Applied Optics*, 1987, 26(16): 3474−3478.

[11] P. Wang, J. Chang, C. Zu, B. Sun, G. Lv, S. Zhang, *et al.*, "A four-element sensor array consisting of asymmetric distributed-feedback fiber lasers," *Photonic Sensors*, 2014, 4(2): 180−187.

[12] H. F. Qi, Z. Q. Song, S. J. Li, J. Guo, C. Wang, and G. D. Peng, "Apodized distributed feedback fiber laser with asymmetrical outputs for multiplexed sensing applications," *Optics Express*, 2013, 21(9): 11309−11314.

[13] F. Favre, "Theoretical analysis of external optical feedback on DFB semiconductor lasers," *IEEE Journal of Quantum Electronics*, 1995, 1(23): 81−88.

[14] K. Yelen, L. M. B. Hickey, and M. N. Zervas, "A new design approach for fiber DFB lasers with improved efficiency," *IEEE Journal of Quantum Electronics*, 2011, 40(6): 711−720.

[15] L. Ma, Y. M. Hu, H. Luo, and Z. L. Hu, "DFB fiber laser hydrophone with flat frequency response and enhanced acoustic pressure sensitivity," *IEEE Photonics Technology Letters*, 2009, 21(17): 1280−1282.

23

Distributed Acoustic Sensing With Michelson Interferometer Demodulation

Xiaohui LIU[1], Chen WANG[1], Ying SHANG[1], Chang WANG[1*], Wenan ZHAO[1], Gangding PENG[1,2], and Hongzhong WANG[3]

[1]*Shandong Key Laboratory of Optical Fiber Sensing Technologies, Laser Institute of Shandong Academy of Sciences, Jinan, Shandong, 250014, China*

[2]*School of Electrical Engineering & Telecommunications, the University of New South Wales, NSW, 2052, Australia*

[3]*Shengli Oilfield Xinsheng Geophysical Technology Co. Ltd., Dongying, Shandong, 257000, China*

[*]Corresponding author: Chang WANG E-mail: wang960100@163.com

Abstract: The distributed acoustic sensing (DAS) has been extensively studied and widely used. A distributed acoustic sensing system based on the unbalanced Michelson interferometer with phase generated carrier (PGC) demodulation was designed and tested. The system could directly obtain the phase, amplitude, frequency response, and location information of sound wave at the same time and measurement at all points along the sensing fiber simultaneously. Experiments showed that the system successfully measured the acoustic signals with a phase-pressure sensitivity about −148 dB (re rad/μPa) and frequency response ripple less than 1.5 dB. The further field experiment showed that the system could measure signals at all points along the sensing fiber simultaneously.

Keywords: Distributed acoustic sensing; phase generated carrier; interferometer; frequency response

1. Introduction

Distributed acoustic sensing (DAS) is a novel technology which offers the capability of measurement at thousands of points simultaneously, using a simple and unmodified optical fiber as the sensing element. Compared with conventional point sensors, DAS needs no special reflectors or fiber Bragg gratings in its optical path, which greatly reduces the operation difficulty in field test. And it can measure thousands of continuous points along the sensing fiber and form acoustic or seismic imaging. These advantages cause DAS to be extensively studied and widely used in the fields of oil and gas exploration, national defense, security, and so on [1–5].

The DAS system operates according to a radar-style process: it sends a series of pulses into the fiber and records the return of the naturally occurring scattered signal against time. One representative of this backscattered sensing is phase sensitive optical time domain reflectometer (φ-OTDR), which inputs a narrow line-width laser to the sensing fiber and monitors the phase changes of Rayleigh backscattered lights [6–9]. The other one is Brillouin optical time domain reflectometer (BOTDR) which uses Brillouin backscattered lights to measure dynamic strain and temperature [10–12].

A major limitation of these distributed sensors above is that they are incapable of determining the full vector acoustic field, namely the amplitude, frequency, and phase, which is a necessity for acoustic or seismic imaging. Measuring the full acoustic field is a much harder technical challenge to be overcome. But, in doing this, it is possible to achieve high resolution seismic imaging and also make other novel systems, for example, a massive acoustic antenna.

In this work, a DAS system based on the Michelson interferometer with phase generated carrier (PGC) demodulation was designed and tested. Experiments showed that the system could successfully acquire the acoustic signal information, including the location, frequency, amplitude, and phase, at all points along the sensing fiber simultaneously.

2. System design

The DAS system diagram is shown in Fig. 1. The light source used in the system is a distributed-feedback laser diode (DFB-LD) which emits a continuous wave light with the power of 10 mW and linewidth of 5 kHz. The light is injected into an acoustic-optic modulator (AOM) to generate a pulsed light. The pulses of the light have the width of 50 ns and repetition rate *R* of 50 kHz. Then the modulated pulses are amplified by an erbium-doped fiber amplifier (EDFA) in which a narrow-band light filter is integrated to remove the spontaneous emission noise. Then the amplified pulses are launched into a single mode sensing fiber (Corning SMF-28e) by a circulator, and the Rayleigh-backscattered light is excited at all positions of the sensing fiber. When an acoustic signal is applied on a position of the sensing fiber, it changes the optical path difference of the two backscattered lights exited ahead and behind of the position. So we can introduce an unbalanced interferometer to make the two neighborhood-position's Rayleigh-backscattered lights interfere and obtain the acoustic signal. The backscattered light is amplified by another EDFA to improve the signal-to-noise-ratio (SNR). Then the amplified light passes through an isolator and is injected into a Michelson interferometer which consists of a 2×2 coupler, a phase modulator (PM), and two Faraday rotation mirrors (FRMs). The light is split into two equal beams by the coupler. One of the beams is modulated by the PM with the 2-kHz sinusoidal wave to generate the phase carrier, which decides that the highest signal frequency measured is no more than 1 kHz theoretically. The length difference of two interferometer arms is 5 m. The final interference signals outputted from the coupler are detected by a photodetector (PD) and digitized by an analog to digital (A/D) converter. Then the digitized signals are arrayed according to the position of the sensing fiber. Then the signals processing scheme is accomplished by a software program of PGC demodulation.

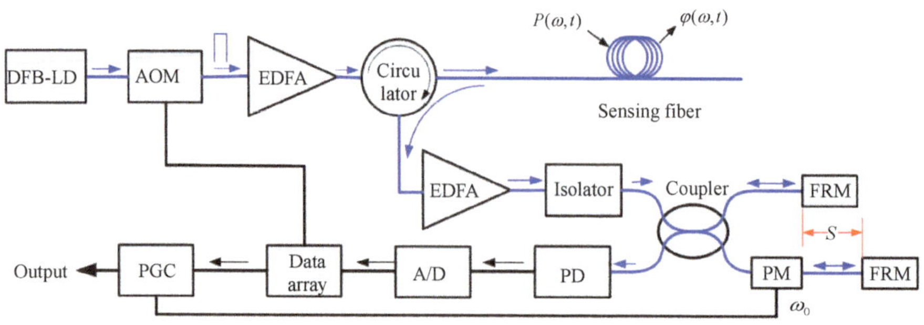

Fig. 1 DAS system block diagram.

The interferometer output signal I can be expressed as

$$I = A + B\cos\left[C\cos\omega_0 t + \psi(t)\right] \quad (1)$$

where A is the average optical power of the interferometer output signal. $B=kA$, $k\leq 1$, and k is the interference fringe visibility. $C\cos\omega_0 t$ is the phase generated carrier. $\psi(t)=\varphi(\omega, t)+\phi(t)$, $\varphi(\omega, t)$ is the phase change induced by the acoustic signal, and $\phi(t)$ is the slow variation of the initial phase caused by the environment.

The PGC demodulation scheme is shown in Fig. 2. The interferometer output signal I is fed into the PGC demodulation scheme, and the final output of the system, which contains the tested signal $\varphi(\omega, t)$, is [13]

$$B^2GHJ_1(C)J_2(C)\phi(\omega,t) \equiv K\phi(\omega,t) \quad (2)$$

where G and H are the amplitude of fundamental and double frequency signal, respectively. $K=B^2GHJ_1(C)J_2(C)$ is the gain factor of the system. Finally, we calibrate the constant K and obtain the acoustic signal from the high-pass filter (HPF).

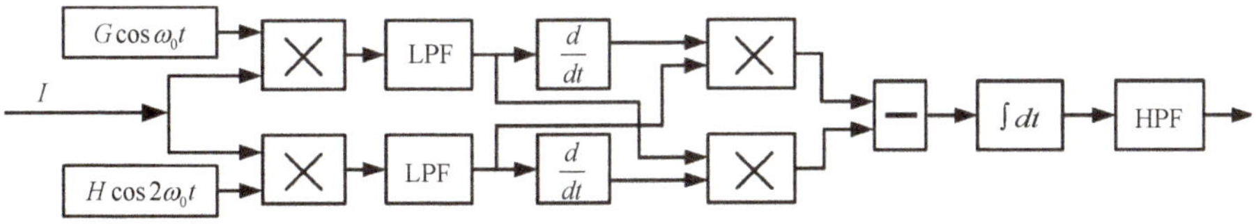

Fig. 2 PGC demodulation scheme.

By introducing the unbalanced interferometer with a arm-length difference of 5 m, the signal we got every time represents the phase change between two points of 5 m apart on the sensing fiber. That is to say, the spatial resolution of the system is 5 m. The repetition rate of the pulses decides the system's maximum detection length. The time interval between two pulses should be longer than the round trip time of the light traveling in the sensing fiber to insure that there is only one pulse inside the fiber at any time. For the 50-kHz repetition rate, we can know that the detection range is around 2 km which is determined by $L<c/2Rn_f$.

3. Lab experiments

A water tank system was used to examine the performance of the DAS system (Fig. 3). An underwater speaker was fixed in water and driven by a function generator. A sensing fiber with a length of 10 m was wrapped into a fiber ring with a diameter of 8 cm to reduce the influence of the acoustic-intensity differences, which was put into water and connected to the DAS instrument. A commercial piezoelectric hydrophone (Type RHSA20 made by 715th Research Institute of China Shipbuilding Industry Corporation) was also placed close to the fiber ring to measure the acoustic pressure amplitude. The sound fields of the fiber ring and the piezoelectric hydrophone were almost the same.

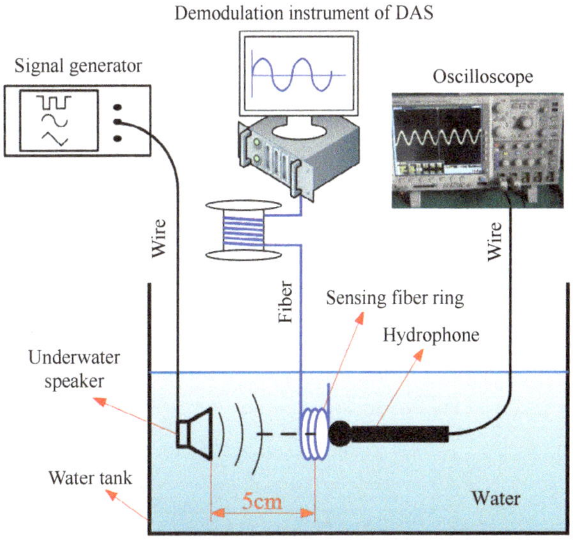

Fig. 3 Schematic diagram of underwater experiment.

Firstly, we fixed the frequency at 200 Hz and changed the amplitude of the sinusoidal wave generated by the function generator. The acoustic pressure amplitudes in different acoustic intensities were measured by the piezoelectric hydrophone, and the phase-pressure sensitivity of the DAS system was calculated. Figure 4 shows the instantaneous frequency of our system demodulated and the spectral analysis via fast Fourier transform (FFT) of the demodulated signal. It indicates that the signals of DAS and hydrophone have almost the same waveform and SNR. It demonstrates that the DAS system can measure the full vector acoustic field, including the amplitude, frequency, and phase. Figure 5 shows the phase-pressure relationship of the DAS system. When the acoustic pressure increases, the demodulated phase variation increases linearly. The phase-pressure sensitivity was calculated through our system demodulated phase amplitude divided by piezoelectric hydrophone detected acoustic pressure amplitude. The achieved phase-pressure sensitivity was 0.0384 rad/Pa (−148 dB, re rad/μPa).

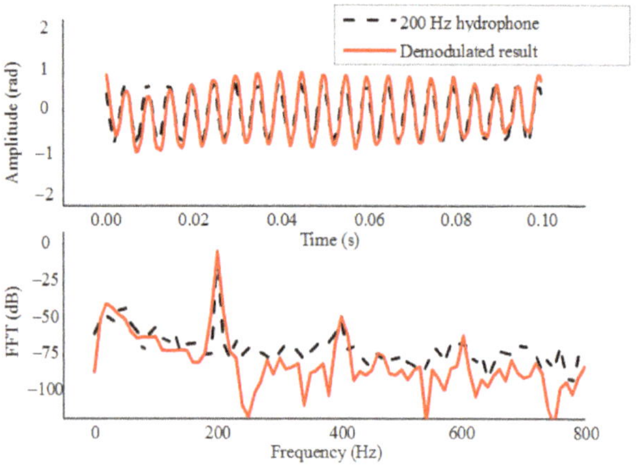

Fig. 4 Signals of DAS sensor and hydrophone.

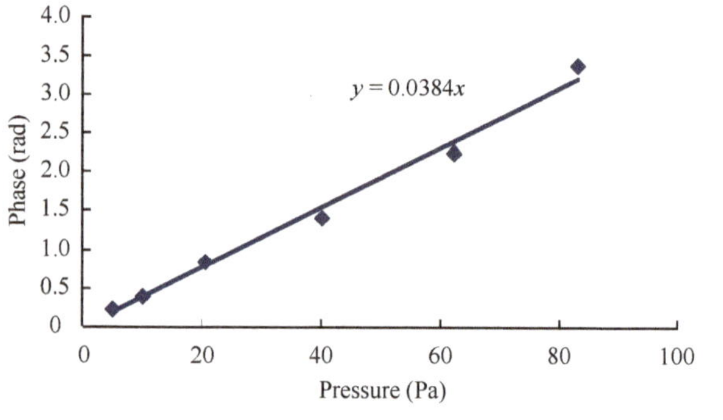

Fig. 5 Phase-pressure relationship.

Secondly, we fixed the amplitude and changed the frequency of the sinusoidal wave generated by the function generator to measure the frequency response of our system. Figure 6 shows the frequency response of the system from 50 Hz to 500 Hz. The response flatness of less than 1.5 dB was obtained.

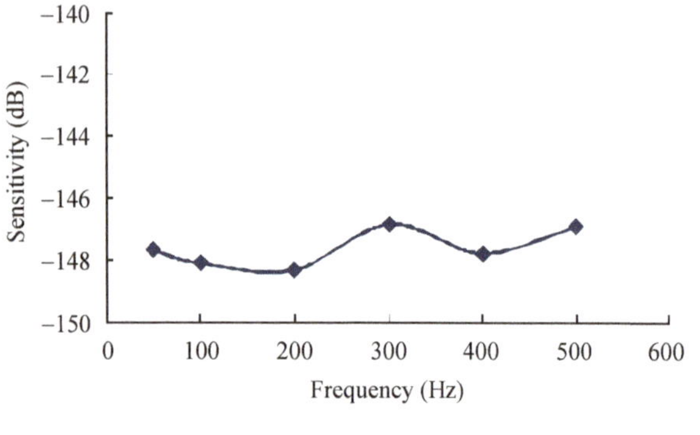

Fig. 6 DAS frequency response curve.

4. Field experiments

Furthermore, a field experiment was introduced to examine the system. We buried a 110-m-long fiber cable with 50 cm underneath the ground, which was marked from 33 m to 143 m shown in Fig. 7. And a man walked on the ground along the buried fiber cable. The fiber cable was connected to the DAS instrument in which there was a monitor program running to process the acoustic signals of all the points on the fiber cable in real time. The output of the program is shown in Fig. 8. The gray level represents the intensity, which is the square of the amplitude of the acoustic signal we tested. According the figure, we can clearly identify the signal of every walking step and infer the man walking 19 steps in 10 s and about 15 m forward. So the signals at all points along the sensing fiber can be measured by the DAS system simultaneously.

Fig. 7 Schematic diagram of walking experiment.

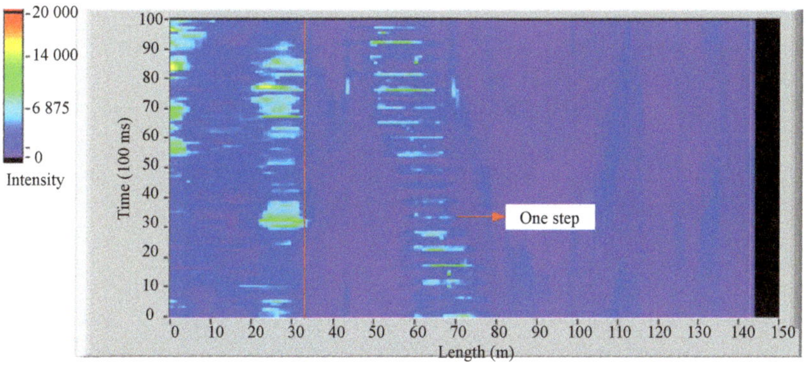

Fig. 8 Output of DAS instrument program.

5. Conclusions

An optical fiber distributed acoustic sensing system based on the unbalanced Michelson interferometer and PGC demodulation is demonstrated. The system can directly obtain the phase, amplitude, frequency response, and location information of sound wave at the same time and measure at all points along the sensing fiber simultaneously. Experiments showed that the system successfully measured the acoustic signals with the phase-pressure sensitivity of about –148 dB (re rad/μPa) and frequency response flatness of less than 1.5 dB. The further field experiment showed that the system could measure signals at all points along the sensing fiber simultaneously. The DAS can be used as geophone for surface, seabed, and downhole measurements, encompassing vertical seismic profiles. It can also offer integrity and surveillance monitoring solutions for pipelines and intrusions by using a single optical fiber cable.

Acknowledgment

This work was supported by Science and Technology Development Projects (Nos. 2014GGX103019 and 2015GSF120001) of Shandong Province, and Independent Innovation Major Project (No. 2014ZZCX04206) of Shandong Province.

References

[1] M. D. Thomas, M. F. Barry, A. Jonathan, and S. Dou, "Field testing of fiber-optic distributed acoustic sensing (DAS) for subsurface seismic monitoring," *Leading Edge*, 2013, 32(6): 699–706.

[2] A. Mateeva, J. Lopez, H. Potters, J. Mestayer, B. Cox, D. Kiyashchenko, *et al.*, "Distributed acoustic sensing for reservoir monitoring with vertical seismic profiling," *Geophysical Prospecting*, 2014, 62(4): 679–692.

[3] M. Mondanos, T. Parkera, C. H. Milnea, J. Yeoa, T. Colemana, and M. Farhadiroushana, "Distributed temperature and distributed acoustic sensing for

remote and harsh environments," *SPIE*, 2015, 9491(0F): 1–8.

[4] M. M. Molenaar, D. Hill, P. Webster, E. Fidan, and B. Birch. "First downhole application of distributed acoustic sensing for hydraulic-fracturing monitoring and diagnostics," *SPE Drilling and Completion*, 2011, 7(2): 193–202.

[5] A. Owen, G. Duckworth, and J. Worsley, "OptaSense: fiber optic distributed acoustic sensing for border monitoring," *Fiber Optic Sensors & Systems*, 2012, 26(8): 362–364.

[6] J. C. Juarez, E. W. Maier, K. N. Choi, and H. F. Taylor, "Distributed fiber-optic intrusion sensor system," *Journal of Lightwave Technology*, 2005, 23(6): 2081–2087.

[7] J. C. Juarez and H. F. Taylor, "Field test of a distributed fiber-optic intrusion sensor system for long perimeters," *Applied Optics*, 2007, 46(11): 1968–1971.

[8] M. Zhang, S. Wang, Y. Zheng, Y. Yang, X. Sa, and L. Zhang, "Enhancement for Φ-OTDR performance by using narrow linewidth light source and signal processing," *Photonic Sensors*, 2016, 6(1): 58–62.

[9] F. Peng and X. Cao, "A hybrid Φ/B-OTDR for simultaneous vibration and strain measurement," *Photonic Sensors*, 2016, 6(2): 121–126.

[10] Y. Dong, L. Chen, and X. Bao, "Time-division multiplexing based BOTDA over 100 km sensing length," *Optics Letters*, 2011, 36(2): 277–279.

[11] A. Minardo, A. Coscetta, R. Bernini, and L. Zeni, "Heterodyne slope-assisted brillouin optical time-domain analysis for dynamic strain measurements," *Journal of Optics*, 2016, 18(2): 1–7.

[12] X. Bao and L. Chen, "Recent progress in optical fiber sensors based on Brillouin scattering at University of Ottawa," *Photonic Sensors*, 2011, 1(2): 102–117.

[13] Z. Sun, J. Wang, J. Chang, J. Ni, L. Min, C. Wang, *et al.*, "Fiber laser sensor interrogation system development and test," *SPIE*, 2012, 8351(1): 73–77.

24

Modeling of All-Optical Even and Odd Parity Generator Circuits Using Metal-Insulator-Metal Plasmonic Waveguides

Lokendra SINGH, Amna BEDI, and Santosh KUMAR*

Photonics Lab, Department of Electronics and Communication Engineering, DIT University, Dehradun-248009, India

*Corresponding author: Santosh KUMAR E-mail: santoshrus@yahoo.com

Abstract: Plasmonic metal-insulator-metal (MIM) waveguides sustain excellent property of confining the surface plasmons up to a deep subwavelength scale. In this paper, linear and S-shaped MIM waveguides are cascaded together to design the model of Mach-Zehnder interferometer (MZI). Nonlinear material has been used for switching of light across its output ports. The structures of even and odd parity generators are projected by cascading the MZIs. Parity generator and checker circuit are used for error correction and detection in an optical communication system. Study and analysis of proposed designs are carried out by using the MATLAB simulation and finite-difference-time-domain (FDTD) method.

Keywords: Plasmonics; MIM waveguides; Mach-Zehnder interferometer; nonlinear process; FDTD method

1. Introduction

In today's scenario of photonic industry, the major challenge is the miniaturization of optical components for wavelength division multiplexing [1]. Despite the deep subwavelength confinement of surface plasmon polaritons (SPPs) with respect to other nanophotonic waveguides such as photonic crystals, SPP waveguides work well beyond the diffraction limit [2–4]. Plasmonics is one of the promising fields of integrated photonics due to its numerous applications in optical clocking, cancer therapy, nano-lithography, solar cells, etc. In designing of building blocks for future generation of integrated optical components and devices, plasmonic waveguides play a significant role. Recently, several types of waveguides were discovered such as slots [5, 6], plasmonic wedge waveguide [7], and dielectric ridge on the metal surface [8], but among all these waveguides the metal-insulator-metal (MIM) waveguide is generally preferred due to its their exceptional property of confining the surface plasmons to a deep subwavelength scale [9–13]. In these waveguides, propagation is achieved by using even modes due to their low-loss confinement profile. Considering the photonic device altogether, directional couplers (DCs) are serves as a prime component to design the elementary part of optical circuits. Recently, several designs of directional couplers have been proposed by using plasmonic waveguides [14–17]. The major phenomenon of coupling of light within the DC is beating of even-odd modes. As the equal splitting of power through the directional coupler is wavelength dependent, for broadband response either wavelength dependency should be removed by making coupler adiabtic, or input power with its coupling length should be controlled. In the proposed structure of DC connected in the Mach-Zehnder interferometer (MZI), equal splitting of power is attained with coupling length and propagation length of 1 μm and 10.2 μm, respectively. Hence, by adjoining the two DCs using two linear MIM waveguides, a design of MZI has been designed within the footprints of (16.2×4) μm. Although interferometric circuit using Fabry-Perot interferometer is also used in various applications, the need of too many sources will lead to an increase in the cost and complexity of devices [18]. One of the arms of MZI is filled with an active material. Here, MEH-PPV [poly (2-methoxy-5-(28-ethylexyloxy)-PPV)] is used as an active material with refractive index and response time of n=1.65 and τ=2.1e^{-15} s, respectively [19]. The presence of active materials in the structure of MZI is used to switch the light across its output ports by changing the refractive index of waveguide with respect to power of input signals. The changes in phase canceling out or adding each other depend on nonlinear processes. If phase of both the signals cancels out each other then it crosses phase modulation (XPM), and if it adds then it is self phase modulation (SPM). Due to XPM, the optical signal arrives at through port and SPM thrives the signal at cross port of MZI. Thus, by cascading plasmonic DCs and MZIs, all-optical structures of even and odd parity generators are modeled within the footprints of (125×10) μm and (205×10) μm, respectively. Recently, some switching circuits have been proposed by using $LiNbO_3$-based MZI, which work on the principle of electro-optic effect [20]. Although the work is excellent, it still lacks advanced applications in terms of device miniaturization up to a deep subwavelength scale (hence overall design of device remains in thousand of micrometers), and due to the presence of electrical signals for altering the refractive index of waveguides, it limits the speed of device. Some switching circuits were also proposed by using semiconductor optical amplifier (SOA) but they were backsitted due to the phenomenon of gain saturation [21]. All optical logic gates were also proposed using photonic crystal but being lack of the nonlinear material inside the structure will enable the switching of optical signals for complex circuits [22].

The proposed design of parity generators are verified by using the finite difference time domain (FDTD) method [23] and MATLAB simulation. Section 2 presents the models of even and odd parity generators with their mathematical formulations. In Section 3, the propagation of light through proposed structures is projected. Finally, the conclusions of complete work are presented in Section 4.

2. Design of parity generator circuits

The combinational circuits are useful in designing various integrated circuits which are very useful in micro-controlling and central processing units. In this work, for designing the all optical circuits of parity generators, MIM plasmonic waveguides are used. To create the MIM geometry, a metal (SiON) having high refractive index of $n_1 = 2.01$ with permittivity of −1+j2.9 is deposited over a low index substrate having refractive index of $n_2 = 1.44$, and air is considered as dielectric between them. The thickness of dielectric is taken as 0.5 μm for the propagation of surface plasmons with a propagation length of 10.2 μm. Thus by using these MIM waveguides, the structure of MZI is designed as shown in Fig. 1, whose second linear arm is filled with nonlinear Kerr or active material. When the optical signals with power of E_{in}=0.014 W/μm (considered as digital logic "0") are given at the first input port, then by obeying the phenomenon of SPM, signals are obtained at the second output port of MZI as shown in Fig. 2(a). When signals with the power E_{in}=0.081 W/μm (considered as digital logic "1") are incident at the first input port of MZI, then due to XPM optical signals arrive at the first output port of it as shown in Fig. 2(b). The extinction ratio (ER) of single MZI is about 26 dB at output power difference of 0.7 dB and 0.8 dB. The analyzed value of ER for single MZI is quite useful and enough for designing the all-optical logic gates [24].

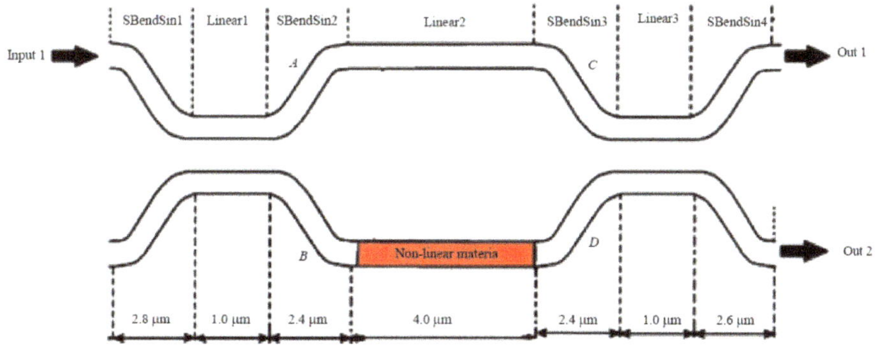

Fig. 1 Schematic of single Mach-Zehnder interferometer.

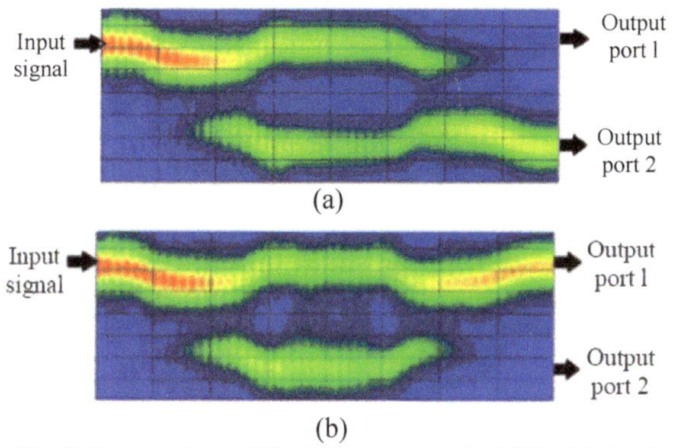

Fig. 2 Propagation of light through single MZI: (a) for low intensity signals and (b) for high intensity signals.

Thus, this toggling of optical signals across output ports of MZI with respect to input power of source is mathematically written as [25–28]

$$T_{\text{out1}} = \sin^2\left(\frac{\Delta\varphi}{2}\right) at E_{\text{in}} = 0.014 \text{ W/μm} \quad (1)$$

$$T_{\text{out2}} = \cos^2\left(\frac{\Delta\varphi}{2}\right) at E_{\text{in}} = 0.081 \text{ W/μm} \quad (2)$$

where $\Delta\varphi$ is the total change in phase in two linear arms of MZI. Equations (1) and (2) represent the mathematical expressions for outputs of MZI when input is fed with high and low powers, respectively.

2.1 Even parity generator

Figure 3 shows the digital circuit and K-map of three bit even parity generators, and Table 1 presents the truth table of the even parity generator. The circuit of the even parity generator is designed by cascading six MZIs and four directional couplers as shown in Fig. 4. For the proper functioning of device, three optical sources are placed in front of the second input ports of MZI1, MZI2, and MZI3. Three directional couplers DC1, DC2, and DC3 for splitting the optical energy into two equal parts are connected at the second output port of all three MZIs, respectively. The second output ports of DC1 and DC2 are combined together with the first output port of DC3 to feed the MZI5. While the first output ports of DC1 and DC2 are combined together to feed the MZI4 at the second input port DC4 is connected at the second output port of same MZI. Then the second output ports of DC3 and DC4 are combined together to feed MZI6 at its first input port. Finally, the second and first output ports of MZI5 and MZI6 are combined together to get the required output of even parity generator.

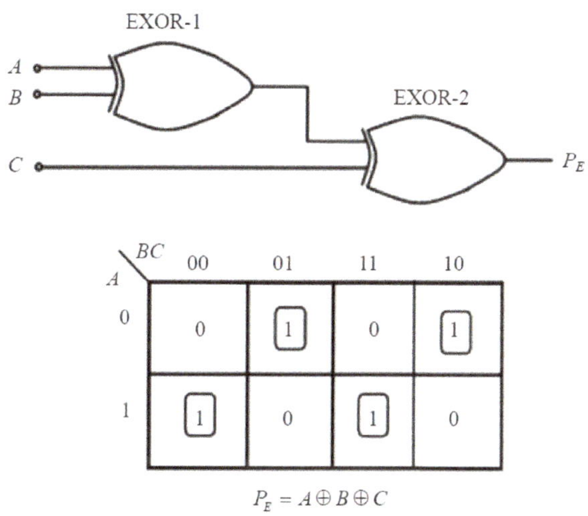

Fig. 3 Digital circuit and K-map of three bit even parity generator circuit.

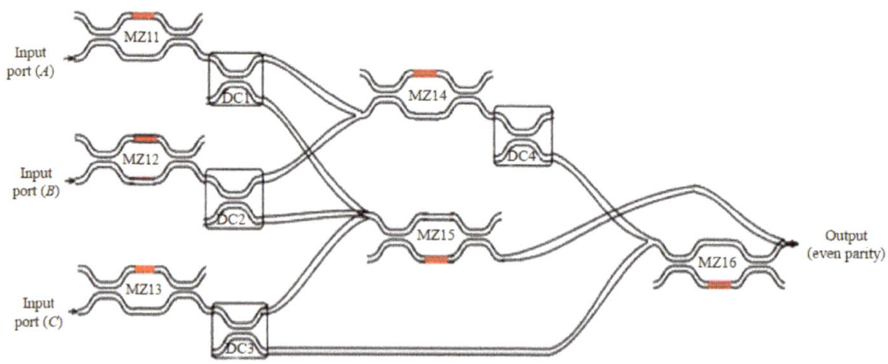

Fig. 4 Schematic diagram of even parity generator.

Table 1 Truth table of even parity generator.

Min-terms	Input signals			Output signals
	A	B	C	P_E
m_1	0	0	0	0
m_2	0	0	1	1
m_3	0	1	0	1
m_4	0	1	1	0
m_5	1	0	0	1
m_6	1	0	1	0
m_7	1	1	0	0
m_8	1	1	1	1

In Fig. 1, the power at the output of directional coupler towards the input side can be written as [24]

$$P_A = \sin^2(\alpha L)$$
$$P_B = \cos^2(\alpha L) \tag{3}$$

where α is the attenuation constant of DC towards input side, and L is its coupling length which is 1 μm. Thus, by using (1)–(3), mathematical expressions for the output of even parity generator circuit can be written as

$$m_1 = \cos^2\left(\frac{\Delta\varphi_{\text{MZI1}}}{2}\right) + \cos^2\left(\frac{\Delta\varphi_{\text{MZI2}}}{2}\right) + \cos^2\left(\frac{\Delta\varphi_{\text{MZI3}}}{2}\right)\Bigg\} \tag{4}$$

$$m_2 = \cos^2\left(\frac{\Delta\varphi_{\text{MZI1}}}{2}\right) + \cos^2\left(\frac{\Delta\varphi_{\text{MZI2}}}{2}\right) + \sin^2\left(\frac{\Delta\varphi_{\text{MZI3}}}{2}\right)\sin^2(\alpha L_{\text{DC3}})\cos^2\left(\frac{\Delta\varphi_{\text{MZI5}}}{2}\right) \tag{5}$$

$$m_3 = \cos^2\left(\frac{\Delta\varphi_{\text{MZI1}}}{2}\right) + \sin^2\left(\frac{\Delta\varphi_{\text{MZI2}}}{2}\right)\cos^2(\alpha L_{\text{DC2}}) \times \cos^2\left(\frac{\Delta\varphi_{\text{MZI5}}}{2}\right) + \cos^2\left(\frac{\Delta\varphi_{\text{MZI3}}}{2}\right) \tag{6}$$

$$m_4 = \cos^2\left(\frac{\Delta\varphi_{\text{MZI1}}}{2}\right) + \left\{\begin{matrix} \sin^2\left(\frac{\Delta\varphi_{\text{MZI2}}}{2}\right)\cos^2(\alpha L_{\text{DC2}}) + \\ \sin^2\left(\frac{\Delta\varphi_{\text{MZI3}}}{2}\right)\sin^2(\alpha L_{\text{DC3}}) \end{matrix}\right\}\sin^2\left(\frac{\Delta\varphi_{\text{MZI5}}}{2}\right) \tag{7}$$

$$m_5 = \sin^2\left(\frac{\Delta\varphi_{\text{MZI1}}}{2}\right)\cos^2(\alpha L_{\text{DC1}})\cos^2\left(\frac{\Delta\varphi_{\text{MZI5}}}{2}\right) + \cos^2\left(\frac{\Delta\varphi_{\text{MZI2}}}{2}\right) + \cos^2\left(\frac{\Delta\varphi_{\text{MZI3}}}{2}\right) \tag{8}$$

$$m_6 = \left\{\begin{matrix} \sin^2\left(\frac{\Delta\varphi_{\text{MZI1}}}{2}\right)\cos^2(\alpha L_{\text{DC1}}) + \\ \sin^2\left(\frac{\Delta\varphi_{\text{MZI3}}}{2}\right)\sin^2(\alpha L_{\text{DC3}}) \end{matrix}\right\}\sin^2\left(\frac{\Delta\varphi_{\text{MZI5}}}{2}\right) + \cos^2\left(\frac{\Delta\varphi_{\text{MZI2}}}{2}\right) \tag{9}$$

$$m_7 = \left\{\begin{matrix} \sin^2\left(\frac{\Delta\varphi_{\text{MZI1}}}{2}\right)\cos^2(\alpha L_{\text{DC1}}) + \\ \sin^2\left(\frac{\Delta\varphi_{\text{MZI2}}}{2}\right)\cos^2(\alpha L_{\text{DC2}}) \end{matrix}\right\}\sin^2\left(\frac{\Delta\varphi_{\text{MZI5}}}{2}\right) + \cos^2\left(\frac{\Delta\varphi_{\text{MZI3}}}{2}\right) \tag{10}$$

$$m_8 = \left[\left\{\begin{matrix} \sin^2\left(\frac{\Delta\varphi_{\text{MZI1}}}{2}\right)\sin^2(\alpha L_{\text{DC1}}) + \\ \sin^2\left(\frac{\Delta\varphi_{\text{MZI2}}}{2}\right)\sin^2(\alpha L_{\text{DC2}}) \end{matrix}\right\} \times \begin{matrix} \sin^2\left(\frac{\Delta\varphi_{\text{MZI4}}}{2}\right)\cos^2(\alpha L_{\text{DC2}}) + \\ \sin^2\left(\frac{\Delta\varphi_{\text{MZI3}}}{2}\right)\cos^2(\alpha L_{\text{DC3}}) \end{matrix}\right]\sin^2\left(\frac{\Delta\varphi_{\text{MZI6}}}{2}\right).$$

Thus, overall output of even parity generator (EPG)

can be written as $\text{OUT}_{\text{EPG}} = m_2 + m_3 + m_5 + m_8$, since there is no signal at output ports for the rest of min-terms.

The timing diagram of even parity generator through MATLAB is presented in Fig. 5, which is verified by its truth table given in Table 1.

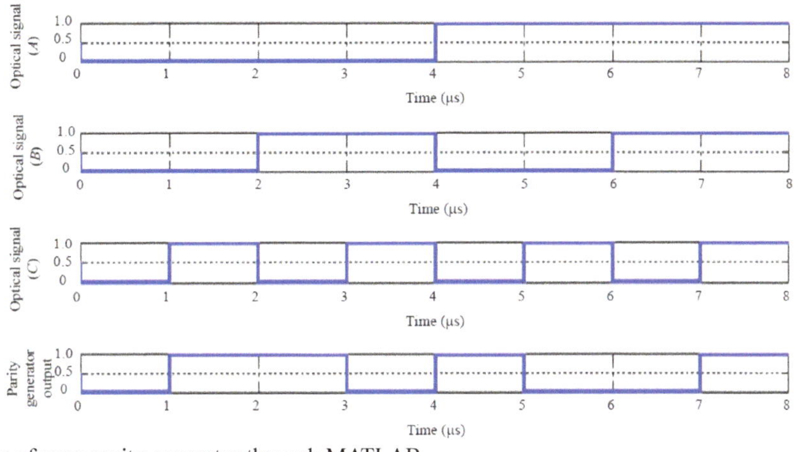

Fig. 5 Timing diagram of even parity generator through MATLAB.

In Fig. 5, the first three rows present the timing sequence for input optical signals, and the fourth row presents the timing signals for the output of parity generator. In the timing diagram, signals in the fourth row with higher magnitudes represent the presence of signal at output, while signals with lower magnitude show absence of signal.

2.2 Odd parity generator

Figure 6 shows a digital circuit and K-map of a three-bit-odd-parity generator, and Table 2 presents the truth table of odd parity generator. The design of an odd-parity generator is obtained by cascading seven MZIs and five DCs as shown in Fig. 7.

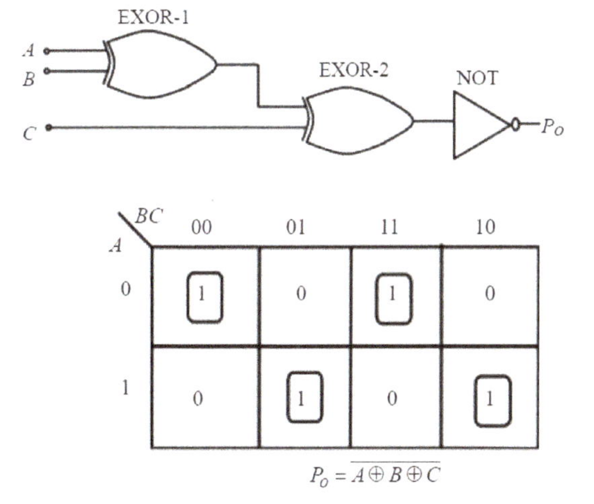

Fig. 6 Digital circuit and K-map of a three-bit-odd-parity generator.

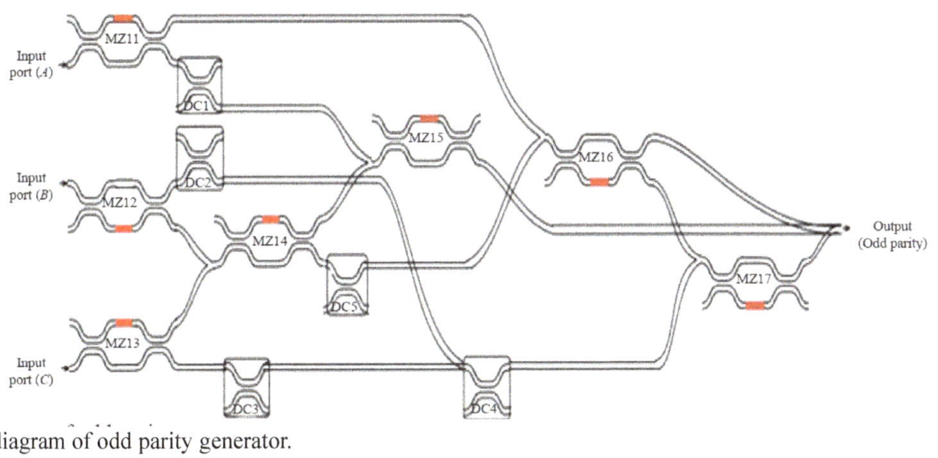

Fig. 7 Block diagram of odd parity generator.

Input optical signals are given at the second, first and second input ports of MZI1, MZI2, and MZI3, respectively. DC1 and DC2 are connected at the second and first output ports of MZI1 and MZI2, respectively. The second and first output ports of MZI2 and MZI3 are cobbled together to feed the MZI4 with optical signals at its second input port, while the second and first output ports of DC1 and MZI4 are combined together to feed MZI5 at its second input port. To feed MZI6 at its first input port, the first output ports of MZI1 and DC5 (which is connected at the second output port of MZI4) are combined together. The second and first output ports of DC2 and DC3 (which is connected at the second output port of MZI3) are combined together to feed DC4, whose first output port is combined with the second output port of MZI6 to feed the MZI7. Finally, to get the output of odd parity generator, the first output ports of MZI6 and MZI7 are combined together with the second output port of MZI5.

The mathematical expression for outputs of odd parity generator circuit using (1)–(3) can be written as

$$m_1 = \left\{\begin{array}{l} \cos^2\left(\frac{\Delta\varphi_{\mathrm{MZI1}}}{2}\right) + \left[\begin{array}{l} \cos^2\left(\frac{\Delta\varphi_{\mathrm{MZI2}}}{2}\right) + \\ \cos^2\left(\frac{\Delta\varphi_{\mathrm{MZI3}}}{2}\right) \end{array}\right] \times \\ \sin^2\left(\frac{\Delta\varphi_{\mathrm{MZI4}}}{2}\right)\sin^2\left(\alpha L_{\mathrm{DC5}}\right) \end{array}\right\} \sin^2\left(\frac{\Delta\varphi_{\mathrm{MZI6}}}{2}\right)$$

$$\begin{aligned} m_2 = {} & \left[\begin{array}{l} \cos^2\left(\frac{\Delta\varphi_{\mathrm{MZI1}}}{2}\right)\cos^2\left(\frac{\Delta\varphi_{\mathrm{MZI6}}}{2}\right) + \\ \sin^2\left(\frac{\Delta\varphi_{\mathrm{MZI3}}}{2}\right)\sin^2\left(\alpha L_{\mathrm{DC3}}\right)\sin^2\left(\alpha L_{\mathrm{DC4}}\right) \end{array}\right] \times \\ & \cos^2\left(\frac{\Delta\varphi_{\mathrm{MZI7}}}{2}\right) + \cos^2\left(\frac{\Delta\varphi_{\mathrm{MZI2}}}{2}\right) \times \\ & \cos^2\left(\frac{\Delta\varphi_{\mathrm{MZI4}}}{2}\right)\cos^2\left(\frac{\Delta\varphi_{\mathrm{MZI5}}}{2}\right) \end{aligned}$$

$$\begin{aligned} m_3 = {} & \left[\begin{array}{l} \cos^2\left(\frac{\Delta\varphi_{\mathrm{MZI1}}}{2}\right)\cos^2\left(\frac{\Delta\varphi_{\mathrm{MZI6}}}{2}\right) + \\ \sin^2\left(\frac{\Delta\varphi_{\mathrm{MZI2}}}{2}\right)\sin^2\left(\alpha L_{\mathrm{DC2}}\right)\sin^2\left(\alpha L_{\mathrm{DC4}}\right) \end{array}\right] \times \\ & \cos^2\left(\frac{\Delta\varphi_{\mathrm{MZI7}}}{2}\right) + \cos^2\left(\frac{\Delta\varphi_{\mathrm{MZI3}}}{2}\right) \times \\ & \cos^2\left(\frac{\Delta\varphi_{\mathrm{MZI4}}}{2}\right)\cos^2\left(\frac{\Delta\varphi_{\mathrm{MZI5}}}{2}\right) \end{aligned}$$

$$m_4 = \left[\left\{\begin{array}{l} \cos^2\left(\frac{\Delta\varphi_{\mathrm{MZI1}}}{2}\right)\cos^2\left(\frac{\Delta\varphi_{\mathrm{MZI6}}}{2}\right) + \\ \left[\begin{array}{l} \sin^2\left(\frac{\Delta\varphi_{\mathrm{MZI2}}}{2}\right)\sin^2\left(\alpha L_{\mathrm{DC2}}\right) + \\ \sin^2\left(\frac{\Delta\varphi_{\mathrm{MZI3}}}{2}\right)\sin^2\left(\alpha L_{\mathrm{DC3}}\right) \end{array}\right] \times \\ \sin^2\left(\alpha L_{\mathrm{DC4}}\right) \end{array}\right\} \sin^2\left(\frac{\Delta\varphi_{\mathrm{MZI7}}}{2}\right)\right]$$

$$\begin{aligned} m_5 = {} & \sin^2\left(\frac{\Delta\varphi_{\mathrm{MZI1}}}{2}\right)\cos^2\left(\alpha L_{\mathrm{DC1}}\right)\cos^2\left(\frac{\Delta\varphi_{\mathrm{MZI5}}}{2}\right) + \\ & \left[\cos^2\left(\frac{\Delta\varphi_{\mathrm{MZI2}}}{2}\right) + \cos^2\left(\frac{\Delta\varphi_{\mathrm{MZI3}}}{2}\right)\right] \times \\ & \sin^2\left(\frac{\Delta\varphi_{\mathrm{MZI4}}}{2}\right)\sin^2\left(\alpha L_{\mathrm{DC5}}\right) \times \\ & \cos^2\left(\frac{\Delta\varphi_{\mathrm{MZI6}}}{2}\right)\cos^2\left(\frac{\Delta\varphi_{\mathrm{MZI7}}}{2}\right) \end{aligned}$$

$$\begin{aligned} m_6 = {} & \left[\begin{array}{l} \sin^2\left(\frac{\Delta\varphi_{\mathrm{MZI1}}}{2}\right)\cos^2\left(\alpha L_{\mathrm{DC1}}\right) + \\ \cos^2\left(\frac{\Delta\varphi_{\mathrm{MZI2}}}{2}\right)\cos^2\left(\frac{\Delta\varphi_{\mathrm{MZI4}}}{2}\right) \end{array}\right] \sin^2\left(\frac{\Delta\varphi_{\mathrm{MZI5}}}{2}\right) + \\ & \sin^2\left(\frac{\Delta\varphi_{\mathrm{MZI3}}}{2}\right)\sin^2\left(\alpha L_{\mathrm{DC3}}\right) \times \\ & \sin^2\left(\alpha L_{\mathrm{DC4}}\right)\cos^2\left(\frac{\Delta\varphi_{\mathrm{MZI7}}}{2}\right) \end{aligned}$$

$$\begin{aligned} m_7 = {} & \left[\begin{array}{l} \sin^2\left(\frac{\Delta\varphi_{\mathrm{MZI1}}}{2}\right)\cos^2\left(\alpha L_{\mathrm{DC1}}\right) + \\ \cos^2\left(\frac{\Delta\varphi_{\mathrm{MZI3}}}{2}\right)\cos^2\left(\frac{\Delta\varphi_{\mathrm{MZI4}}}{2}\right) \end{array}\right] \times \\ & \sin^2\left(\frac{\Delta\varphi_{\mathrm{MZI5}}}{2}\right) + \sin^2\left(\frac{\Delta\varphi_{\mathrm{MZI2}}}{2}\right)\sin^2\left(\alpha L_{\mathrm{DC2}}\right) \times \\ & \sin^2\left(\alpha L_{\mathrm{DC4}}\right)\cos^2\left(\frac{\Delta\varphi_{\mathrm{MZI7}}}{2}\right) \end{aligned}$$

$$\begin{aligned} m_8 = {} & \sin^2\left(\frac{\Delta\varphi_{\mathrm{MZI1}}}{2}\right)\cos^2\left(\alpha L_{\mathrm{DC1}}\right)\cos^2\left(\frac{\Delta\varphi_{\mathrm{MZI5}}}{2}\right) + \\ & \left[\begin{array}{l} \sin^2\left(\frac{\Delta\varphi_{\mathrm{MZI2}}}{2}\right)\sin^2\left(\alpha L_{\mathrm{DC2}}\right) + \\ \sin^2\left(\frac{\Delta\varphi_{\mathrm{MZI3}}}{2}\right)\sin^2\left(\alpha L_{\mathrm{DC3}}\right) \end{array}\right] \sin^2\left(\alpha L_{\mathrm{DC3}}\right) \times \\ & \cos^2\left(\frac{\Delta\varphi_{\mathrm{MZI7}}}{2}\right) + \sin^2\left(\frac{\Delta\varphi_{\mathrm{MZI2}}}{2}\right)\sin^2\left(\alpha L_{\mathrm{DC2}}\right) \times \\ & \sin^2\left(\alpha L_{\mathrm{DC4}}\right)\cos^2\left(\frac{\Delta\varphi_{\mathrm{MZI7}}}{2}\right) \end{aligned}$$

Table 2 Truth table of odd parity generator.

Min-terms	Input signals			Output signals
	A	B	C	P_O
m_1	0	0	0	1
m_2	0	0	1	0
m_3	0	1	0	0
m_4	0	1	1	1
m_5	1	0	0	0
m_6	1	0	1	1
m_7	1	1	0	1
m_8	1	1	1	0

Thus, the overall output of odd parity generator (OPG) is written as $\mathrm{OUT}_{\mathrm{OPG}} = m_1 + m_4 + m_6 + m_7$, since there is no optical signal at the output port for the rest of min-terms.

The timing diagram of the odd parity generator through MATLAB is presented in Fig. 8, which is verified by its truth table given in Table 2. In Fig. 8, the first three rows represent the timing signals for input signals, and the fourth row represents the timing signals for the generated output of the odd parity generator, which is truly matched with its truth table. In timing diagram at the fourth row, timing signals with low and high magnitudes only represent the presence and absence of optical signals at the desired output.

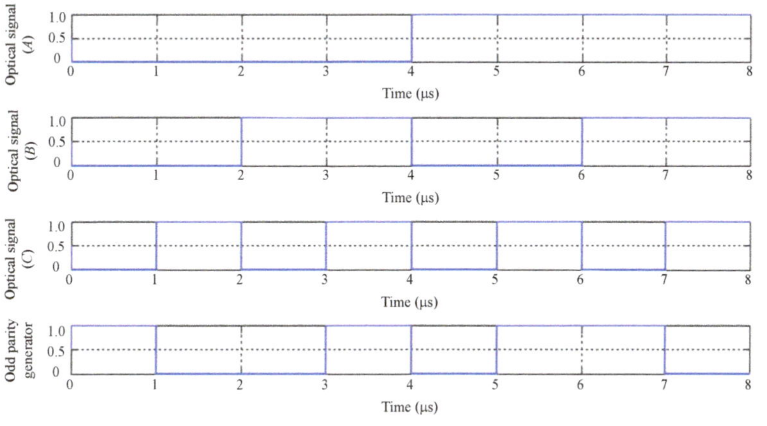

Fig. 8 Timing diagram of odd parity generator.

3. Simulation results using FDTD

The finite difference time domain (FDTD) method has been used for analysis of propagation of optical signals through the design of parity generator circuits, and their verifications are done by using MATLAB. The continuous wave (CW) source is used to provide the input signals with Gaussian distribution under the transverse magnetic (TM) polarization at the wavelength of 1550 nm. The Gaussian distribution of signals spectrum is very useful because the interference occurs with side lobes of spectrum, while the maximum information is transmitted on the peak lobe which remains unaffected. For the simulation of the proposed device, perfect matched layers are considered as boundary layers for all interfaces of metal and dielectric with $1e^{-12}$ as reflection coefficient. The half width of input source is 0.39 μm within the mesh size of Δx=0.021 μm and Δz=0.051 μm, under the perfect matched layer (PML) as boundary conditions for all interfaces of metal and dielectric.

3.1 Simulation results of even parity generator

The simulation result of propagation of light through the even parity generator with combination of all possible input signals is presented in Fig. 9.

(1) Case I: A=0, B=0, C=0

In this case, A=B=C=0 means all three input ports are fed with optical signals having low input power, and thus by following the phenomenon of SPM, these signals are obtained at the first output ports of MZI1, MZI2, and MZI3. Hence, there is no signal at the output port of the even parity generator as shown in Fig. 9, which is truly matched with its truth table given in Table 1.

(2) Case II: A=0, B=0, C=1

Here, A=B=0 means low-power signals are

incident at the second input ports of MZI1 and MZI2, and by obeying the principle of SPM, they arrive at the first output ports of the same MZIs. While C=1 means optical high-power signals of high power are provided at the second input port of MZI3 and due to XPM, they arrive at the second output port, where they propagate through DC3 and get split into two equal parts. Then MZI5 gets input optical signals at its first input port from the first output port of DC3, where they obey the laws of SPM due to its low power and are obtained at the second output port of MZI5, which is considered as the output port of EPG. Signals from the second output port of DC3 propagate through MZI6 by obeying the laws of SPM and arrive at its second output port. Thus, in this case, optical signals arrive at the desired output port of EPG due to optical signals of high power at input of MZI3 as shown in Fig. 9.

(3) Case III: A=0, B=1, C=0

Here, A=C=0 means the low-power optical signala are given at the second input ports of MZI1 and MZI3. B=1 means optical high-power signals are incident at the second input port of MZI2. Thus, output optical signals arrive at the desired output port of EPG due to the presence of optical high-power signals at MZI2, which is truly matched with the truth table given in Table 1.

(4) Case IV: A=0, B=1, C=1

In this case, A=0 means low-power signals are given at the second input port of MZI1. B=C=1 means signals of high power are incident at second input ports of MZI2 and MZI3. Hence, in this case output of EPG is zero as shown in Fig. 9.

(5) Case V: A=1, B=0, C=0

Here, A=1 means high-power signals are given at the second input port of MZI1. B=C=0 means low-power signals are given at second input ports of MZI2 and MZI3. In this case, optical signals arrive at the desired output port of EPG, due to the presence of high-powers signal at the second input port of MZI1 (as shown in Fig. 9).

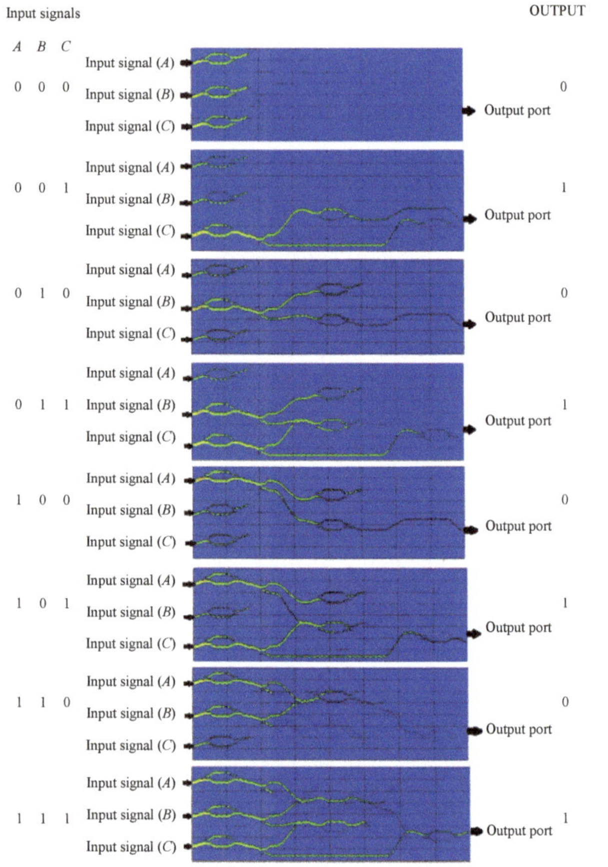

Fig. 9 Propagation of the optical signals through the even parity generator circuit for all possible combinations of input signals obtained through the FDTD method.

(6) Case VI: A=1, B=0, C=1

Here, A=C=1 means high-power signals are given at the second input ports of MZI1 and MZI3. B=0 means low-power signals are given at the second input port of MZI2. In this case, there is no optical signal at the desired output port of EPG, which is truly matched with the truth table shown in Table 1.

(7) Case VII: A=1, B=1, C=0

In this case, A=B=1 means high-power signals are given at the second input ports of MZI1 and MZI2. C=0 means low-power signals are given at the second input port of MZI3. Thus, in this case there is no optical signal at the desired output port of EPG.

(8) Case VIII: A=1, B=1, C=1

In this case, A=B=C=1 means high-power signals are given at the second input ports of MZI1, MZI2, and MZI3. Hence, in this case optical signals are obtained at the desired port of EPG, which is truly matched with its truth table.

3.2 Simulation results odd parity generator

The propagation of light through its model with all possible combinations of input signals is presented in Fig. 10.

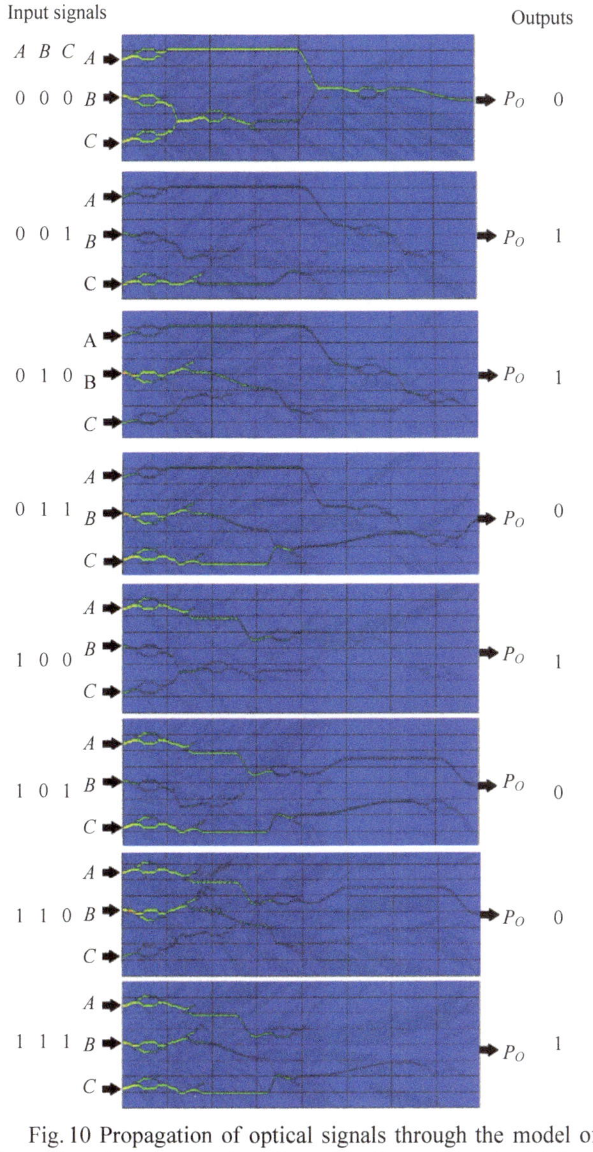

Fig. 10 Propagation of optical signals through the model of the odd parity generator for all combinations of inputs obtained through the FDTD method.

(1) Case I: A=0, B=0, C=0

In this case, A=B=C=0 means low-power signals are given at the second, first, and second input port of MZI1, MZI2, and MZI3, respectively, and by obeying the phenomenon of SPM the signals are obtained at the first, second, and first output ports of same MZIs. Then signals from the second and first output port of MZI2 and MZI3 are combined together to feed the MZI4 at its second input port, and by following the laws of XPM they arrive at the second output port of same MZI, where they propagate through DC5 and get equally split. Then signals from the first output ports of MZI1 and DC5 are combined together to feed MZI6 at its first input port, and due to its high power they obey the laws of XPM and arrive at the first output port of MZI6, which is considered as the output port for odd parity generator. Thus, in this case an optical signal arrives at the desired output port of OPG, which is truly matched with its truth table given in Table 2.

(2) Case II: A=0, B=0, C=1

Here, A=B=0 means low-power signals are given at the second and first input ports of MZI1 and MZI2, where they obey the phenomenon of SPM and arrive at the first and second output ports of same MZI. Then signals from the first output port of MZI1 propagate through MZI6 and due to the low power they arrive at the second output port due to SPM. While signals from the second output port of MZI2 are given to the second input port of MZI4, and after travelling through it signals arrive at its first output port due to the low power. Then MZI5 gets optical signals at its second input port from the first output port of MZI4, and after propagating through MZI5, the signals arrive at first output ports of same MZI. C=1 means high power signals are given at the second input port of MZI3 and arrive at the second output port due to XPM, where they propagate through DC3 and get equally split. Further, signals from the first output port of DC3 is again split by DC4, and then signals from the second and first output ports of MZI6 and DC4 are cobbled together to feed MZI7 at its first input port. After propagating through MZI7, signals will appear at its second output port due to the low power. Thus in this case, there is no signal at desired output port of OPG.

(3) Case III: A=0, B=1, C=0

Here, A=C=0 means signals with low powers are

incident at the second input ports of MZI1 and MZI3. *B*=1 means high-power signals are given at the first input port of MZI2. Thus, in this case there is no optical signal at the desired output port of OPG.

(4) Case IV: *A*=0, *B*=1, *C*=1

Here, *A*=0 means the second input port of MZI1. *B*=*C*=1 means high intensity signals are given at the first and second input ports of MZI2 and MZI3. Hence, in this case optical signals are obtained at the desired output port of OPG, which is truly matched with its truth table.

(5) Case V: *A*=1, *B*=0, *C*=0

In this case, *A*=1 means signals with high power are given at the second input port of MZI1. *B*=*C*=0 means low-power signals are given at the first and second input ports of MZI2 and MZI3. Thus, in this case there is no optical signal at desired output port of OPG.

(6) Case VI: *A*=1, *B*=0, *C*=1

Here, *A*=*C*=1 means the second input ports of MZI1 and MZI3 are fed with high-power signals. *B*=0 means low-power signals are incident at the first input port of MZI2. Thus, in this case optical signals arrive at the desired output port of OPG, which is truly matched with the truth table given in Table 2.

(7) Case VII: *A*=1, *B*=1, *C*=0

Here, *A*=*B*=1 means high-power signals are given at the second and first input ports of MZI1 and MZI2. *C*=0 means low-power signals are fed at the second input port of MZI3. Finally, output signals approach to the port which is assigned as output port for OPG.

(8) Case VIII: *A*=1, *B*=1, *C*=1

Here, as *A*=*B*=*C*=1 means high-power signals are given at the second, first, and second input ports of MZI1, MZI2, and MZI3. Thus, in this case there is no optical signal at desired output port of OPG.

4. Conclusions

In this paper, models of even and odd parity generator are projected by using MIM plasmonic waveguides due to their enormous capability of confining the surface plasmons up to a deep subwavelength scale. The concept of nonlinear MZI using MIM plasmonic waveguides has been used to cascade the desired structures. The proposed concept of cascading the MZIs is useful for integrating the all-optical devices. The circuits of parity generators are mainly used for error detection and correction in optical communication networks.

References

[1] W. Wei, X. Zhang, and X. Ren, "Asymmetric hybrid plasmonic waveguide with centimetric scale propagation length under subwavelength confinement for photonic components," *Nanoscale Research Letters*, 2014, 9(1): 1–8.

[2] W. L. Barnes, A. Dereux, and W. E. Thomas, "Surface plasmon subwavelength optics," *Nature*, 2003, 424(6950): 824–830.

[3] D. K. Gramotnev and S. I. Bozhelvonyi, "Plasmonics beyond the diffraction limit," *Nature Photonics*, 2010, 4(2): 83–91.

[4] Y. Chen and H. Ming, "Review of surface plasmon resonance and localized surface plasmon resonance sensor," *Photonic Sensors*, 2012, 2(1): 37–49.

[5] G. Veronis and S. H. Fan, "Guided subwavelength plasmonic mode supported by a slot in a thin metal film," *Optics Letters*, 2005, 30(24): 3359–3361.

[6] L. Liu, H. Zhanghua, and H. Sailing, "Novel surface plasmon waveguide for high integration," *Optics Express*, 2015, 13(17): 6645–6650.

[7] A. Boltasseva, S. V. Valentyn, B. N. Rasmus, M. Esteban, G. R. Sergio, and I. B. Sergey, "Triangular metal wedges for subwavelength plasmon polariton guiding at telecom wavelengths," *Optics Express*, 2008, 16(8): 5252–5260.

[8] A. Kumar, J. Gosciniak, V. S. Volkov, S. Papaioannou, D. Kalavrouziotis, K. Vyrsokinos, *et al.*, "Dielectric-loaded plasmonic waveguide components: going practical," *Laser & Photonics Reviews*, 2013, 7(6): 938–951.

[9] W. L. Barnes, A. Dereux, and T. W. Ebbeen, "Surface plasmon subwavelength optics," *Nature*, 2003, 424(6950): 824–830.

[10] R. Charbonneau, N. Lahoud, G. Mattiussi, and P. Berini, "Demonstration of integrated optics elements based on long-ranging surface plasmon polaritons," *Optics Express*, 2005, 13(3): 977–984.

[11] L. Liu, Z. Han, and S. He, "Novel surface plasmon

waveguide for high integration," *Optics Express*, 2005, 13(17): 6645–6650.

[12] J. A. Dionne, L. A. Sweatlock, H. A. Atwater, and A. Polman, "Plasmon slot waveguides: towards chip-scale propagation with subwavelength scale localization," *Physical Review B*, 2006, 73(3): 035407-9.

[13] G. Veronis and S. Fan, "Modes of subwavelength plasmonic slot waveguides," *Journal of Lightwave Technology*, 2007, 25(9): 2511–2521.

[14] F. Lou, Z. Wang, D. Dai, L. Tylen, and L. Wousnski, "Experimental demonstration of ultra-compact directional couplers based on silicon hybrid plasmonic waveguides," *Applied Physics Letters*, 2012, 100(24): 241105-1–241105-4.

[15] M. Z. Alam, J. N. Caspers, J. S. Aitchison, and M. Mojahedi, "Compact low loss and broadband hybrid plasmonic directional coupler," *Optics Express*, 2013, 21(13): 16029–16034.

[16] D. K. Gramotnev, K. C. Vernon, and D. F. P. Pile, "Directional coupler using gap plasmon waveguides," *Applied Physics B*, 2008, 93(1): 99–106.

[17] Q. Li, Y. Song, G. Zhou, Y. Su, and M. Qiu, "Asymmetric plasmonic-dielectric coupler with short coupling length, high extinction ratio, and low insertion loss," *Optics Letters*, 2010, 35(19): 3153–3155.

[18] P. Jia, G. Fang, and D. Wang, "Characterization of miniature fiber-optic Fabry-Perot interferometric sensors based on hollow silica tube," *Photonic Sensors*, 2016, 6(3): 193–198.

[19] M. A. Bader, G. Marowsky, A. Bahtiar, K. Koynov, C. Bubeck, H. Tillmann, *et al.*, "Poly (*p*-phenylenevinylene) derivatives: new promising materials for nonlinear all-optical waveguide switching," *Journal of the Optical Society of America B*, 2002, 19(9): 2250–2262.

[20] S. Kumar, Chanderkanta, and A. Amphawan, "Design of parity generator and checker circuit using electro-optic effect of Mach-Zehnder interferomenters," *Optics Communications*, 2016, 364: 195–224.

[21] L. Wang, Y. Wang, C. Wu, and F. Wang, "All-optical flip-flop based on coupled SOA-PSW," *Photonic Sensors*, 2016, 6(4): 366–371.

[22] X. S. Christina and A. P. Kabilan, "Design of optical logic gates using self-collimated beams in 2d photonic crystal," *Photonic Sensors*, 2012, 2(2): 173–179.

[23] J. A. Pereda, A. Vegas, and A. Prieto, "An improved compact 2D full-wave FDTD method for general guided wave structures microwave," *Microwave & Optical Technology Letters*, 2003, 38(4): 331–335.

[24] S. Kumar, L. Singh, S. K. Raghuwanshi, and N. K. Chen, "Design of full-adder and full-subtractor using metal-insulator-metal plasmonic waveguides" *Plasmonics*, 2016, 11: 1–11.

[25] S. Kumar, L. Singh, and S. Swarnakar, "Design of one bit magnitude comparator using nonlinear plasmonic waveguide," *Plasmonics*, 2016, 11: 1–7.

[26] S. Kumar and L. Singh, "Proposed new approach to design all optical AND gate using plasmonic based Mach-Zehnder interferometer for high speed communication," *SPIE*, 2016, 9884: 98842D.

[27] S. K. Raghuwanshi, A. Kumar, and S. Kumar, "1×4 signal router using three Mach-Zhender interferometers," *Optical Engineering*, 2013, 52(3): 035002-1–035002-9.

[28] G. P. Agrawal, *Nonlinear fiber optics*. Salt Lake City: Academic Press, 2006.

25

Effect of Radio Frequency Magnetron Sputtering Power on Structural and Optical Properties of Ti6Al4V Thin Films

Mohammed K. KHALAF[1*], H. F. Al-TAAY[2], and Dawood S. ALI[3]

[1] *Ministry of Science and Technology, Center of Applied Physics, Baghdad, Iraq*

[2] *Department of Physics, College of Science for Women, University of Baghdad, Baghdad, Iraq*

[3] *Department of Physics, College of Education for Pure Sciences, University of Anbar, Anbar, Iraq*

[*] Corresponding author: Mohammed K. KHALAF E-mail: mohammedkhkh@yahoo.com

Abstract: In this research, the effects of target sputtering power on the structure and optical properties of radio frequency (RF) sputtered Ti6Al4V films were investigated. Different sputtering RF powers were used to produce different thicknesses of Ti6Al4V thin films. From the X-ray diffraction, it was found that the Ti6A14V films had polycrystalline cubic and hexagonal structures and increased films crystallinity and crystalline size with increasing the sputtering power. Atomic forces microscopy (AFM) gave us a nanometric film character, films homogeneity, and surfaces roughness. A higher degree of roughness and average grain size with increasing RF power was exhibited. Band gap and refractive index of Ti6Al4V thin films varied with sputtering RF powers.

Keywords: RF magnetron sputtering; Ti6Al4V; structural properties; optical properties.

1. Introduction

Due to the interesting chemical and mechanical properties, the titanium alloy is far used in many applications [1, 2]. One particular alloy, the Ti6Al4V, has the most performance among the different grades of titanium. The Ti6Al4V thin films have attracted large scientific and practical interest since their specific properties enable various applications as microstructure materials in surgical appliances [3, 4]. The pure titanium is a monophasic, physiologically inert, and non-toxic metal. Ternary titanium alloys containing Al and V exhibit α and β phases structure that has attractive mechanical properties, high wear resistance, hardness, tenacity, resistance to fatigue, and high corrosion resistance [5]. Besides having a low density, it has an excellent biocompatibility of permitting its use in the fabrication of medical implants [6]. There are many deposition methods used to prepare Ti6Al4V, by means of chemical vapor deposition (CVD) [7], electrochemical method [8], selective laser melting [9], and physical vapor deposition (PVD) such as RF-DC sputtering [10]. In this work, Ti6Al4V thin films have been deposited on glass substrates by the radio frequency (RF) magnetron sputtering method with five different deposition power conditions. The crystallographic properties and surface morphology of the films were studied by the X-ray diffraction (XRD) and atomic forces microscopy (AFM) techniques. The optical property measurements for Ti6Al4V thin films are obtained by using the UV (ultraviolet) -visible recording spectrometer.

2. Experiment

In our study, Ti6Al4V thin films were prepared by using the RF magnetron sputtering technique (CRC600 CO.USA-made). The thin films were deposited on the glass substrate with different powers. The glass slides were sequentially cleaned in an ultrasonic bath with acetone and ethanol. Finally, they were rinsed with distilled water and dried. The sputtering chamber was evacuated to 5×10^{-5} mbar base pressure using diffusion and mechanical booster pump combination prior to the deposition. Before the deposition of Ti6Al4V films, a Ti6Al4V target (99.99% pure and 5 cm diameter) was pre-sputtered in pure argon atmosphere for 10 min in order to remove oxide on the surface of the target. Ti6Al4V films were deposited by the RF sputtering system in pure argon gas (99.9%) with pressure of 5×10^{-2} mbar. The X-ray diffraction measurements of thin films were performed by using the diffract meter type (SHIMADZU-6000). The AFM was a contact mode used to analyze the morphological feature on Angstrom Inc. (AA3000). The optical properties measurements for Ti6Al4V thin films were obtained by using the UV-Visible recording spectrometer (UV-2601 PC Shimadzu software 1700 1650). The thickness of the films has been calculated by using the Device FT-650 Film Thickness (FT) Probe System.

3. Results and discussion

3.1 X-ray diffraction (XRD)

It can be expected that low sputtering power exhibits a low deposition rate which is due to less energetic argon over the target species and less ejected atoms from the target material. However, when the sputtering power was increased, the sputtering yield of the Ti6Al4V films markedly was increased. Figure 1 shows the XRD analysis for Ti6Al4V thin films deposited on glass substrates with different powers (50 W, 75 W, 100 W, 125 W, and 150 W, respectively). The XRD pattern illustrates that the Ti6Al4V films had a polycrystalline structure with peaks attributed to (110) diffractions for cubic structure or (002) diffractions for hexagonal structure and (102) diffractions for cubic structure, identified with standard peaks (card No. 96-900-8555 and 96-900-8518). Also, note that an increase in the RF power led to an increase in the peak intensity (i.e. an increase in films crystallinity). The mobility improvement of adatoms sputtered on the surface, which was required to form highly crystalline films. Because it is believed that high DC sputtering power in the magnetron sputtering system energizes inert argon gas to provide sufficient kinetic energy to adatoms, the surface diffusion of these adatoms was then expected to enhance with the momentum transfer to the nucleation and growth of the Ti6Al4V films. Increasing the RF power will make an increase in the grain size, as shown in Table 1. This may be due to the enhancement of crystallinity in the films. The films of crystalline was improved which led to a decrease in the number of grain boundaries. A significant line was broadened which is a characteristic of nanoparticles [1, 11].

Table 1 Comparison between the Exp. and Std. the values of d_{hkl} for the (Ti6Al4V) peaks shown in XRD for different RF powers on a glass substrate.

Power (W)	2θ (deg.)	β (FWHM) (deg.)	d_{hkl} Exp.(Å)	G.S (nm)	d_{hkl} Std.(Å)	Phase	hkl
50	38.4010	1.7020	2.3422	4.9	2.3380	Cubic	(110)
					2.3430	Hex.	(002)
	53.8120	1.6503	1.7022	5.4	1.7268	Hex.	(102)
75	38.5902	0.9230	2.3312	9.1	2.3380	Cubic	(110)
					2.3430	Hex.	(002)
	53.7980	1.4120	1.7026	6.3	1.7268	Hex.	(102)
100	38.4320	0.6013	2.3404	14.0	2.3380	Cubic	(110)
					2.3430	Hex.	(002)
	53.8210	0.9340	1.7020	9.5	1.7268	Hex.	(102)
125	38.4610	0.5780	2.3387	14.6	2.3380	Cubic	(110)
					2.3430	Hex.	(002)
	53.7650	1.1043	1.7036	8.1	1.7268	Hex.	(102)
150	38.4426	0.5431	2.3398	15.5	2.3380	Cubic	(110)
					2.3430	Hex.	(002)
	53.7630	0.5312	1.7037	16.8	1.7268	Hex.	(102)

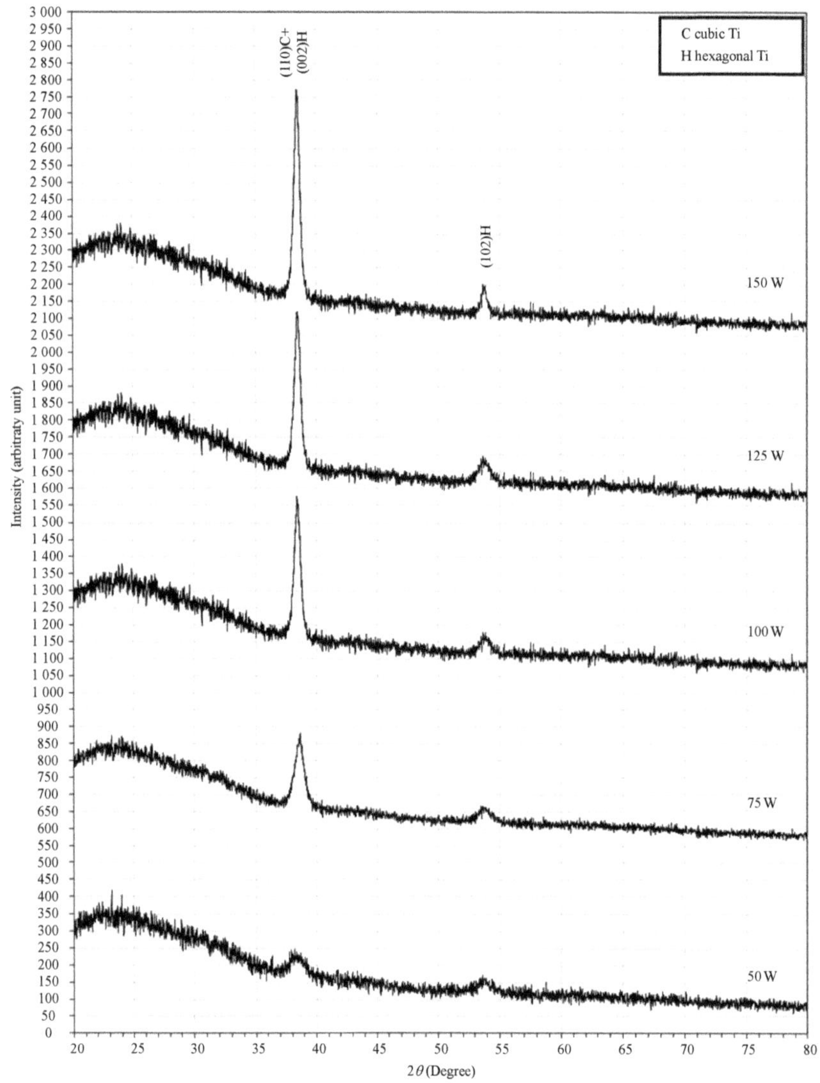

Fig. 1 X-ray diffraction patterns of the (Ti6A14V) films deposited on the glass substrate at different sputtering powers.

3.2 Atomic force microscope

The surface morphology of Ti6Al4V films deposited on the glass substrate was studied by AFM to monitor the growth of nanostructure under the influence of different deposition powers. Figure 2 shows 2D and 3D AFM images of Ti6Al4V thin film deposited at different working sputtering powers. The images have light and dark regions. From the colors, brightness is used to specify the vertical profile of the thin film surface, where light regions represent the highest points and the dark points are the depressions. This figure confirms that the films are uniform, and the substrate surface is well covered with grains that are nearly uniformly distributed. From these images, it is observed that the surfaces of the films exhibit more degree of roughness with increasing the RF power. In addition, an increase in the average grain size leads to an increase in the root mean square roughness (RMS), as shown in Table 2. The Ti6Al4V film deposited at higher sputtering power exhibits profound large grains with orientations. These morphologies are due to the fact that sputtering power helps increase the surface mobility of adatoms, which is required to form continuous films. The surface diffusion of these adatoms is then enhanced by the higher sputtering power, which results in a provision of the momentum transfer to the growing surface.

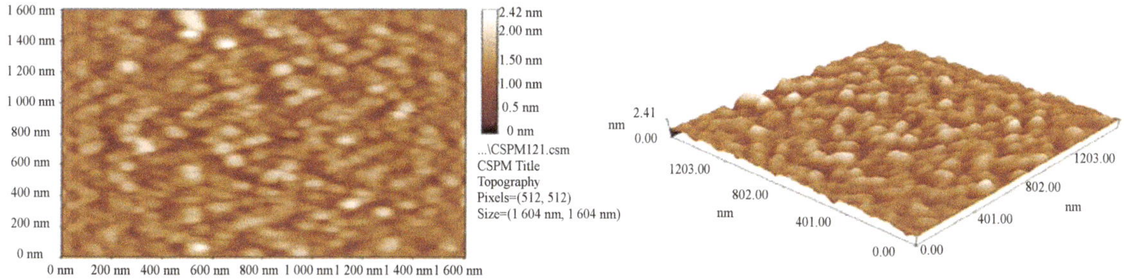

(a)

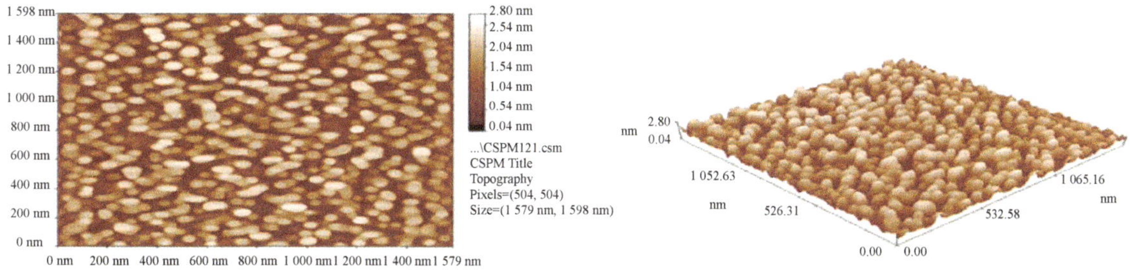

(b)

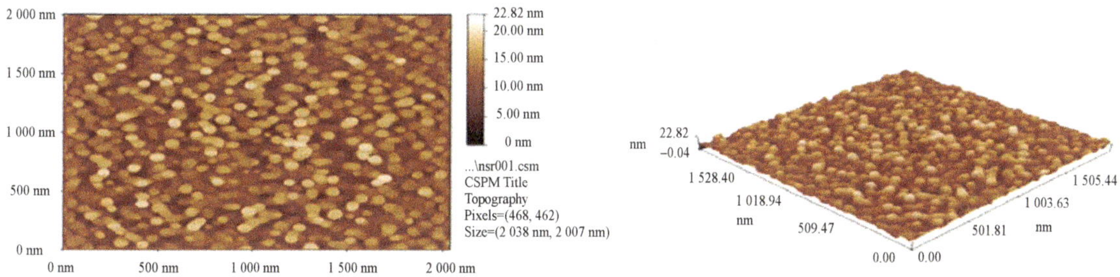

(c)

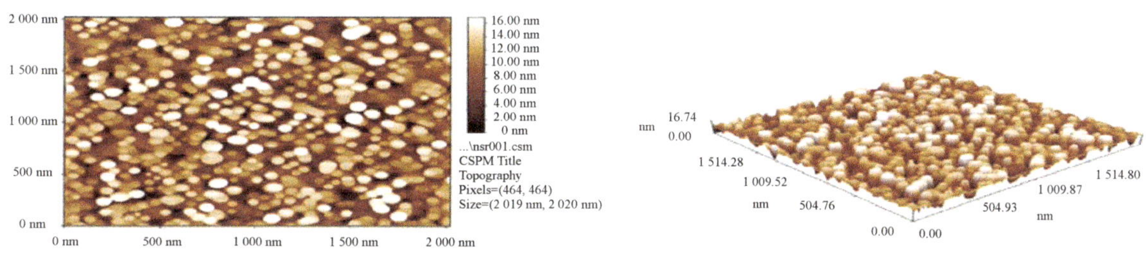

(d)

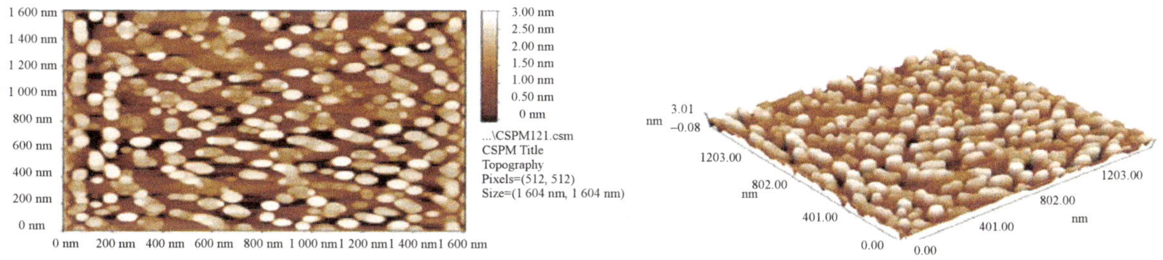

(e)

Fig. 2 Two-dimensional (2D) and three-dimensional (3D) AFM images of (Ti6Al4V) thin films deposited on glass substrate at different RF powers: (a) 50 W, (b) 75 W, (c) 100 W, (d) 125 W, and (e) 150 W.

Table 2 AFM data of (Ti6Al4V) thin films deposited on the glass substrate at different RF powers.

Power (W)	Average diameter of granularity (nm)	Root mean square roughness (nm)	Average roughness (nm)
50	50.21	0.255	0.190
75	57.38	0.569	0.487
100	75.38	0.848	0.730
125	81.70	3.320	2.710
150	92.85	4.000	3.340

3.3 Optical measurements

The optical properties of the Ti6Al4V thin films deposited by RF magnetron sputtering were analyzed by UV-visible spectroscopy in the wavelength range of 400 nm–1100 nm as shown in Fig. 3. The transmission spectra of Ti6Al4V thin film at different RF powers (50 W, 75 W, 100 W, 125 W, and 150 W) decrease with an increase in RF power when films thicknesses increase (244.89 nm, 435.26 nm, 598.98 nm, 866.23 nm, and 910.46 nm, respectively). The transmittance patterns of all deposited thin films on glass increase with an increase in wavelength (λ). A decrease in the transmittance spectra is caused by an increase in the loss of light scattering as the grain size increases [12].

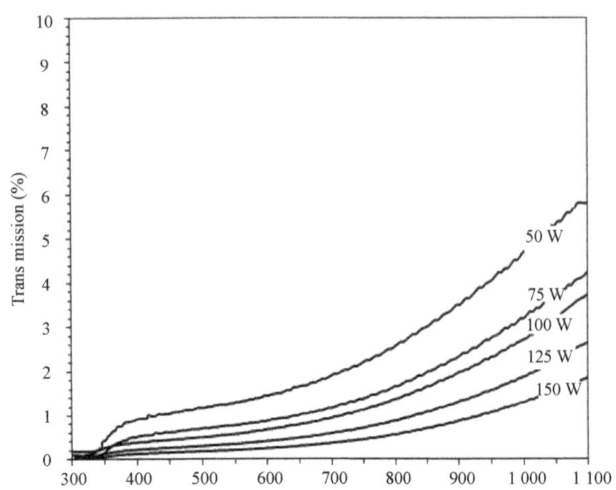

Fig. 3 Transmittance as a function of wavelength for (Ti6Al4V) thin film at different working RF powers.

Absorbance spectra of Ti6Al4V thin films at different RF powers are shown in Fig. 4. An increase in absorbance due to a decrease in transmission associates with a change in the thickness. The figure shows that the optical absorption in the UV region is high. The absorption patterns of all deposited thin films on glass decrease with an increase in the wavelength.

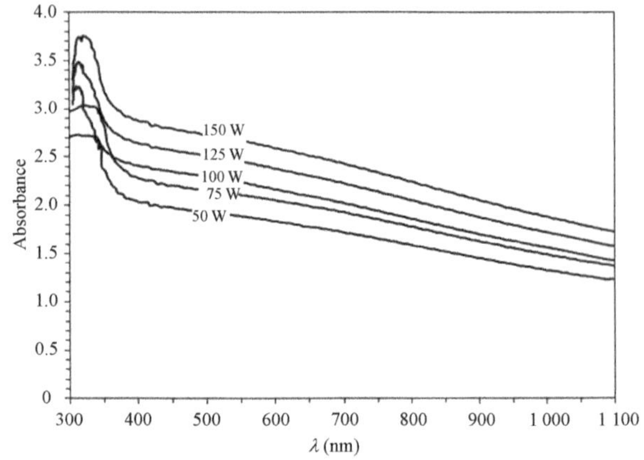

Fig. 4 Absorbance as a function of wavelength for (Ti6Al4V) thin film at different working RF powers.

Figure 5 displays the absorption coefficient as a function of wavelength for Ti6Al4V thin films deposited at different RF powers. The absorption coefficient increases with an increase in the film thickness which is proportional to RF power. Also, the absorption coefficient of deposited thin films on glass decreases with an increase in the wavelength.

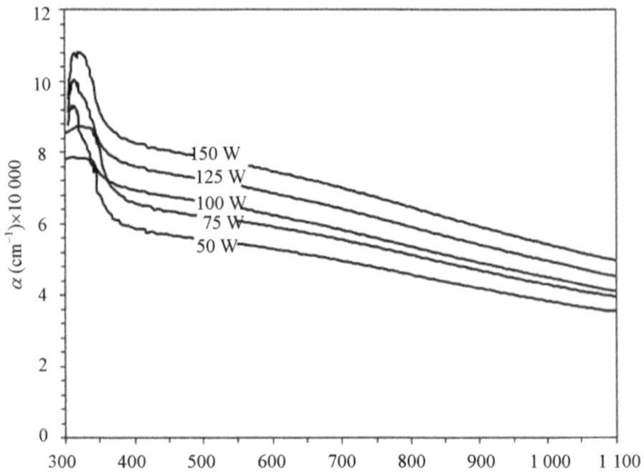

Fig. 5 Absorption coefficient as a function of wavelength for (Ti6Al4V) thin film at different working RF powers.

Figure 6 illustrates the variation of extinction coefficient with wavelength in the range of 400 nm–1100 nm for Ti6Al4V films deposited on the glass substrate at different RF powers. The extinction coefficient depends mainly on the absorption coefficient, and we notice that the extinction coefficient decreases with an increase in the wavelength because of the increment in the absorption coefficient [13].

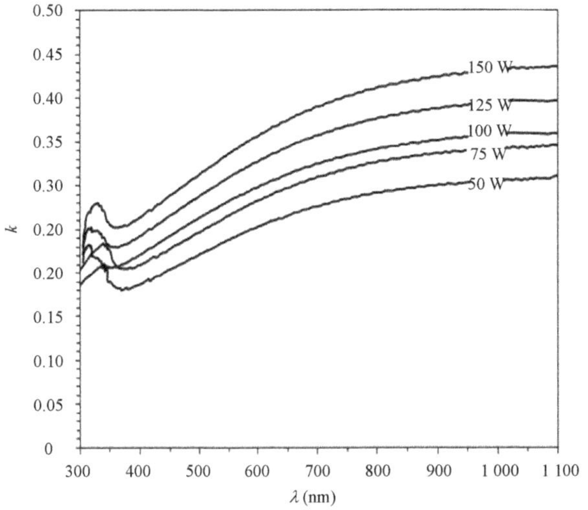

Fig. 6 Extinction coefficient as a function of wavelength for (Ti6Al4V) thin film at different working RF powers.

Figures 7 and 8 show the dielectric constant and dielectric loss as a function of wavelength for Ti6Al4V thin films deposited at different RF powers. The dielectric constants decrease with an increase in RF power and increase with an increase in the wavelength (λ). The variation of the dielectric constant depends on the value of the refractive index. By contrast, the dielectric loss depends mainly on the extinction coefficient values which are related to the variations of absorption [13].

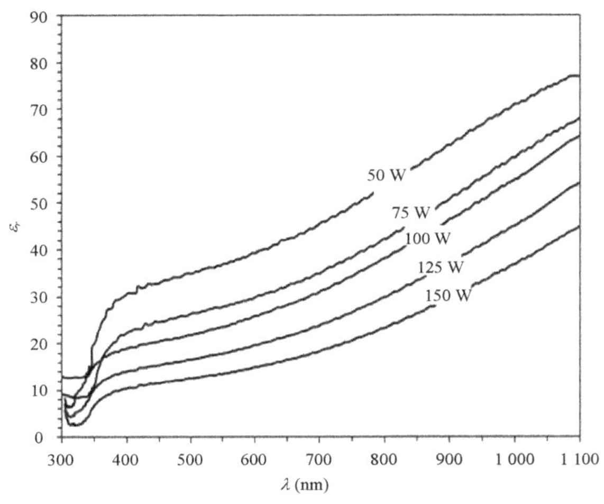

Fig. 7 Dielectric constant real as a function of wavelength for (Ti6Al4V) thin film at different working RF powers.

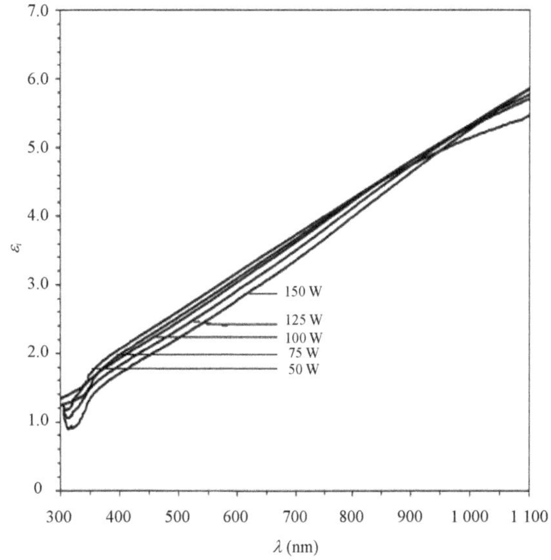

Fig. 8 Dielectric constant imaginary as a function of wavelength for (Ti6Al4V) thin film at different working RF powers.

The variations of the refractive index are versus wavelength in the range 400 nm–1100 nm. It is clear from this figure that the refractive index in general increases with an increase in the thickness, due to the different deposited thicknesses, as shown in Fig. 6.

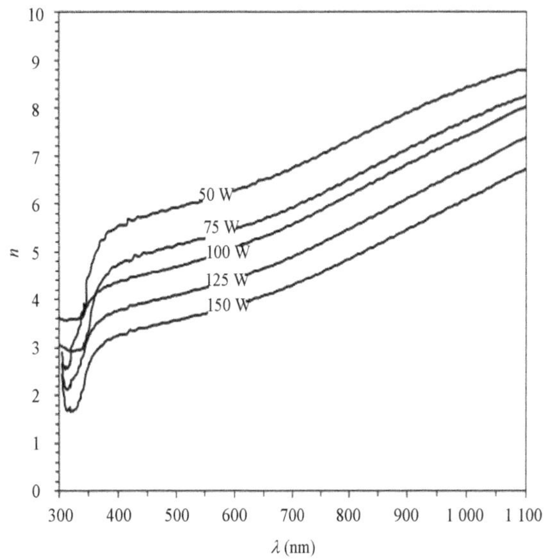

Fig. 9 Refractive index as a function of wavelength for (Ti6Al4V) thin film at different working RF powers.

Table 3 shows the variation of optical parameters at 500 nm wavelength for (Ti6Al4V) films at different RF powers. This table illustrates that T, n, ε_r, ε_i, and E_g decrease with an increase in the RF power while α and K decrease.

Table 3 Optical parameters for (Ti6Al4V) films at the different thicknesses.

Sample	$T\%$	α (cm^{-1})	K	n	ε_r	ε_i	E_g(eV)	Thickness (nm)
50	1.17	55621	0.22	5.92	34.95	2.63	1.54	244.89
75	0.71	61801	0.25	5.16	26.58	2.55	1.44	435.26
100	0.50	66222	0.26	4.68	21.85	2.47	1.39	598.98
125	0.31	72400	0.29	4.09	16.62	2.36	1.34	866.23
150	0.19	78578	0.31	3.57	12.61	2.23	1.23	910.46

4. Conclusions

We investigated the effects of target sputtering power on the structure and mechanical properties of Ti6Al4V film on the glass substrate as deposited by the RF magnetron sputtering technique. The results showed that the structure of the Ti6Al4V films deposited at all the target sputtering powers was polycrystalline with dual structure phases cubic and hexagonal. Films crystallinity and crystalline size increased with an increase in the RF power. AFM data indicated that film roughness was less for samples deposited on the glass substrate at lower sputtering power. It was observed from UV-visible measurements that the absorbance and the extinction coefficient (k) for deposited thin films increased with an increase in the RF power, while other parameters such as dielectric constants and refractive index decreased. The present work can be a guideline for obtaining good quality Ti6Al4V thin films for biomedical applications.

Acknowledgment

We acknowledge the support of the Dept. of Physics, College of Education for Pure Sciences, University of Anbar and Ministry of Science and Technology, Center of Applied Physics, Baghdad, Iraq. We acknowledge Aqeel F. Hasan, Jamal. F. Hamodi, and Mohammed Odaa for their precious help.

References

[1] C. M. Garzón, J. E. Alfonso, and E. C. Corredor, "Characterization of adherence for Ti6Al4V films RF magnetron sputter grown on stainless steels," *D. Reidel*, 1981, 81(185): 175–181.

[2] A. C. Faria, R. C. Rodrigues, A. P. Claro, M. D. G. C. de Mattos, and R. F. Ribeiro, "Wear resistance of experimental titanium alloys for dental applications," *Journal of the Mechanical Behavior of Biomedical Material*, 2011, 4(8): 1873–1879.

[3] S. Roessler, R. Zimmermann, D. Scharnweber, C. Werner, and H. Worch, "Characterization of oxide layers on Ti6Al4V and titanium by streaming potential and streaming current measurements," *Colloids and Surfaces B Biointerfaces*, 2002, 26(4): 387–395.

[4] H. Copete, F. Vargas, and A. Echavarría, "Mediante proyección térmica por combustión oxiacetilénica sobre un sustrato de ti6al4v in vitro evaluation of oxy-fuel thermal sprayed hydroxyapatite coatings onto ti6al4v substrates," *Dyna*, 2013, 10(177): 101–107.

[5] A. E. B. Torres, S. B. Neves, J. C. N. Abreu, C. L. Cavalcante, and D. M. Ruthven, "Single-and multi-component liquid phase adsorption measurements by headspace chromatography," *Brazilian Journal of Chemical Engineering*, 2001, 18(1): 121–125.

[6] S. Roessler, R. Zimmermann, D. Scharnweber, C. Werner, and H. Worch, "Characterization of oxide layers on Ti6Al4V and titanium by streaming potential and streaming current measurements," *Colloids and Surfaces B Biointerfaces*, 2002, 26(4): 387–395.

[7] T. Falcade, L. M. Antonini, T. E. Schmitzhaus, and C. D. F. Malfatti, "Tribological study of ti6al4v coated with amorphous carbon films obtained by electrodeposition," in *15th International Conference on Experimental Mechanics*, Porto, Portugal, pp. 1–11, 2012.

[8] D. H. He, P. Wang, P. Liu, X. K. Liu, F. C. Ma, and J. Zhao, "HA coating fabricated by electrochemical deposition on modified Ti6Al4V alloy," *Surface and Coatings Technology*, 2015, 277: 203–209.

[9] B. Song, S. Dong, B. Zhang, H. Liao, and C. Coddet, "Effects of processing parameters on microstructure and mechanical property of selective laser melted Ti6Al4V," *Materials and Design*, 2012, 35: 120–125.

[10] S. Xu, J. Long, L. Sim, C. H. Diong, and K. Ostrikov, "RF plasma sputtering deposition of hydroxyapatite bioceramics: synthesis, performance, and biocompatibility," *Plasma Processes and Polymers*, 2005, 2(5): 373–390.

[11] K. V. Dijk, H. G. Schaeken, J. C. G. Wolke, C. H. M.

Maree, F. H. P. M. Habraken, J. Verhoeven, *et al.*, "Influence of discharge power level on the properties of hydroxyapatite films deposited on Ti6A14V with RF magnetron sputtering," *Journal Biomedical Materials Research*, 1995, 29(2): 269–276.

[12] N. Muslim, Y. W. Soon, C. M. Lim, and N. Y. Voo, "Influence of sputtering power on properties of titanium thin films deposited by rf magnetron sputtering," *ARPN Journal of Engineering and Applied Sciences*, 2015, 10(16): 7184–7189.

[13] M. F. A. Alias, K. M. Rashid, and K. A. Adem, "Optical properties for Ti doped thin ZnO films prepared by PLD," *International Journal of Innovative Research in Science, Engineering and Technology*, 2014, 3(8): 15538–15544.

Enhancement of FSO Communications Links Under Complex Environment

Satea H. ALNAJJAR*, Ammar A. NOORI*, and Arwa A. MOOSA*

Department of Network Engineering, Al-iraqia University, Baghdad, Iraq

*Corresponding author: Satea H. ALNAJJAR E-mail: dr.satea@gmail.com
Ammar A. NOORI E-mail: eng.ammar.noori@gmail.com
Arwa A. MOOSA E-mail: eng.arwa.amir@gmail.com

Abstract: Free space optical communication is a line-of-sight (LOS) technology that uses lasers to provide optical bandwidth connections. Potential disturbance arising from the weather condition is one of the most effective factors that influence the bi-directional free space optics (FSO) performance. The complex weather condition in the Middle East region and Arabian Gulf has been dominated by dust storms activities. Dust storms directly affect the characteristics of FSO and consequently lead to an increase in the bit error rate (BER) and deterioration *Q*-factor to bad levels due to the high attenuation factor. In this research, the authors compare the differences between two bi-directional FSOs. One is the traditional link, and the other has been developed to enhance the system performance under the dust storms condition. The proposed design consists of dual FSO channels, and each one includes erbium-doped fiber amplifier (EDFA) optical amplifiers. This design has demonstrated the proficiency in addressing the attenuation that occurs due to weather stickers. The results prove there is an improvement in performance by measuring the *Q*-factor. In addition, BER can be significantly improved, and further communicating distance can be achieved by utilizing 1550 nm with multiple channels and EDFA.

Keywords: FSO; atmospheric attenuation; laser; dust attenuation; EDFA

1. Introduction

The weather condition is one of the most effective factors that have a big impact on the free space optics (FSO) link performance. Link range, the bit error rate (BER), quality factor, and optical power will experience fatal degradation in the complex weather environment. Atmospheric attenuation of the FSO system can be triggered by weather fluctuation such as fog, dust storm, snow, rain, and haze since they are the most dominant outage sources in the FSO system [1]. It can affect the system by means of absorption and scattering of laser photons by variety of aerosols and gaseous molecules in the atmosphere. Absorption occurs due to reciprocal action between the photons of the optical signal and the molecules. Scattering also has the same interaction but without affecting the wave length [2]. Studies have shown that optical attenuation can go up to 340 dB/km in fog with a visibility of 50 m [1]. Different studies of FSO link availability and parameters optimization of the transmission link were proposed in the past for various application scenarios. The dedicated indoor laboratory atmospheric chamber to investigate the effects of fog and dust storm on the FSO link performance was developed in [3]. A new laboratory structure to replicate the sandstorm based on fluid dynamics simulations has been addressed and reported in [4]. The optical amplifying-and-forward (OAF) relaying technique using the optical amplifier was first proposed in [5], and in [6] the problem of beam divergence caused by atmospheric turbulence through the implementation of FSO system over atmosphere for a long distance was avoided by using wave length division multiplexing (WDM) and parameters tuning of FSO link to increase the overall system performance.

In this paper, a new FSO link design has been proposed to enhance the system performance under the dust storms condition, extend the link range, and calculate the attenuation caused by dust storms using the Kim model for the laser beam propagation with an approximate visibility of 70 m [1, 4]. The new design upgraded by using two FSO channels with erbium-doped fiber amplifier (EDFA) optical amplifiers at the sending and receiving ends to ensure a low BER with an extended distance of up to 2.5 km.

The rest of this paper is organized as follows: Section 2 presents the dust storms attenuation for the FSO and the application of the Kim model to approximate the atmospheric attenuation, Section 3 devotes to the description of the proposed FSO design with a performance comparison among one FSO link channel, two FSO link channels, and two channels with EDFAs, Section 4 shows the simulation results and discussion, and Section 5 is the research conclusions.

2. Dust storms attenuation for FSO link

One of the challenges of the FSO channel in which the signal loss and link failure are caused is the atmospheric attenuation due to absorption, scattering, and scintillation [1, 7]. All these effects vary with time and depend on the current local conditions, weather, and distance. Water particles and carbon dioxide mainly cause the absorption of optical signals, whereas the fog, rain, snow, and cloud cause the scattering of optical signals transmitting in the free space. This scattering causes portion of the light beam travelling from a source to deflect away from the intended receiver [1], and the atmospheric attenuation is given by the following Beer's law equation, which represents the relationship between the power of the transmitted signal and the received signal in the presence of the atmospheric attenuation [1, 7]:

$$\tau = P_L = P_0 \times \mathrm{EXP}(-\alpha \times L) \quad (1)$$

where P_L is the laser power at the range L, P_0 is the laser power of the source, and α is the atmospheric attenuation coefficient. The atmospheric attenuation (τ) in dB can be calculated as follows:

$$\tau_R = 4.3429 \times \alpha \times L[\mathrm{dB}] \quad (2)$$

and the atmospheric attenuation coefficient according to the Kim model equals to

$$\alpha = \frac{3.91}{V} \times \left(\frac{550\mathrm{nm}}{\lambda}\right)^{-q} \quad (3)$$

where λ is the wavelength in nm, V is the visibility, and q is the size distribution of the scattering particles. The attenuation of the transmitted signal can be obtained from the previous model to various weather conditions using q [1, 6, 7]:

$q = 1.6$ for visibility ($V > 50$ km)

$q = 1.3$ for visibility (6 km $< V <$ 50 km)

$q = 0.16\,V + 0.34$ for visibility(1 km $< V <$ 6 km)

$q = V - 0.5$ for visibility (0.5 km $< V <$ 1 km)

$q = 0$ for visibility ($V <$ 0.5 km).

This form of calculation is very handy because for a given wavelength, the amount of the attenuation only depends on the visibility [1]. The visibility is used in the equation above as in [1, 4].

In this research, three different wavelengths are adopted to compare the attenuation effect on each one of them, and the calculations of the atmospheric attenuation coefficient are shown in Table 1.

Table 1 FSO link attenuation in dB/km according to the Kim model.

Visibility (km)	dB/km (450 nm)	dB/km (785 nm)	dB/km (1550 nm)	Weather
0.05	340	340	340	Heavy fog
0.07	242	242	242	Heavy dust
0.2	85	85	85	Moderate
0.5	34	34	34	Dust-fog
1	19	14	10	Rain
2	10	7	4	Haze
4	5	3	2	
10	2	1	0.4	Clear
23	1	0.5	0.2	

3. Design of the proposed FSO system

FSO is a line-of-sight technology that uses lasers to provide optical bandwidth connections, and it is also an optical communications technique that propagates the light in the free space such as air, outer space, vacuum, or something similar to wirelessly transmit data for telecommunications and computer networking. In recent years, the demand for FSO communications has increased considerably due to its ability to carry high bandwidth data in the long distance up to 8 km with low bit error rates [3, 4].

Figure 1 shows the essential concepts and the devices that are used to design the FSO system utilizing the optical communications system design software "Optisystem™" [8].

The receiver component of the FSO on both sides has a specific minimum sensitivity at a given data rate, and the received power must remain above the lowest sensitivity to guarantee the reliable operation of the system. The connectivity of the FSO channel is affected by the carrier medium (air) between the sender and the receiver, thus, the weather conditions impact the signal-transmission process. Therefore, the weather conditions must be considered when using the FSO system. Thus, the system is analyzed under a range of weather conditions, depending on the attenuation coefficient affecting the channel transmitter.

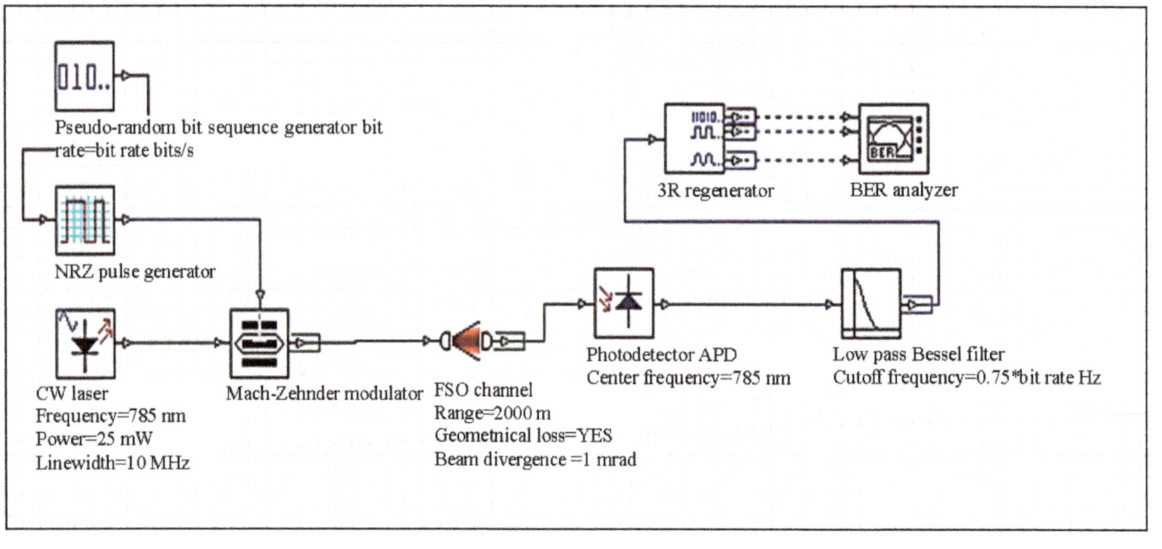

Fig. 1 One-channel FSO system components in "Optisystem™".

The FSO transmission part in the system includes a pseudo-random bit generator, a non-return-to-zero (NRZ) pulse generator, continuous wave (CW) laser diodes where it acts as the carrier source with the wavelength (450 nm, 750 nm, and 1550 nm) and a Mach-Zehnder modulator. An avalanche photo diode (APD) and a low-pass Gaussian filter are used in the receiving side, which causes 40% timing jitter reduction and reduces the switch window loss of the traveling-wave electron absorption modulator (TW-EAM) [9].

In this research, the performance comparison is handled for the previous design (Fig. 1) that has one FSO channel with another design (Fig. 2) having two channels.

Figure 2 shows an enhancement of the system performance in dust storms condition and extends the link range by using two channels enforced with two optical amplifiers EDFAs. The new system design is consisted of two FSO channels with the fork power splitter and power combiner in order to take advantage of the constrictive combination of the light power, and an optical amplifier is used to increase the performance of the system. The gain of the amplifier is set to be 20 dB. The amplified signal is then directly sent to the receiver through the free space optical channel, and another one is suited in the reversed side as shown in Fig. 3.

The parameters in Table 2 are used to simulate the channel transmission in the FSO systems via the optical simulator software, "Optisystem™".

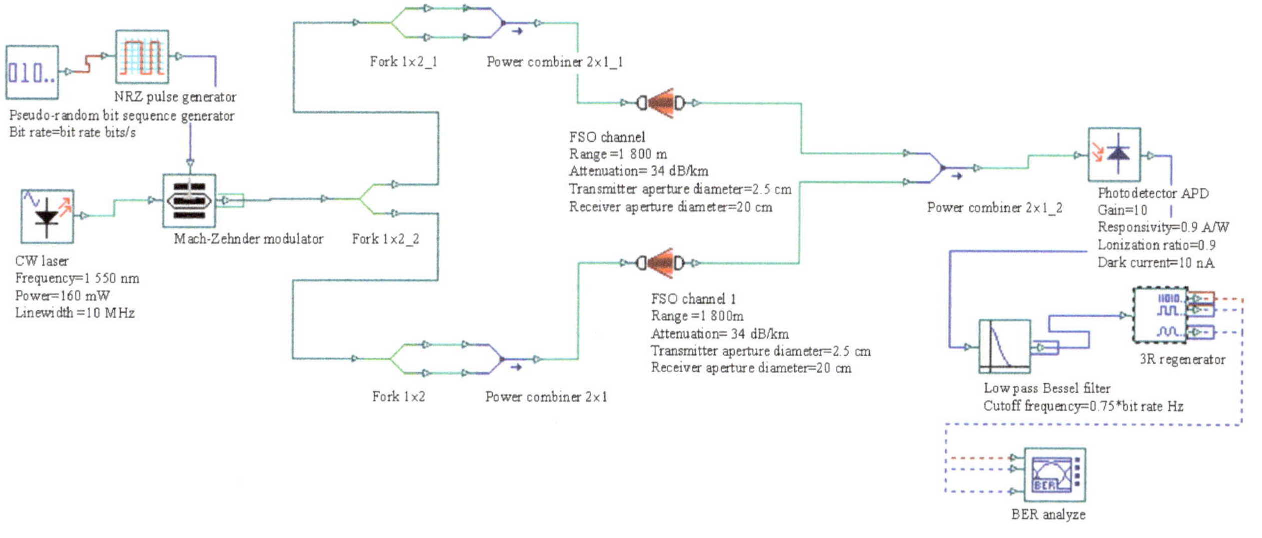

Fig. 2 Two-channel FSO system.

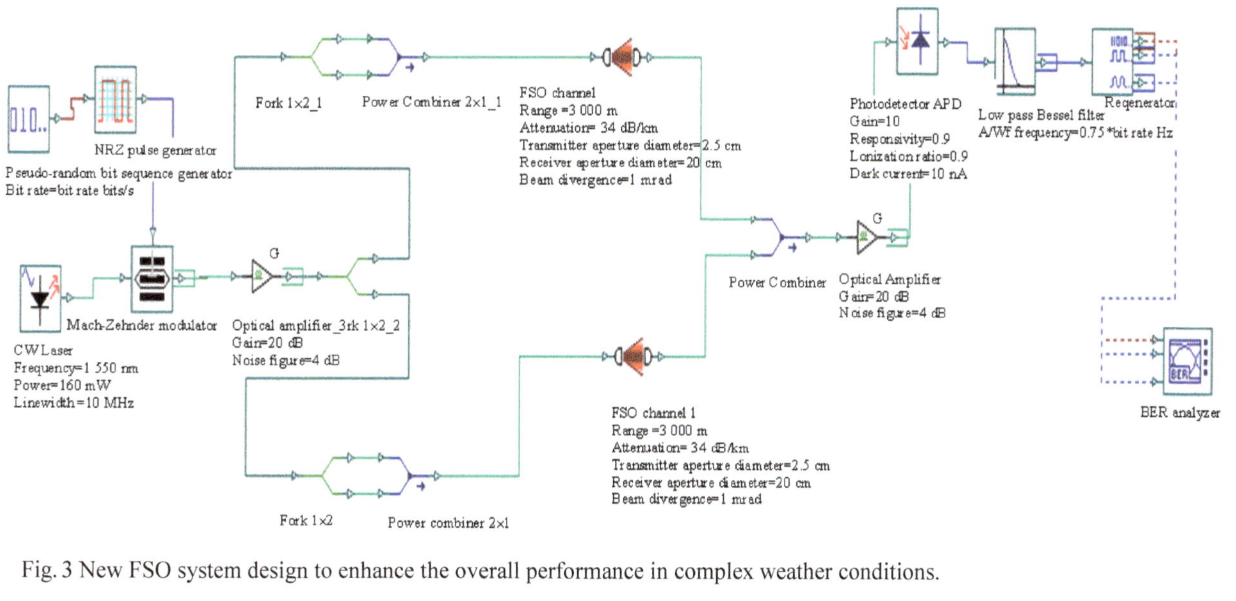

Fig. 3 New FSO system design to enhance the overall performance in complex weather conditions.

Table 2 FSO link parameters.

Channel configurations	Parameter
Dynamic range	3000 m
Attenuation for dust storms	34 dB/km–242 dB/km
Diameter of the aperture of the receiver	20 cm
Diameter of the aperture of the transmitter	2.5 cm
Transmitter and receive loss	1 dB
Transmission power	160 mV
Wavelength	1550 nm
Beam divergences	1 mrad

4. Results and discussion

In this section, numerical results for the BER performance of FSO links based on the above system model will be presented for various number links designs with different wavelengths and distances. The FSO system is considered with transmitting and receiving apertures with the size of 2.5 cm and 20 cm, respectively. The attenuations are measured with different attenuations ranges, according to the Kim model, which is illustrated at 1550 nm as shown in Fig. 4 and at 850 nm as shown in Fig. 5. It is obvious that using multiple channels with EDFA amplifiers at 1550 nm could have the best result for different attenuations. Figure 6 shows the system's quality factor while Fig. 7 shows the BER comparison among one channel, two channels, and two channels with EDFA under the moderate dust scenario.

The BER performance is illustrated in Fig. 7, in which the systems with one and two channels in general are affected by the weather attenuation from 1 km and BER becomes approximately 1.8E–13 at the distance of 1250 m.

Figure 8 illustrates the eye diagram distortion utilizing the BER analyzer in Optisystem™. The enhancement in the bit error rate is clearly shown by using the new design of two channels with EDFAs. The approximate BER is 2.12E–10 at the distance of 2300 m.

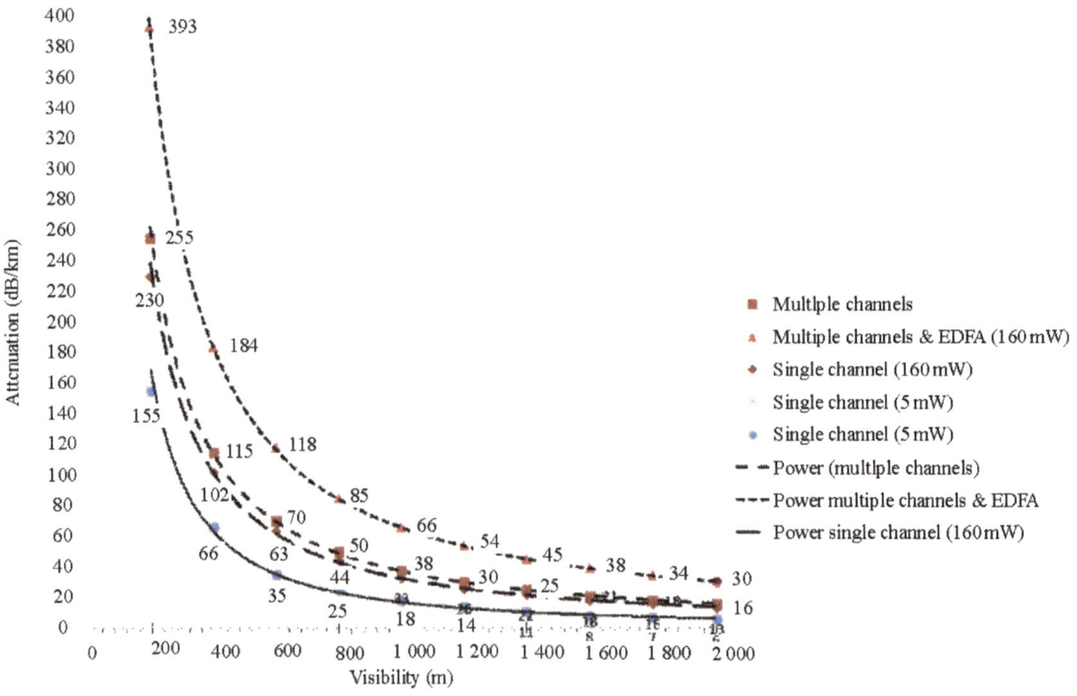

Fig. 4 Variation of the attenuation at 1550 nm in the function of visibility, compared, using the Kim model.

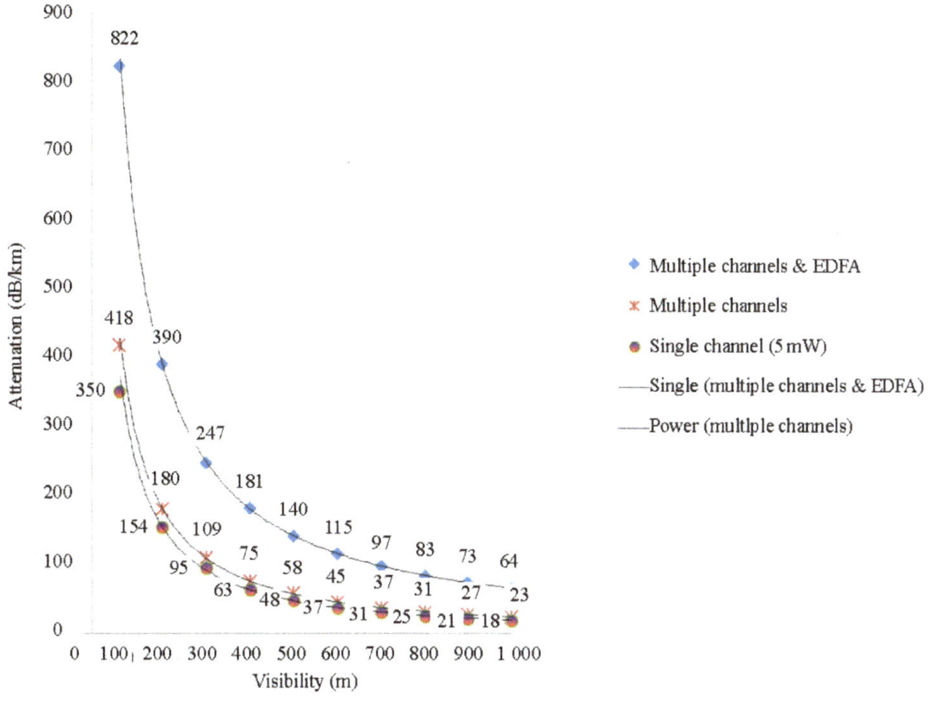

Fig. 5 Variation of the attenuation at 850 nm in the function of visibility, compared, using the Kim model.

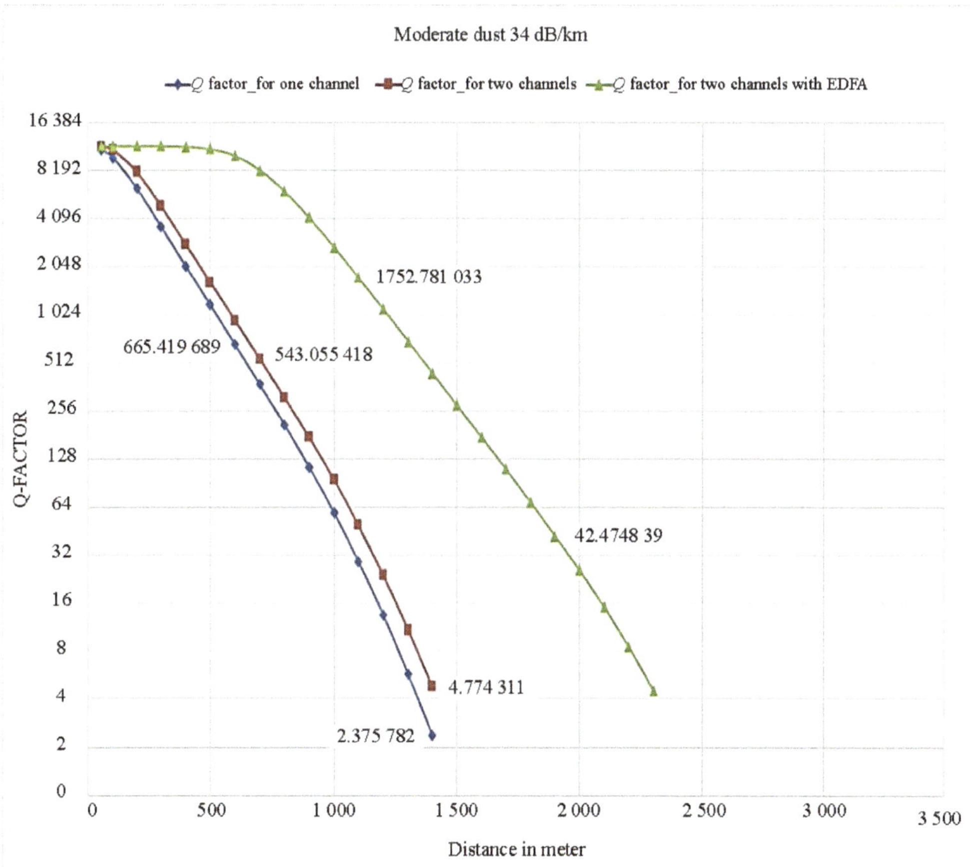

Fig. 6 Quality factor at 34 dB/km according to the distance.

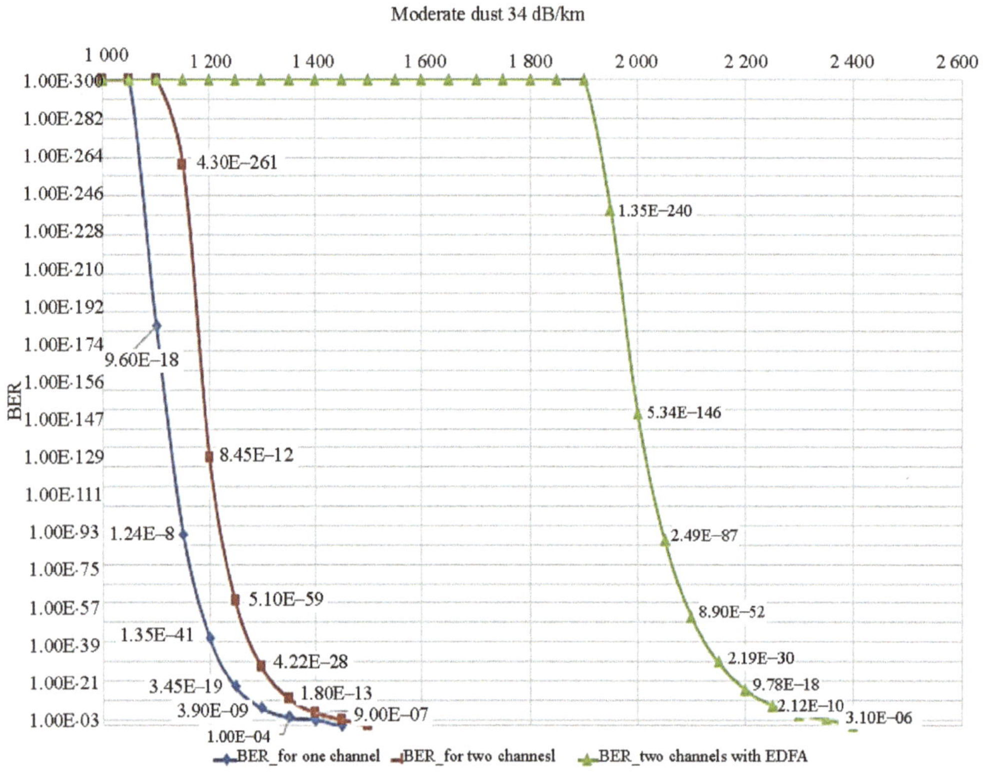

Fig. 7 BER performance comparison according to the distance at 34 dB/km.

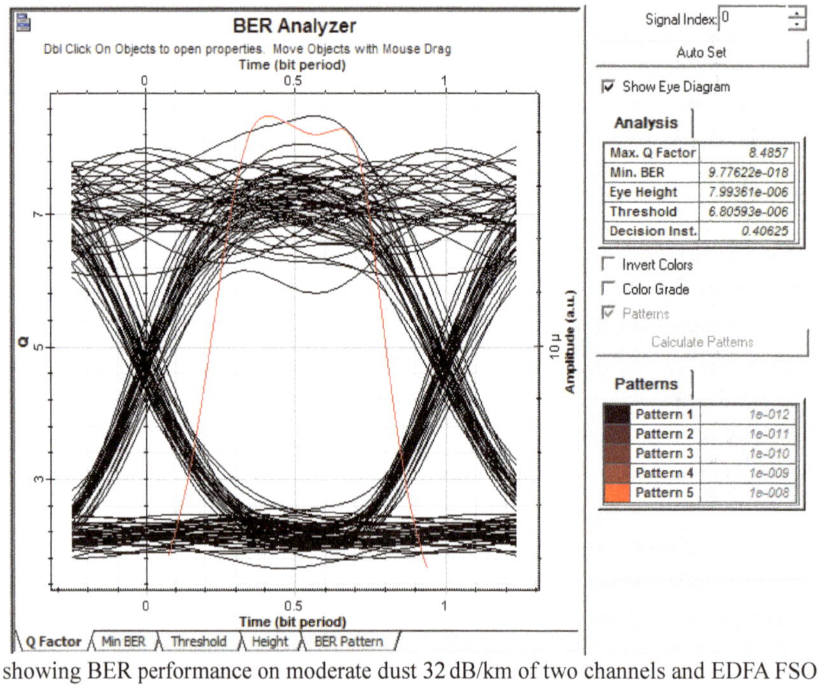

Fig. 8 Eye diagram showing BER performance on moderate dust 32 dB/km of two channels and EDFA FSO system at 2.3 km.

The heavy dust storms are considered as a design's challenge since its attenuation is calculated using the Kim model with a visibility of 70 m to be around 242 dB/km. Figure 9 shows the system's quality factor, and Fig. 10 shows the BER comparison among one channel, two channels, and two channels with EDFA under the heavy dust scenario.

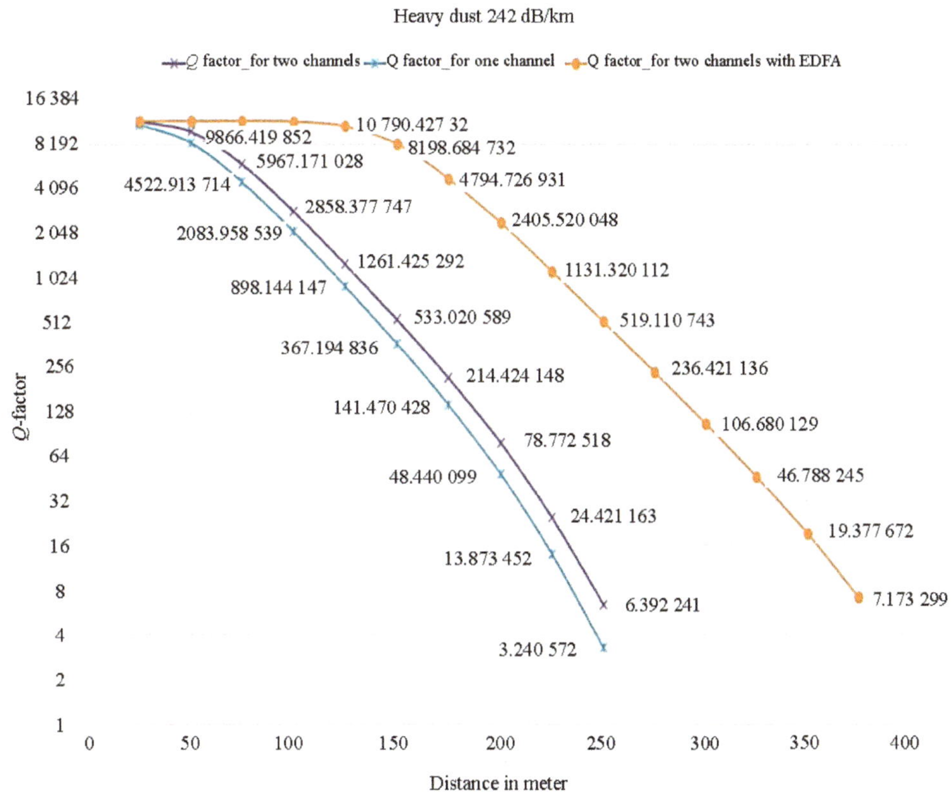

Fig. 9 Quality factor at 242 dB/km according to the distance.

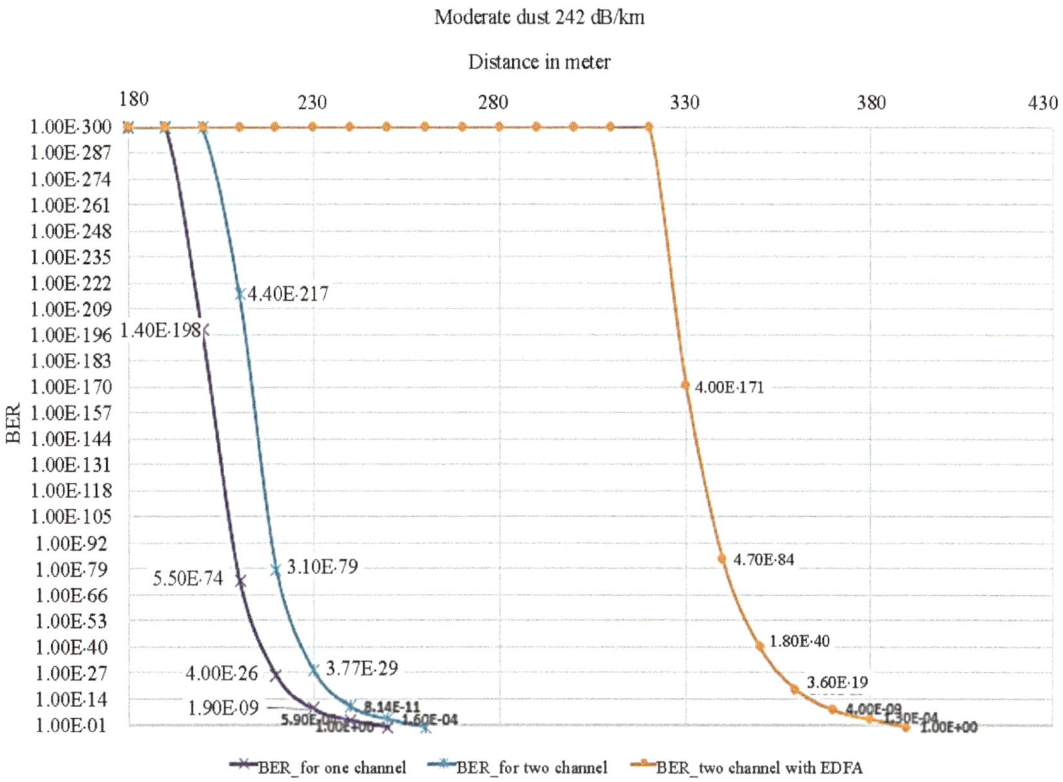

Fig. 10 BER performance comparison according to the distance at 242 dB/km.

The BER performance in Fig. 10 and the eye diagram in Fig. 11 clearly show that the performances are improved significantly with the increasing number of transmitting links, which in effect reduces the effect of the weather turbulence variance for the delivering channel, for example, the BER performance for 242 dB/km is zero using the proposed design of two channels and EDFAs at the distance of 280 m. In this point, the other designs are at the highest BER. Further, the EDFA amplifiers will also increase the channels performance as BER for 242 dB/km at the distance up to 38.

In distance 0 m, all wavelengths have equivalent BER, knowing that the minimum acceptable BER is 1.0E–9 [10].

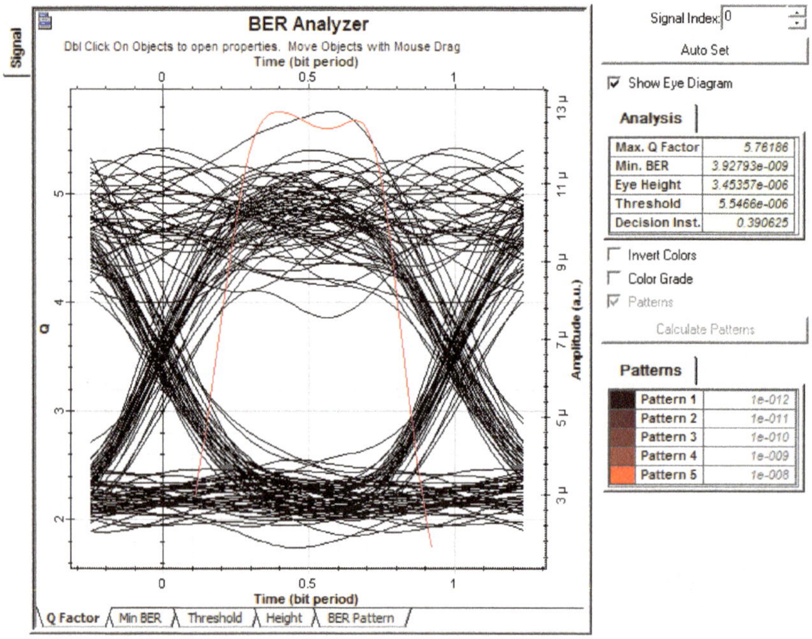

Fig. 11 Eye diagram showing BER performance on heavy dust 242 dB/km of two channels with EDFA FSO system at 380 m.

Although there is some discrepancy observed in these different designs, the performance will still be limited to some distance, however an increase in the number of links and the power of the amplifiers will increase the performance of the FSO system.

5. Conclusions

In this paper, the FSO system BER performance has been investigated over different turbulence atmospheric weathers. Results demonstrate that FSO links could be efficiently effective for higher attenuations such as in the heavy fog and dust by using multiple channels with EDFA amplifiers. Moreover, the effect of the attenuation will be decreased, and the BER performance will be increased via the rising number of transmitting links plus the power of amplifiers.

In addition, the performance of the links bounded by the wavelength that was observed, however the numerical result shows that at higher attenuation of different wavelengths the BER have the same performance. As a result, the BER performance could be significantly improved, and further communicating distance could be achieved at 1550 nm with multiple channels and EDFA amplifiers. Finally, this improvement in performance has led to enhance the quality of the service.

References

[1] I. I Kim, B. McArthur, and E. J. Korevaar, "Comparison of laser beam propagation at 785 nm and 1550 nm in fog and haze for optical wireless

communications," *SPIE*, 2001, 4214(2): 26–37.

[2] A. Raja, Q. Kagalwala, T. Landolsi, and M. El-Tarhuni, "Free-space optics channel characterization under UAE weather conditions," in *2007 IEEE International Conference on Signal Processing and Communications*, Dubai, pp. 856–859, 2007.

[3] J. Perez, Z. Ghassemlooy, S. Rajbhandari, M. Ijaz, and H. Minh, "Ethernet FSO communications link performance study under a controlled fog environment," *IEEE Communications Letters*, 2012, 16(3): 408–410.

[4] Z. Ghassemlooy, J. Perez, and E. Leitgeb, "On the performance of FSO communications links under sandstorm conditions," in *International Conference on Telecommunications*, 2013, pp. 53–58.

[5] M. Karimi and M. Nasiri-Kenari, "Free space optical communications via optical amplify-and-forward relaying," *Journal Lightwave Technology*, 2011, 29(2): 242–248.

[6] H. A. Fadhil, A. Amphawan, H. A. B. Shamsuddin, T. H. Abd, H. M. R. Al-Khafaji, S. Aljunid, *et al.*, "Optimization of free space optics parameters: an optimum solution for bad weather conditions," *Optik – International Journal for Light and Electron Optics*, 2013, 124(19): 3969–3973.

[7] N. A. Mohammed, A. S. El-Wakeel, and M. H. Aly, "Performance evaluation of FSO link under NRZ-RZ line codes, different weather conditions and receiver types in the presence of pointing errors," *Open Electrical & Electronic Engineering Journal*, 2014, 6(1): 28–35.

[8] S. H. Alnajjar, M. S. R. F. Malek, and M. S. Ahmad, "Low-altitude platform to enhance communications reliability in disaster environments," *Journal of Advances in Information Technology*, 2014, 5(1): 21–30.

[9] Z. Hu, H. F. Chou, J. E Boors, and D. J. Blumenthal, "40-gb/s optical clock recovery using a compact traveling-wave electroabsorption modulator-based ring oscillator," *IEEE Photonics Technology Letters*, 2004, 16(5): 1376–1378.

[10] B. Wandernoth, "5 photon/bit low complexity 2 MBit/s PSK transmission breadboard experiment with homodyne receiver applying synchronization bits and convolutional coding," in *20th European Conference on Optical Communications, Firenze*, pp. 59–62, 1994.

Permissions

All chapters in this book were first published in PS, by Springer International Publishing AG.; hereby published with permission under the Creative Commons Attribution License or equivalent. Every chapter published in this book has been scrutinized by our experts. Their significance has been extensively debated. The topics covered herein carry significant findings which will fuel the growth of the discipline. They may even be implemented as practical applications or may be referred to as a beginning point for another development.

The contributors of this book come from diverse backgrounds, making this book a truly international effort. This book will bring forth new frontiers with its revolutionizing research information and detailed analysis of the nascent developments around the world.

We would like to thank all the contributing authors for lending their expertise to make the book truly unique. They have played a crucial role in the development of this book. Without their invaluable contributions this book wouldn't have been possible. They have made vital efforts to compile up to date information on the varied aspects of this subject to make this book a valuable addition to the collection of many professionals and students.

This book was conceptualized with the vision of imparting up-to-date information and advanced data in this field. To ensure the same, a matchless editorial board was set up. Every individual on the board went through rigorous rounds of assessment to prove their worth. After which they invested a large part of their time researching and compiling the most relevant data for our readers.

The editorial board has been involved in producing this book since its inception. They have spent rigorous hours researching and exploring the diverse topics which have resulted in the successful publishing of this book. They have passed on their knowledge of decades through this book. To expedite this challenging task, the publisher supported the team at every step. A small team of assistant editors was also appointed to further simplify the editing procedure and attain best results for the readers.

Apart from the editorial board, the designing team has also invested a significant amount of their time in understanding the subject and creating the most relevant covers. They scrutinized every image to scout for the most suitable representation of the subject and create an appropriate cover for the book.

The publishing team has been an ardent support to the editorial, designing and production team. Their endless efforts to recruit the best for this project, has resulted in the accomplishment of this book. They are a veteran in the field of academics and their pool of knowledge is as vast as their experience in printing. Their expertise and guidance has proved useful at every step. Their uncompromising quality standards have made this book an exceptional effort. Their encouragement from time to time has been an inspiration for everyone.

The publisher and the editorial board hope that this book will prove to be a valuable piece of knowledge for researchers, students, practitioners and scholars across the globe.

List of Contributors

Fei Peng and Xuli Cao
Key Laboratory of Optical Fiber Sensing & Communications (Ministry of Education), University of Electronic Science and Technology of China, Chengdu, 611731, China

Wolfgang Wildner and Dietmar Drummer
Institute of Polymer Technology, Friedrich-Alexander-University Erlangen-Nürnberg, Erlangen, Germany

H. Moayyed, I. T. Leite, L. Coelho and J. L. Santos
CAP / INESC TEC - Technology and Science and Faculty of Sciences from the University of Porto, Rua do Campo Alegre 687, 4169-007 Porto, Portugal

D. Viegas
INL - International Iberian Nanotechnology Laboratory, Av. Mestre José Veiga, 4715-310 Braga, Portugal

Jingsheng Lv, Haifeng Qi, Zhiqiang Song, Jian Guo, Jiasheng Ni and Chang Wang
Shandong Key Laboratory of Optical Fiber Sensing Technologies, Laser Institute of Shandong Academy of Sciences, Jinan, 250014, China

Gangding Peng
Shandong Key Laboratory of Optical Fiber Sensing Technologies, Laser Institute of Shandong Academy of Sciences, Jinan, 250014, China
School of Electrical Engineering & Telecommunications, The University of New South Wales, NSW, 2052, Australia

Oday A. Hammadi
Department of Physics, College of Education, Al-Iraqia University, Baghdad, IRAQ

Yuchen Shao, Xiuyou Han, Xiaonan Han, Zhili Lu, Zhenlin Wu and Mingshan Zhao
School of Physics and Optoelectronic Engineering, Dalian University of Technology, Dalian, 116024, China

Jie Teng
No. 38 Research Institute, China Electronics Technology Group Corporation, Hefei, 230088, China

Jinyan Wang
School of Chemical Engineering, Dalian University of Technology, Dalian, 116024, China

Geert Morthier
Photonics Research Group, Department of Information Technology, Ghent University-IMEC, 9000 Ghent, Belgium

Zezheng Qiu and Liping Pang
School of Aeronautic Science and Engineering, Beijing University of Aeronautics and Astronautics, Beijing, 100191, China

Tong Zheng and Hongquan Qu
College of Electrical & Information Engineering, North China University of Technology, Beijing, 100144, China

Chuntong Liu, Zhengyi Zhang, Hongcai Li, Zhenxin He, Xiaofeng Zhao and Rongjing Wu
Department Two, Rocket Force University of Engineering, Xi'an 710025, China

Lina Wang, Yongjun Wang, Chen Wu and Fu Wang
School of Electronic Engineering, Beijing University of Posts and Telecommunications, Beijing, 100876, China

Shijie Deng Michael Ap Mcauliffe and Liam Lewis
Centre for Advanced Photonics & Process Analysis, Cork Institute of Technology and Tyndall National Institute, Cork, Ireland

Guillaume Huyet
Centre for Advanced Photonics & Process Analysis, Cork Institute of Technology and Tyndall National Institute, Cork, Ireland
Department of Physical Science, Cork Institute of Technology, Cork, Ireland

William Doherty
Department of Physical Science, Cork Institute of Technology, Cork, Ireland

Urszula Salaj-Kosla
Department of Chemical and Environmental Science, University of Limerick, Limerick, Ireland

Xiaonan Han, Xiuyou Han, Yuchen Shao, Zhenlin Wu and Mingshan Zhao
School of Physics and Optoelectronic Engineering, Dalian University of Technology, Dalian, 116024, China

Jie Teng
No. 38 Research Institute, China Electronics Technology Group Corporation, Hefei, 230088, China

Shuhui Bo
Technical Institute of Physics and Chemistry, Chinese Academy of Sciences, Beijing, 100190, China

Yuxin Liang
School of Physics and Optoelectronic Engineering, Dalian University of Technology, Dalian, 116024, China
Photonics Research Group, Department of Information Technology, Ghent University-IMEC, 9000 Ghent, Belgium

Geert Morthier
Photonics Research Group, Department of Information Technology, Ghent University-IMEC, 9000 Ghent, Belgium

Xuejun Zhang, Xin Zhang, Haifei Hu and Xuefeng Zeng
Changchun Institute of Optics, Fine Mechanics and Physics, Chinese Academy of Sciences, Changchun, 130033, China
Key Laboratory of Optical System Advanced Manufacturing Technology, Chinese Academy of Sciences, Changchun, 130033, China

Chi Song
Changchun Institute of Optics, Fine Mechanics and Physics, Chinese Academy of Sciences, Changchun, 130033, China
University of Chinese Academy of Sciences, Beijing, 100049, China

Georgios N. Tsigaridas
Department of Physics, School of Applied Mathematical and Physical Sciences, National Technical University of Athens, Zografou Campus, GR-15780 Zografou, Athens, Greece

Abdellatif Cherifi and Benamar Bouhafs
University of Tlemcen, Theoretical Physics Laboratory, Physics of Department, Tlemcen, 13000, Algeria

Ming Li
State Key Laboratory of Crystal Materials, Shandong University, Jinan, 250100, China
Jinan Research Center, Laser Institute of Shandong Academy of Sciences, Jinan, 250014, China

Xiangang Xu
State Key Laboratory of Crystal Materials, Shandong University, Jinan, 250100, China

Zhihui Sun, Xiaolei Zhang, Shujuan Li, Zhiqiang Song, Meng Wang, Jian Guo, Jiasheng Ni and Chang Wang
Jinan Research Center, Laser Institute of Shandong Academy of Sciences, Jinan, 250014, China

Gangding Peng
School of Electronic Engineering and Telecommunication, University of New South Wales, Sydney, 2052, Austrilia

Lei Zhang, Zhang Zhang, Yichuan Wang, Wei Fang and Limin Tong
State Key Laboratory of Modern Optical Instrumentation, College of Optical Science and Engineering, Zhejiang University, Hangzhou, 310027, China

Meiying Ye
College of Material, Chemistry and Chemical Engineering, Hangzhou Normal University, Hangzhou, 310036, China

Md. Shadidul Islam
Department of Information and Communication Technology, Mawlana Bhashani Science and Technology University Santosh, Tangail-1902, Bangladesh

Bikash Kumar Paul and Kawsar Ahmed
Department of Information and Communication Technology, Mawlana Bhashani Science and Technology University Santosh, Tangail-1902, Bangladesh
Group of Bio-Photomatriχ, Tangail-1902, Bangladesh

Sayed Asaduzzaman
Department of Information and Communication Technology, Mawlana Bhashani Science and Technology University Santosh, Tangail-1902, Bangladesh
Group of Bio-Photomatriχ, Tangail-1902, Bangladesh
Department of Software Engineering, Daffodil International University, Sukrabad, Dhaka, Bangladesh

Hang Heng
Center for Analysis and Testing, Nanjing Normal University, Nanjing, 210097, China

Rong Wang
Department of Neurosurgery, Nanjing Drum Tower Hospital, Nanjing, 210008, China
The Affiliated Hospital of Nanjing University Medical School, Nanjing, 210008, China

Lei Wei and Xiaoxue Gong
University of Chinese Academy of Sciences, Beijing, 100039, China
Changchun Institute of Optics, Fine Mechanics and Physics, Chinese Academy of Sciences, Changchun, 130033, China

Lei Zhang
Changchun Institute of Optics, Fine Mechanics and Physics, Chinese Academy of Sciences, Changchun, 130033, China
Chang Guang Satellite Technology LTD.Co, Changchun, 130000, China

Asama N. Naje, Russul R. Ibraheem and Fuad T. Ibrahim
University of Baghdad, Collage of Science, Department of Physics, Baghdad, Iraq

Zhouzhou Yu, Qi Jiang, Hao Zhang and Junjie Wang
School of Control Science and Engineering, Shandong University, Jinan, 250061, China
State Key Lab of Microbial Technology, School of Life Science, Shandong University, Jinan, 250100, China

Wen Liu, Ying Feng and Huayong Yang
College of Optoelectronic Science and Engineering, National University of Defense Technology, Changsha, 410073, China

Lina Ma and Zhengliang Hu
Academy of Ocean Science and Engineering, National University of Defense Technology, Changsha, 410073, China

Xiaohui Liu, Chen Wang, Ying Shang, Chang Wang and Wenan Zhao
Shandong Key Laboratory of Optical Fiber Sensing Technologies, Laser Institute of Shandong Academy of Sciences, Jinan, Shandong, 250014, China

Gangding Peng
Shandong Key Laboratory of Optical Fiber Sensing Technologies, Laser Institute of Shandong Academy of Sciences, Jinan, Shandong, 250014, China
School of Electrical Engineering & Telecommunications, the University of New South Wales, NSW, 2052, Australia

Hongzhong Wang
Shengli Oilfield Xinsheng Geophysical Technology Co. Ltd., Dongying, Shandong, 257000, China

Lokendra Singh, Amna Bedi and Santosh Kumar
Photonics Lab, Department of Electronics and Communication Engineering, DIT University, Dehradun-248009, India

Mohammed K. Khalaf
Ministry of Science and Technology, Center of Applied Physics, Baghdad, Iraq

H. F. Al-Taay
Department of Physics, College of Science for Women, University of Baghdad, Baghdad, Iraq

Dawood S. Ali
Department of Physics, College of Education for Pure Sciences, University of Anbar, Anbar, Iraq

Satea H. Alnajjar, Ammar A. Noori and Arwa A. Moosa
Department of Network Engineering, Al-iraqia University, Baghdad, Iraq

Index